Primary Explosives

Robert Matyáš · Jiří Pachman

Primary Explosives

Springer

Robert Matyáš
Faculty of Chemical Technology
University of Pardubice
Pardubice
Czech Republic

Jiří Pachman
Faculty of Chemical Technology
University of Pardubice
Pardubice
Czech Republic

ISBN 978-3-642-28435-9   ISBN 978-3-642-28436-6 (eBook)
DOI 10.1007/978-3-642-28436-6
Springer Heidelberg New York Dordrecht London

Library of Congress Control Number: 2012945904

© Springer-Verlag Berlin Heidelberg 2013
This work is subject to copyright. All rights are reserved by the Publisher, whether the whole or part of the material is concerned, specifically the rights of translation, reprinting, reuse of illustrations, recitation, broadcasting, reproduction on microfilms or in any other physical way, and transmission or information storage and retrieval, electronic adaptation, computer software, or by similar or dissimilar methodology now known or hereafter developed. Exempted from this legal reservation are brief excerpts in connection with reviews or scholarly analysis or material supplied specifically for the purpose of being entered and executed on a computer system, for exclusive use by the purchaser of the work. Duplication of this publication or parts thereof is permitted only under the provisions of the Copyright Law of the Publisher's location, in its current version, and permission for use must always be obtained from Springer. Permissions for use may be obtained through RightsLink at the Copyright Clearance Center. Violations are liable to prosecution under the respective Copyright Law.
The use of general descriptive names, registered names, trademarks, service marks, etc. in this publication does not imply, even in the absence of a specific statement, that such names are exempt from the relevant protective laws and regulations and therefore free for general use.
While the advice and information in this book are believed to be true and accurate at the date of publication, neither the authors nor the editors nor the publisher can accept any legal responsibility for any errors or omissions that may be made. The publisher makes no warranty, express or implied, with respect to the material contained herein.

Printed on acid-free paper

Springer is part of Springer Science+Business Media (www.springer.com)

Preface

Primary explosives, or initiators, represent a small subset of those chemical compounds called explosives and are used for the purpose of initiating explosions. Simply said, they can be found in various initiation devices where it is necessary to use a mechanical, thermal, or electrical stimulus to prime the combustion or detonation of the main explosive.

We are currently confronted with an overwhelming volume of negative news about the use and abuse of explosives for criminal activities or terrorist bomb attacks, such that the word *explosive* itself has acquired a rather pejorative connotation and is often replaced with the more neutral, but rather meaningless, term *energetic material*. Leaving aside the Orwellian Newspeak, it is fair to say—without any exaggeration—that, although not generally known, explosives are part and parcel of everyday life, much more so now than in the past. If one just considers a small group of explosives used for initiation—primary explosives—we find that most of us are in contact with them on a daily basis, in the form of initiators or pyrotechnic devices in air bag systems, for example. Given this, it seems rather surprising that there has been nothing published in English summarizing the vast knowledge of their properties, preparation, or usage.

Of course, claiming that no information is available on primary explosives would not be entirely true. However, it is common for primary explosives to be discussed in a chapter or a paragraph in publications dealing with explosives in general or in a wider picture (e.g., in T. Urbański's "Chemistry and Technology of Explosives" or David's "The Chemistry of Powder and Explosives"). Another significant source of information are the hundreds or thousands of specialized articles in technical periodicals and papers published in conference publications, which are far too focused on a particular substance or phenomenon and do not allow an easy grasp of the function of primary explosives in the wider context. A fair amount of information is also included in encyclopedias on explosives (of which the most famous and comprehensive are Fedoroff, Shefield and Kaye's "Encyclopaedia of Explosives and Related Items" or Meyer's "Explosives"). However, information included in these publications is usually rather brief and generally omits many essential details. Very comprehensive and valuable sources of information on primary explosives are the many publications written in non-English languages, which, despite today's translation capabilities, often remain inaccessible for the majority of potential readers due to the language barrier. This is really regrettable as there are numerous comprehensive and high-quality sources produced in Russian or the Czech language. Here, one should note Bagal's 500-page publication "Khimiya i tekhnologiya initsiiruyushchikh vzryvshchatykh veshchestv" (Chemistry and Technology of Primary Explosives) written in Russian or Hanus's Czech publication "Méně známé třaskaviny" (Lesser Known Primary Explosives).

Given all the above, the authors of this publication decided to summarize the information on primary explosives included in the sources mentioned above. The information given is not restricted to the traditional primary explosives used in numerous applications, but there is also information on the latest trends and substances which are currently the subject of research and development projects, which are regarded as being potentially useful in the future. The publication also contains some information on primary explosives which are frequently produced illegally by nonprofessionals. Their inclusion represents the authors' intention to provide accurate information on this group of substances, for which the general public, and even the experts, find rather fanciful, distorted, and false information in the mass media and on the Internet. Being chemical engineers, we could not resist the temptation to include also a few substances which are unlikely to be used practically in the future but whose properties are so interesting that omitting them would be unforgiveable.

The publication has been written for the general public interested in the field of explosive chemical compounds, but especially for chemistry students and teachers, researchers working in the field of explosives, police officers, criminologists, forensic analysts, soldiers, engineers working in the production of initiators, rock blasters, and others who are likely to come across primary explosives in their various forms in their professional lives. A basic knowledge of chemistry at a secondary school level may be of great benefit to the reader; however, even a complete layman may learn a lot about the properties, methods of preparation, or use of individual substances. Ideally, the publication should assist in achieving a deeper understanding of the role of primary explosives, helping to demystify their extreme sensitivity and dangerousness, and providing precise definitions, enhancing understanding of the historical context of their development and outlining the potential future use of this group of substances.

With the exception of the first two chapters dealing with general performance and sensitivity properties of primary explosives, the structure of the publication is rather simple, each chapter covering a group of substances. The chapters which follow are consistently subdivided and include information on the discovery of the substance, sometimes including a few anecdotal or historical pieces of information, followed by a summary of physical, chemical, and explosive properties, a brief description of methods of preparation and a final part giving information on its usage. We have not sought to discuss special properties in great detail. We aimed to summarize what we consider to be most important and we paid special attention to thorough referencing to enable the reader easily to find detailed information in the available literature sources. Detailed instructions are not provided for individual syntheses; methods resulting in individual substances are highlighted. In places, where necessary, the topic has been discussed in more detail; in others, it has been simplified. However, it is always rather easy to find the original source, including all details, using the references. To make the rather technical text more user-friendly, we decided to accompany it with photographs of a number of the substances discussed. In most cases, these are the authors' unpublished photographs of products made or supplied by the authors and which cannot therefore be found in the available literature.

In conclusion, let us express our gratitude and thanks to all who supported the development of this project. Without their help, its creation would not have been possible. We would like to thank especially Prof. How-Ghee Ang, director of the Energetics Research Institute, Nanyang Technological University in Singapore, who had the original idea to create such a work and who generously supported the writing of the book throughout the first year of its preparation. We would also like to thank Prof. Svatopluk Zeman, director of the Institute of Energetic Materials, University of Pardubice, Czech Republic for his support in the second stage of its development, especially for creating an inspiring environment for completing the publication. Further, thanks to Mr. Jiří Strnad, in memoriam, for providing valuable and unpublished information, which significantly enhanced the chapter on Explosive Properties of Primary Explosives.

We would also like to express our thanks to Jiří Nesveda and Pavel Valenta for providing expert consultation and for supplying samples of a number of primary explosives, to Lenka Murcková for her ability to obtain even the unobtainable, to Monika Šubrtová for her infinite patience when producing graphs, and to Jakub Šelešovský, Zdeněk Jalový and John Svoboda for the first review of this publication. We would especially like to thank Prof. Stanisław Cudziło from Wojskowa Akademia Techniczna in Warszawa for reviewing the manuscript.

This work was created as part of two projects: project of the Ministry of Education, Youth and Sports of the Czech Republic No. MSM-0021627501 and project of the Ministry of Interior of the Czech Republic No. VG 20102014032.

Contents

1 **Introduction to Initiating Substances** 1
   1.1 Primary Explosives 4
   1.2 Priming Compositions 5
   1.3 Environmental Hazards: Emergence of Green Initiating Substances 6
   References 10

2 **Explosive Properties of Primary Explosives** 11
   2.1 Influence of Density on Detonation Parameters 11
   2.2 Initiating Efficiency 13
       2.2.1 Influence of Density and Compacting Pressure 13
       2.2.2 Influence of Specific Surface 19
       2.2.3 Influence of the Charge Diameter 20
       2.2.4 Influence of Confinement 20
       2.2.5 Influence of Secondary Charge Type 22
       2.2.6 Mixtures 23
   2.3 Sensitivity 23
       2.3.1 Impact Sensitivity 25
       2.3.2 Friction Sensitivity 29
       2.3.3 Sensitivity to Electrostatic Discharge 31
       2.3.4 Sensitivity to Flame 32
   References 33

3 **Fulminates** 37
   3.1 Introduction 37
       3.1.1 Fulminic Acid 37
       3.1.2 Mercury Fulminate 39
       3.1.3 Silver Fulminate 58
       3.1.4 Other Fulminates 62
   References 66

4 Azides

- 4.1 Azoimide ... 71
- 4.2 Lead Azide ... 72
  - 4.2.1 Physical and Chemical Properties ... 72
  - 4.2.2 Chemical Reactivity ... 74
  - 4.2.3 Sensitivity ... 77
  - 4.2.4 Explosive Properties ... 78
  - 4.2.5 Preparation ... 80
  - 4.2.6 Spontaneous Explosions During Crystal Growth ... 85
  - 4.2.7 Uses ... 86
- 4.3 Other Substances Derived from Lead Azide ... 87
  - 4.3.1 Basic Lead Azide ... 87
  - 4.3.2 Lead (IV) Azide ... 88
- 4.4 Silver Azide ... 89
  - 4.4.1 Physical and Chemical Properties ... 89
  - 4.4.2 Sensitivity ... 91
  - 4.4.3 Explosive Properties ... 92
  - 4.4.4 Preparation ... 93
  - 4.4.5 Uses ... 96
- 4.5 Copper Azides ... 96
  - 4.5.1 Physical and Chemical Properties ... 97
  - 4.5.2 Explosive Properties ... 98
  - 4.5.3 Preparation ... 100
  - 4.5.4 Undesired Formation of Copper Azides ... 102
  - 4.5.5 Uses ... 104
- 4.6 Other Metallic Azides ... 105
  - 4.6.1 Physical and Chemical Properties ... 105
  - 4.6.2 Explosive Properties ... 106
  - 4.6.3 Preparation ... 108
  - 4.6.4 Uses ... 110
- 4.7 Organic Azides ... 110
- 4.8 Cyanuric Triazide ... 111
  - 4.8.1 Physical and Chemical Properties ... 112
  - 4.8.2 Explosive Properties ... 112
  - 4.8.3 Preparation ... 115
  - 4.8.4 Uses ... 115
- 4.9 4,4′,6,6′-Tetra(azido)hydrazo-1,3,5-triazine and 4,4′,6,6′-Tetra(azido)azo-1,3,5-triazine ... 116
  - 4.9.1 Physical and Chemical Properties ... 116
  - 4.9.2 Explosive Properties ... 116
  - 4.9.3 Preparation ... 117
  - 4.9.4 Uses ... 118
- 4.10 1,3,5-Triazido-2,4,6-trinitrobenzene ... 118
  - 4.10.1 Physical and Chemical Properties ... 118
  - 4.10.2 Explosive Properties ... 119

		4.10.3	Preparation	120
		4.10.4	Uses	121
	4.11	2,3,5,6-Tetraazido-1,4-benzoquinone		121
		4.11.1	Physical Properties	121
		4.11.2	Explosive Properties	122
		4.11.3	Preparation	123
		4.11.4	Uses	123
	References			123
5	**Salts of Polynitrophenols**			131
	5.1	Salts of Picric Acid		131
		5.1.1	Normal Lead Picrate	132
		5.1.2	Basic Lead Picrate	133
	5.2	Salts of Dinitroresorcinol		133
		5.2.1	Lead salts of 2,4-Dinitroresorcinol	133
		5.2.2	Lead Salts of 4,6-Dinitroresorcinol	135
	5.3	Salts of Trinitroresorcine		138
		5.3.1	Lead Styphnate	138
		5.3.2	Basic Lead Styphnate	145
		5.3.3	Double Salts of Lead Styphnate	148
		5.3.4	Barium Styphnate	149
		5.3.5	Other Salts of Styphnic Acid	152
	References			152
6	**Diazodinitrophenol**			157
	6.1	Introduction		157
	6.2	Structure		157
	6.3	Physical and Chemical Properties		159
	6.4	Explosive Properties		160
	6.5	Preparation		162
	6.6	Use		164
	References			165
7	**Salts of Benzofuroxan**			167
	7.1	Introduction		167
	7.2	Salts of 4,6-Dinitrobenzofuroxan		168
		7.2.1	Physical and Chemical Properties	169
		7.2.2	Explosive Properties	169
		7.2.3	Preparation of 4,6-Dinitrobenzofuroxan	172
		7.2.4	Preparation of 4,6-Dinitrobenzofuroxan Salts	173
		7.2.5	Uses	174
	7.3	Potassium Salt of 7-Hydroxylamino-4,6-dinitro-4,7-dihydrobenzofuroxan		175
	7.4	Potassium Salt of 7-Hydroxy-4,6-dinitrobenzofuroxan		176
		7.4.1	Physical and Chemical Properties	176
		7.4.2	Explosive Properties	176

		7.4.3 Preparation	177
		7.4.4 Uses	179
	7.5	Salts of Bis(furoxano)-2-nitrophenol	179
		7.5.1 Physical and Chemical Properties	180
		7.5.2 Explosive Properties	180
		7.5.3 Preparation	181
		7.5.4 Uses	182
	References		183
8	**Tetrazoles**		187
	8.1	Tetrazene	189
		8.1.1 Physical and Chemical Properties	190
		8.1.2 Explosive Properties	191
		8.1.3 Preparation	192
		8.1.4 Uses	193
	8.2	5-Aminotetrazole Salts	194
		8.2.1 Physical and Chemical Properties	194
		8.2.2 Explosive Properties	195
		8.2.3 Preparation	195
		8.2.4 Uses	197
	8.3	5-Nitrotetrazole Salts	197
		8.3.1 Physical and Chemical Properties	197
		8.3.2 Explosive Properties	199
		8.3.3 Preparation	203
		8.3.4 Uses	206
	8.4	5-Chlorotetrazole Salts	207
		8.4.1 Physical and Chemical Properties	207
		8.4.2 Explosive Properties	207
		8.4.3 Preparation	208
		8.4.4 Uses	209
	8.5	5-Azidotetrazole Salts	209
		8.5.1 Physical and Chemical Properties	209
		8.5.2 Explosive Properties	209
		8.5.3 Preparation	210
		8.5.4 Uses	211
	8.6	5,5′-Azotetrazole Salts	212
		8.6.1 Physical and Chemical Properties	213
		8.6.2 Explosive Properties	214
		8.6.3 Preparation	216
		8.6.4 Uses	217
	8.7	Tetrazoles with Organic Substituent	217
		8.7.1 5-Picrylaminotetrazole	218
		8.7.2 1-(1*H*-Tetrazol-5-yl)guanidinium Nitrate	219

8.8	Organic Derivatives of 5-Nitrotetrazole	221
8.9	Organic Derivatives of 5-Azidotetrazole	222
	8.9.1 Explosive Properties	222
	8.9.2 Preparation	222
References		223

9 Tetrazole Ring-Containing Complexes ... 227

9.1	Cobalt Perchlorate Complexes	228
	9.1.1 Pentaamine(5-cyano-2H-tetrazolato-N^2)cobalt(III) perchlorate (CP)	228
	9.1.2 CP Analogs	235
	9.1.3 1,5-Cyclopentamethylenetetrazole Complexes	238
	9.1.4 Tetraammine-cis-bis(5-Nitro-2H-tetrazolato-N^2) cobalt(III) perchlorate (BNCP)	241
	9.1.5 BNCP Analogs	244
	9.1.6 Perchlorate Complexes of 1,5-Diaminotetrazole	245
	9.1.7 Other Perchlorate-Based Complexes	247
9.2	Perchlorate-Free Complexes	247
	9.2.1 Iron- and Copper-Based 5-Nitrotetrazolato-N^2 Complexes	247
	9.2.2 Perchlorate-Free CP Analogs	250
9.3	Other Transition Metal-Based 5-Nitrotetrazolato-N^2 Complexes	251
References		252

10 Organic Peroxides ... 255

10.1	Peroxides of Acetone	255
	10.1.1 Diacetone Diperoxide	256
	10.1.2 Triacetone Triperoxide	262
	10.1.3 Tetraacetone Tetraperoxide	274
10.2	Hexamethylene Triperoxide Diamine	275
	10.2.1 Physical and Chemical Properties	275
	10.2.2 Explosive Properties	278
	10.2.3 Preparation	279
	10.2.4 Uses	280
10.3	Tetramethylene Diperoxide Dicarbamide	280
	10.3.1 Physical and Chemical Properties	281
	10.3.2 Preparation	281
	10.3.3 Use	281
References		282

11 Nitrogen Halides ... 289
11.1 Nitrogen Trichloride ... 289
11.1.1 Physical and Chemical Properties ... 289
11.1.2 Explosive Properties ... 290
11.1.3 Preparation ... 291
11.1.4 Use ... 293
11.2 Nitrogen Tribromide ... 293
11.2.1 Physical and Chemical Properties ... 294
11.2.2 Preparation ... 294
11.3 Nitrogen Triiodide ... 295
11.3.1 Structure ... 296
11.3.2 Physical and Chemical Properties ... 297
11.3.3 Explosive Properties ... 298
11.3.4 Preparation ... 299
11.3.5 Use ... 300
References ... 300

12 Acetylides ... 303
12.1 Silver Acetylides ... 303
12.1.1 Silver Acetylide ... 304
12.1.2 Silver Acetylide–Silver Nitrate ... 308
12.1.3 Silver Acetylide–Silver Hexanitrate ... 312
12.1.4 Other Salts of Silver Acetylide–Silver Nitrate ... 312
12.2 Cuprous Acetylide ... 313
12.2.1 Physical and Chemical Properties ... 313
12.2.2 Explosive Properties ... 314
12.2.3 Preparation ... 315
12.2.4 Uses ... 316
12.3 Cupric Acetylide ... 316
12.3.1 Physical and Chemical Properties ... 316
12.3.2 Explosive Properties ... 317
12.3.3 Preparation ... 317
12.3.4 Uses ... 318
12.4 Mercuric Acetylide ... 318
12.4.1 Physical and Chemical Properties ... 318
12.4.2 Explosive Properties ... 319
12.4.3 Preparation ... 319
12.4.4 Uses ... 320
12.5 Mercurous Acetylide ... 320

12.6	Aurous Acetylide	321
	12.6.1 Physical and Chemical Properties	321
	12.6.2 Explosive Properties	321
	12.6.3 Preparation	321
	References	322
13	**Other Substances**	**325**
13.1	Salts of Nitramines	325
13.2	Organophosphates	328
13.3	Hydrazine Complexes	331
	References	333
Index		**335**

List of Abbreviations

2,4-DNR	2,4-Dinitroresorcinol; 2,4-dinitrobenzene-1,3-diol
2,4-LDNR	Lead salt of 2,4-dinitroresorcinol; lead salt of 2,4-dinitrobenzene-1,3-diol
4,6-DNR	4,6-Dinitroresorcinol; 4,6-dinitrobenzene-1,3-diol
4,6-LDNR	Lead salt of 4,6-dinitroresorcinol; lead salt of 4,6-dinitrobenzene-1,3-diol
5-ATZ	5-Amino-1H-tetrazole
5-PiATZ	5-(2,4,6-Trinitrophenylamino)-1H-tetrazole
Å	Ångström (10^{-10} m)
APCP	Pentaammin-aqua cobalt(III) perchlorate
ATZ	5-Amino-1H-tetrazole
AzTZ	5,5′-Azo-1H-tetrazole
b.p.	Boiling point
BAM	Bundesanstalt für Materialforschung und prüfung
BaS	Barium styphnate; barium salt of 2,4,6-trinitrobenzene-1,3-diol
BFNP	Bis(furoxano)-2-nitrophenol
BNCP	Tetraammine-cis-bis(5-nitro-2H-tetrazolato-N^2)cobalt(III) perchlorate
CL-14	5,7-Diamino-4,6-dinitrobenzofuroxan
CL-20	2,4,6,8,10,12-Hexanitro-2,4,6,8,10,12-hexaazaisowurtzitan (HNIW)
CMC	Carboxymethyl cellulose
CoHN	Cobalt hydrazine nitrate
CP	Pentaamine(5-cyano-2H-tetrazolato-N^2)cobalt(III) perchlorate
CPCN	Pentaammine carbonato cobalt(III) nitrate
CTCN	Tetraamine-carbonato cobalt(III) nitrate
D	Detonation velocity
DADP	Diacetone diperoxide; 3,3,6,6-tetramethyl-1,2,4,5-tetroxane
DAHA	4,4,6,6,8,8-Hexaazido-1,3,5,7,2λ^5,4λ^5,6λ^5,8λ^5-tetrazatetraphosphocine-2,2-diamine
DANT	4,6-Diazido-N-nitro-1,3,5-triazine-2-amine
DDNP	2-Diazo-4,6-dinitrophenol

DDT	Deflagration-to-detonation transition
Dinol	2-Diazo-4,6-dinitrophenol
DNBF	4,6-Dinitrobenzofuroxan
DNR	Dinitroresorcinol; 1,3-dihydroxybenzene
DSC	Differential scanning calorimetry
DTA	Differential thermal analysis
DTG	Differential thermogravimetry
E	Energy
E_{50}	Energy causing initiation in 50 % of the trials
EDNA	N,N-dinitromethylenediamide
E_{min}	Minimal energy causing initiation
en	1,2-Ethylenediamine; ethane-1,2-diamine
ENTA	7,7,9,9-Tetraazido-1,4-dinitro-1,4,6,8,10-pentaaza-$5\lambda^5,7\lambda^5,9\lambda^5$-triphospha-spiro[4.5]deca-5,7,9-triene
ERDE	Explosives Research and Development Establishment
ESD	Electrostatic discharge
Et_2O	Diethylether
F of I	Figure of insensitivity (explosives having a 50 % initiation drop height equal to that of RDX are given a F of I of 80)
FOX-7	2,2-Dinitroethene-1,1-diamine
GNGT	Tetrazene
h_{10}	Impact height causing initiation in 10 % of the trials with particular hammer mass
h_{50}	Impact hight causing initiation in 50 % of the trials with particular hammer mass
h_{min}	Minimal impact hight causing initiation with particular hammer mass
HMTD	Hexamethylene triperoxide diamine; 1,6-diaza-3,4,8,9,12,13-hexaoxabicyclo[4,4,4]tetradecane
HMX	1,3,5,7-Tetranitro-1,3,5,7-tetrazocane (octogen)
HNT	5-Nitro-1H-tetrazole
IR	Infrared spectroscopy
IUPAC	International Union of Pure and Applied Chemistry
KBFNP	Potassium salt of bis(furoxano)-2-nitrophenol
KDNBF	Potassium salt of 2,4-dinitrobenzofuroxan
KDNP	7-Hydroxy-4,6-dinitrobenzofuroxan
LA	Lead azide
LD_{50}	Median lethal dose
LDDS	Double lead salts of styphnic acid with 1,3-di(5-tetrazoyl)triazene
LS	Lead styphnate; lead salt of 2,4,6-trinitrobenzene-1,3-diol
m-	Meta
m.p.	Melting point
Me	Metal
MF	Mercury fulminate
MO	Molecular orbital
n_c/n_a	Molar ratio of catalyst to acetone

List of Abbreviations

NC	Nitrocellulose
NG	Nitroglycerine
NHN	Nickel hydrazine nitrate
NKT	Pentaamine (5-nitro-2H-tetrazolato-N^2)cobalt(III) perchlorate
NMR	Nuclear magnetic resonance
NT	5-Nitro-1H-tetrazole
o-	Ortho
P	Pressure
p-	Para
PAC	Pentaamine(1,5-cyclopentamethylene-tetrazolato-N^3) cobalt(III) perchlorate
PbNATNR	Double lead salt of styphnic acid and 5-nitraminotetrazole
PBX	Plastic bonded explosives
PDT	Plate dent test
PETN	Pentaerytritol tetranitrate
PiC	Picrylchloride; 1-chloro-2,4,6-trinitrobenzene
pK_a	The negative logarithm of the acid dissociation constant
prob.	Probability
PVA	Polyvinylalcohol
PVP-1	Detonation transition element in oil well perforators
RC	Reinforcing cap
RDX	1,3,5-Trinitro-1,3,5-triazinane (hexogen)
SA	Silver azide
SEM	Scanning electron microscopy
SF	Silver fulminate
SII	Simple initiating impulse
STANAG	NATO abbreviation for Standardization Agreement
T	Temperature
TAAT	4,4′,6,6′-Tetra(azido)azo-1,3,5-triazine
TAHT	4,4′,6,6′-Tetra(azido)hydrazo-1,3,5-triazine
TATNB	1,3,5-Triazido-2,4,6-trinitrobenzene
TATP	Triacetone triperoxide; 3,3,6,6,9,9-hexamethyl-1,2,4,5,7,8-hexaoxonane
T_{dec}	Decomposition temperature
TeATeP	3,3,6,6,9,9,12,12-Octamethyl-1,2,4,5,7,8,10,11-octaoxacyclododecane
TeAzQ	2,3,5,6-Tetraazido-1,4-benzoquinone
Tetryl	N-methyl-N,2,4,6-tetranitroaniline
TMD	Theoretical maximum density
TMDD	Tetramethylene diperoxide dicarbamide; 1,2,8,9-tetraoxa-4,6,11,13-tetraazacyclotetradecane-5,12-dione
TNR	Trinitroresorcinol; 2,4,6-trinitrobenzene-1,3-diol
TNT	2,4,6-Trinitrotoluene
UV	Ultraviolet
ΔH_f	Heat of formation
ρ	Density

Chapter 1
Introduction to Initiating Substances

Initiating substances (primaries) are chemical compounds/mixtures used in igniters or detonators to bring about burning or detonation of energetic material. The vocabulary regarding initiators is rather vast, often divided by the type of input and output of the element or its application. In general, igniters (containing a priming composition) are used to provide flame, while detonators (containing the primary explosive) are designed to create a shock wave which starts the detonation reaction in a secondary explosive. Combinations of both are seen, for example, in detonators in which the igniter is first initiated and the resulting flame starts the detonator [1].

The successful initiation of explosives begins with a small external stimulus called the simple initiating impulse (SII). The SII is a nonexplosive type of impulse such as flame, heat, impact, stab, friction, electric spark, etc. The choice depends on the type of material to be initiated and on the desired output effect. The energy provided by this impulse is sufficient enough to initiate the burning of a priming mixture—a sensitive mixture sometimes called prime (in pyrotechnics the term prime is used for the first layer enabling successful ignition). The energy from the combustion of the priming composition needs to be magnified or delayed by a sequence of elements, such as delay, transfer, and output charges. The entire sequence is called an explosive train and is simplified in Fig. 1.1.

The term primer in Fig. 1.1 represents the first element in the explosive train with flame as a desired output (e.g., percussion primer, pyrotechnic prime, electric fusehead, safety fuse, etc.). It needs to be stressed here that consistency in terminology does not exist regarding the term primer which is often used for primary explosives and boosters, but also for a charge consisting of detonating cord, detonator, and transfer charge made of secondary explosive, for detonator fused booster charges or more commonly for percussion-sensitive pyrotechnic mixtures loaded in small metal caps used in ammunition cartridges. Regarding Fig. 1.1, the primer is, as stated, the first element in the explosive train sensitive to the SII. All such elements contain a priming composition, which is typically a pyrotechnic mixture designed specifically for each particular application from a variety of substances (the topic is briefly described in later parts of this chapter). The transfer

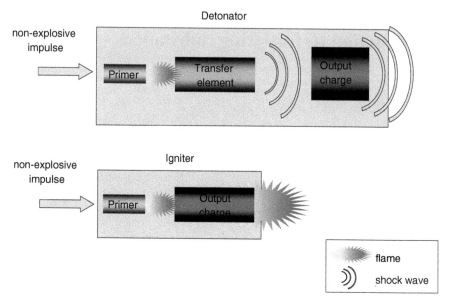

Fig. 1.1 Schematic representation of the two most common types of explosive trains

element in Fig. 1.1 represents any element used in the detonation train to augment or delay the initial impulse. The materials for such an element may range from a primary explosive (column of lead azide or 2-diazo-4,6-dinitrophenol—dinol) or pyrotechnic mixtures (e.g., FeSiCr/Pb_3O_4) to black powder. The output charges are secondary explosives, pyrotechnic mixtures, or propellants depending on the type of application.

In the following text, we have decided to adopt an approach in which we divide the vast group of sensitive condensed energetic materials according to various criteria (Fig. 1.2). The first criterion we used is their usefulness in real applications. We further divide those that have found a use depending upon the type of the desired function into two groups: (a) detonating substances (primary explosives) and deflagrating substances or mixtures (priming mixtures, primes). The term primary explosive is generally used for individual substances (not mixtures) which are further divided based on their response to flame into those which detonate practically immediately after ignition and those that require some distance/time for developing stable detonation. The former are sometimes referred to as detonants in Anglo American literature, whereas in Czech literature these are called primary explosives belonging to the group of lead azide, while the latter belong to the group of mercury fulminate. In Russian literature, the latter are denominated as pseudo primary explosives because they must undergo so-called deflagration to detonation transition, which is a process of accelerating the burning reaction zone to a point where the propagation mechanism changes from being heat transfer driven to being shock wave driven.

1 Introduction to Initiating Substances

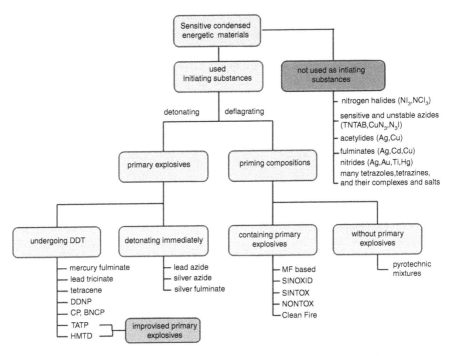

Fig. 1.2 Subdivisions of sensitive energetic materials with some common examples

It is difficult, if not impossible, to draw a clear line between substances used for igniters and those used for detonators. These two groups overlap in many compositions. Mercury fulminate, for example, is a well-known primary explosive used in the past, capable of detonation, but which found a use in priming compositions much earlier than it did in detonators. It should therefore be understood that any type of partitioning of initiating substances is rather artificial and cannot be considered rigorous or definitive. Whether a substance finds a use as a primary explosive in a detonator, as part of a priming mixture or does not find any use at all depends on a large number of criteria which we have tried to introduce in Chap. 2. Many substances are disqualified from practical application due to their inability to meet these stringent criteria. Some examples include substances that are too sensitive for practical application (e.g. fulminating silver, silver acetylide, nitrogen triiodide), substances with low initiating efficiency (e.g. nitrogen triiodide, tetrazene), chemically unstable substances (e.g. 1,6-diaza-3,4,8,9,12,13-hexaoxabicyclo[4,4,4]tetradecane —HMTD), etc. Further, there are substances that are sensitive but are not expected to explode in their typical application (diazonium intermediate products in the dye industry, dibenzoyl and other peroxides used as catalysts of radical reactions, etc.), substances used in the entertainment world for creating sound or light effects, etc.

Further we have included some of those that we believe may attract attention in the future although they are not used at the moment (tetrazole salts and complexes).

The main focus of this monograph is on the individual primary explosives that have been used or reported as promising candidates for potential use. The only exception we included is nitrogen triiodide, which we believe will never find any use. The priming compositions are covered only briefly in this chapter to give more space to discussing the use of primary explosives in a broader perspective.

1.1 Primary Explosives

Primary explosives are sensitive mixtures/compounds that can be easily detonated by an SII—nonexplosive means such as flame, heat, impact, friction, electric spark, etc. They are mainly used in applications where it is desired to produce shock for initiation of a less sensitive explosive (secondary explosive). The ability to initiate detonation of secondary explosives is the characteristic that makes primary explosives useful in detonators and blasting cups or caps (detonators used for mining). In order for this to work, primary explosives must detonate shortly after the nonexplosive initiation. The fast transition from SII to detonation (sometimes referred to as acceleration) is the most important distinction between primary and secondary explosives.

Another important use of primary explosives lies in their application as sensitizers in priming compositions. The purpose of such mixtures is to provide reliable ignition and the role of the primary explosives in such mixtures is to make them sensitive enough to be easily ignited. The detonation of such a mixture is undesirable and must be prevented by careful selection of the ingredients and their quantities for a particular application.

The first substance ever used as a primary explosive was mercury fulminate (MF). It was patented for use in priming mixtures in 1807 and it took another 60 years before Nobel realized that it was possible to use MF in "Fulminate Blasting Caps"—practically the first detonators—for initiation of detonation of nitroglycerine. The price of mercury and the toxicity of decomposition products led to the search for suitable replacements. Even though lead azide (LA) was tested before 1900 as a replacement for MF in detonators, production of MF continued to grow and peaked during World War I. The toxicity of MF, which was even at that time considered a problem in priming mixtures, was not seen as a big issue in the case of detonators. The real replacement of MF in detonators was a lengthy process closely related to improvements in manufacturing of a suitable form of LA.

Apart from the already mentioned lead azide and mercury fulminate, the classic examples of primary explosives include lead styphnate (LS), silver azide (SA), dinol (DDNP), and tetrazene (GNGT). Primary explosives (individual components) are often mixed and used in the form of compositions rather than as single component energetic materials. Mixtures may consist of either individual primary explosives (astryl-MF/SA [2]) or primary explosives plus some nonexplosive additive (LA/LS/dextrine or ASA composition—LA/LS/Al).

Table 1.1 Example of typical historical priming composition [1]

Substance	Amount (wt%)
Mercury fulminate	13.7
Potassium chlorate	41.5
Antimony sulfide	33.4
Powdered glass	10.7
Gelatin glue	0.7

Table 1.2 Example of typical historical priming compositions without potassium chlorates [3]

Substance	Amount (wt%)		
MF	40	50	36
Barium nitrate	40	–	–
Lead chromate	–	24	–
Lead dioxide	–	20	–
Mercury chromate	–	–	40
Antimony sulfide	14	–	20
Glass powder	6	6	4

1.2 Priming Compositions

Priming compositions or priming mixtures are sensitive explosive mixtures that are designed to produce a flame in a particular application. They are most often mixtures containing a primary explosive as one of the components in such a form and amount that ensures the inability to initiate detonation of the mixture. The role of the primary explosive is to sensitize the mixture to external stimuli, not to make it detonatable. Priming mixtures are used in percussion caps for ignition of gun powder, on electric fuseheads, delay elements, etc.

The typical examples of priming mixtures used almost exclusively in the nineteenth century include mercury fulminate, potassium chlorate, antimony sulfide, glass powder, and gum Arabic. The priming mixtures widely used for small arms in the USA in the early 1900s were based on MF. The composition of the most typical one is summarized in Table 1.1.

At the time of black powder, weapons had to be carefully cleaned after use. The combustion products of priming mixtures therefore did not present such a crucial problem since they were in any case removed together with the black powder residues during cleaning. The problem of corrosive products became apparent when black powder was replaced by smokeless powder which meant much less careful cleaning of weapons. Potassium chlorates as a part of the priming composition have therefore been abandoned and replaced by other oxidizers. Some of the compositions from that time can be seen in Table 1.2 [3].

The use of alternative oxidizers led to mixtures that were not corrosive but contained MF, which liberated toxic mercury vapors during firing. To overcome the problem of toxicity, Edmund von Herz proposed replacing MF in priming compositions with lead hypophosphite nitrate (1913) and later by the lead salt of trinitroresorcine (1914). The more important of these two MF substitutes was the

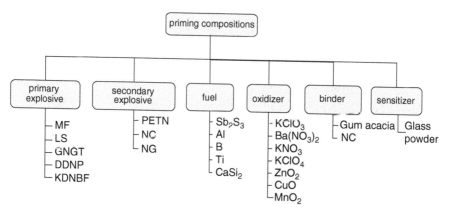

Fig. 1.3 Typical examples of components found in priming compositions

latter—the lead salt of trinitroresorcine; so-called lead styphnate, lead tricinate, or simply just tricinate. This substance, however, does not possess sufficient percussion sensitivity and therefore did not find wider use until the introduction of tetrazene by Rathsburg in 1921 [3]. Some patented compositions contained both MF and tricinate [4].

Typical primary mixtures contain some of the following: lead styphnate, tetrazene, aluminum, antimony sulfide, calcium silicate, lead peroxide, boron, metals, barium nitrate, secondary explosive, binder, sensitizer, etc. (Fig. 1.3). Variations in the ingredients and their relative amounts result in compositions which possess sensitivity and ignition properties tailored to specific requirements.

The replacement of mercury was a significant improvement with respect to the toxicity of the combustion products. It led, however, to the introduction of tricinate which also presents health problems due to its lead content. Concerns over environmental hazards resulted in the search for lead-free or, more generally, heavy metal-free compositions which would eliminate the health threat while maintaining performance and other important properties at least to the level of lead-containing predecessors [5].

1.3 Environmental Hazards: Emergence of Green Initiating Substances

The problem of toxicity, as it is seen today, is caused mainly by lead which is present in primary explosives in form of lead azide and in priming mixtures mostly as lead styphnate. The evolution in the initiating materials is therefore heading toward lead-free, environmentally friendly, "green" initiating substances or more specifically "green primary explosives" and "green priming compositions."

1.3 Environmental Hazards: Emergence of Green Initiating Substances

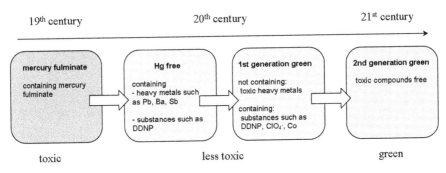

Fig. 1.4 Evolution of primary explosives and priming mixtures from the perspective of toxicity

The term "green" is "science marketing" more than anything else and is quite confusing as it describes compositions not containing lead, as well as compositions without heavy metals such as Hg, Pb, Ba, and Sb or more recently also compositions that do not contain any toxicologically potentially hazardous substances such as DDNP (often found in green compositions today). Further, the term "green initiating substances" or sometimes "green primaries" usually covers substances used as primary explosives as well as substances used in priming mixtures. This may be even more confusing and we will therefore try to distinguish only between "green primary explosives" and "green priming mixtures" with the first ones being the replacement for LA in blasting caps and detonators while the second is a replacement for LS in priming compositions.

Historically the oldest toxic initiating substance is MF. It has to all intents and purposes been replaced in both igniting and detonating initiating devices by lead alternatives. Today these lead-based alternatives coexist with the more modern lead-less compounds or compositions containing other, less toxic, heavy metals. The efforts in finding suitable priming compositions were more successful and led to mixtures not containing lead such as SINTOX, NONTOX, WINCLEAN [5], CCI Clean Fire [6], etc. [7–9]. Future research is therefore aimed toward completely nontoxic alternatives. Results from the search for lead-free primary explosives were not particularly good, and a suitable replacement of lead azide practically does not exist today. The emergence in recent years of a new class of transition metal complexes is a direct response to the toxicity of heavy metals and environmental concerns. From recently published works, one of the more promising possibilities is found in complexes with high nitrogen content such as tetrazoles with nontoxic cations such as sodium, ammonium, and iron [10]. The evolution of the initiating substances and compositions with respect to toxicity is schematically summarized in Fig. 1.4.

In Fig. 1.4 "toxic" refers to the classical compositions with toxic heavy metals, "less toxic" substances relate to those not containing toxic heavy metals but still containing some problematic ingredients, and "green" to nontoxic substances.

The search for suitable replacements of LA and LS should not be seen as a search for one universal substance. It would obviously be nice to have one substance

Table 1.3 Target properties of green explosives and priming mixtures

Criteria	Green primary explosive (blasting caps, detonators)	Green priming mixture (electric pills, percussion caps)
Chemically stable under light, moisture, and in presence of carbon dioxide	X	X
Sufficient thermal stability	X	X
Low toxicity (no heavy metals, DDNP, ClO_4^-)	X	X
Sensitivity within useful limits	X	X
High initiation efficiency	X	
Resistant to dead-pressing good compatibility with construction materials	X	
Fast DDT	X	Unwanted
High flame temperature		X
Low barrel erosion/corrosion		X
Low antilubricating effect		X
Low cost	X	X

replacing both lead salts in both applications but with respect to their different function it should not be the ultimate goal due to different requirements for suitable replacements. High initiation efficiency, very important in primary explosives, is not at all important in primers and suitable properties with respect to barrel corrosion, very important in primers, are practically meaningless in detonators.

Six criteria for "green primaries" have been proposed including [10]:

1. Lack of sensitivity to light and moisture.
2. Sensitivity to initiation in suitable limits (sensitive enough to be initiated by nonexplosive stimulus but not too sensitive to make handling and transport unnecessarily risky).
3. Thermally stable to at least 200 °C.
4. Chemically stable for an extended period.
5. Devoid of toxic metals such as lead, mercury, silver, barium, or antimony.
6. Free of perchlorate which may be a possible teratogen and have an adverse effect on thyroid function.

These criteria are in our opinion relatively broad and cover both primary explosives as well as priming mixtures but do not include some important properties specific to each group. We are therefore proposing criteria summarized in Table 1.3.

It should be also pointed out that the toxicity of combustion products of classical priming mixtures today is mainly concerned with air born lead. It must therefore be stressed that using a lead-free priming mixture would have a somewhat questionable effect if a lead bullet without at least an enclosed bottom were to be used. More

1.3 Environmental Hazards: Emergence of Green Initiating Substances

Table 1.4 Some compositions of priming compositions without heavy metals

Substance	Composition (%) Reported by Kusak			FA956	SINTOX	NONTOX
DDNP	35.0	29.8	40.0	37.0 ± 5	Unreported amount	–
Tetrazene	8.0	19.8	20.0	4.0 ± 1	Unreported amount	Unreported amount
PETN	–	–	–	5.0 ± 1	–	Unreported amount
CuO	–	29.8	–	–	–	–
MnO$_2$	–	–	10.0	–	–	–
Ba(NO$_3$)$_2$	38.0	–	–	32.0 ± 5	–	–
KNO$_3$	–	–	–	–	–	Unreported amount
ZnO$_2$	–	–	–	–	Unreported amount	–
Sb$_2$S$_3$	12.0	–	–	15.0 ± 2	–	–
B	–	–	–	–	–	Unreported amount
Al	7.0	9.8	–	7.0 ± 1	–	–
Ti	–	–	–	–	Unreported amount	–
Glass	–	9.8	28.0	–	–	Unreported amount
Binder	–	1.0	2.0	0.2	Unreported amount	Unreported amount
References	[11]	[11]	[11]	[1]	[3]	[8]

suitable in this case are either fully encapsulated bullets or bullets made from metals other than lead.

Various new energetic materials have been designed and tested as potential candidates for replacement of the lead substances. The green priming mixtures for percussion caps of small arms ammunition for indoor shooting usually use DDNP or tetrazene (or both) as a replacement for tricinate. DDNP, although better than lead salts, is reported to cause some allergic reactions [10]. Example compositions without heavy metals are summarized in Table 1.4.

A suitable replacement for lead azide in detonators still does not exist even though new candidates, mostly based on tetrazole complexes, have been proposed and to a certain limited extent even employed. There are also some alternative approaches which eliminate toxic metals from the detonator by eliminating the primary explosive but they have not yet succeeded in replacing lead azide detonators.

Environmental friendliness is of course only one important aspect to be considered. Explosives that show extraordinary properties such as stability, sensitivity, initiation efficiency, compatibility, or other properties can find their application in some special cases even though they will not meet the green criteria. Typical examples of such explosives may be the mercury salt of tetrazole or cirkon (cadmium(II)tris-carbonohydrazide) perchlorate used in Russia [12, 13].

An enormous amount of work has been done in Russia on perchlorate coordination complexes. A very interesting initiative to find "green" primary explosives among the cobalt perchlorate complexes has been reported by Ilyushin et al. [14]. The differences in perception of toxicity in various parts of the world make direct comparison misleading. Perchlorates, which are not seen as a problem in many countries, are considered unsuitable from the long-term perspective in the USA [10].

References

1. Fedoroff, B.T., Sheffield, O.E., Kaye, S.M.: Encyclopedia of Explosives and Related Items. Picatinny Arsenal, Dover, NJ (1960–1983)
2. Krauz, C.: Technologie výbušin. Vědecko-technické nakladatelství, Praha (1950)
3. Hagel, R., Redecker, K.: Sintox – a new, non-toxic primer composition by Dynamit Nobel AG. Propell. Explos. Pyrotech **11**, 184–187 (1986)
4. Burns, J.E.: Priming mixture. US Patent 1,880,235, 1932
5. Mei, G.C., Pickett, J.W.: Lead free priming mixture for percussion primer. US Patent 5,417,160, 1995
6. Bjerke, R.K., Ward, J.P., Ells, D.O., Kees, K.P.: Primer composition. US Patent 4,963,201, 1990
7. Brede, U., Hagel, R., Redecker, K.H., Weuter, W.: Primer compositions in the course of time: from black powder and SINOXID to SINTOX compositions and SINCO booster. Propell. Explos. Pyrotech **21**, 113–117 (1996)
8. Nesveda, J., Brandejs, S., Jirásek, K.: Non toxic and non-corrosive ignition mixtures. Patent WO 01/21558, 2001
9. Oommen, Z., Pierce, S.M.: Lead-free primer residues: a qualitative characterization of Winchester WinClean, Remington/UMC LeadLess, Federal BallistiClean, and Speer Lawman CleanFire handgun ammunition. J. Forensic Sci. **51**, 509 (2006)
10. Huynh, M.V., Coburn, M.D., Meyer, T.J., Wetzer, M.: Green primaries: environmentally friendly energetic complexes. Proc. Natl. Acad. Sci. USA **103**, 5409–5412 (2006)
11. Kusák, J., Klecka, J., Lehký, L., Svachouček, V., Pechouček, P.: Základy konstrukce munice. Ediční středisko University Pardubice, Pardubice (2003)
12. Danilov, J.N., Ilyushin, M.A., Tselinskii, I.V.: Promyshlennye vzryvchatye veshchestva; chast I Iniciiruyushchie vzryvshchatye veshchestva. Sankt-Peterburgskii gosudarstvennyi tekhnologicheskii institut, Sankt-Peterburg (2001)
13. Andreev, V.V., Neklyudov, A.G., Pozdnyakov, S.A., Fogelzang, A.E., Sinditskii, V.P., Serushkin, V.V., Egorshev, V.Y., Kolesov, B.I.: Kapsyul-detonator (varianty). Patent RU 2,104,466, 1996
14. Ilyushin, M.A., Tselinskii, I.V., Sudarikov, A.M.: Razrabotka komponentov vysokoenergeticheskikh kompozitsii. SPB:LGU im. A. S. Pushkina - SPBGTI(TU), Sankt-Peterburg (2006)

Chapter 2
Explosive Properties of Primary Explosives

The main requirements for primary explosives are sensitivity within useful limits, high initiating efficiency, reasonable fluidity, resistance to dead-pressing, and long-term stability. Useful limits mean that the substance must be sensitive enough to be initiated by an SII but not too sensitive as to be unsafe for handling or transportation. The initiating efficiency, perhaps the most important parameter, determines the ability of a primary explosive to initiate secondary explosives. The reasonable free flowing properties are important for manufacturing where the primary explosives are often loaded volumetrically. Primary explosives must not undergo desensitization when pressed thereby yielding a dead-pressed product. The long-term stability and compatibility with other components, even at elevated temperatures, are essential because primary explosives are often embedded inside more complex ammunition and are not expected to be replaced during their service life. They must also be insensitive to moisture and atmospheric carbon dioxide. Parameters important for secondary explosives such as brisance, strength, detonation velocity, or pressure are of lesser importance to primary explosives although they are of course related to the above properties.

2.1 Influence of Density on Detonation Parameters

Primary explosives are generally prepared in the form of crystalline or powdery material with low bulk densities and large specific surface. This form is hardly ever suitable for direct application and therefore it has to be further processed. For use in detonators they must be compacted by pressing to the detonator cups in a way that assures the best initiation properties.

When higher pressures are used to achieve higher densities, a phenomenon called "dead-pressing" may occur, leading to a material which is hard to ignite and which, if ignited, only burns without detonation [1]. Pressing a primary explosive to a point where it loses its capability to detonate is therefore not desirable.

The phenomenon of dead pressing is not common to all primary explosives. Many azides, including lead azide, cannot be easily dead-pressed. On the other

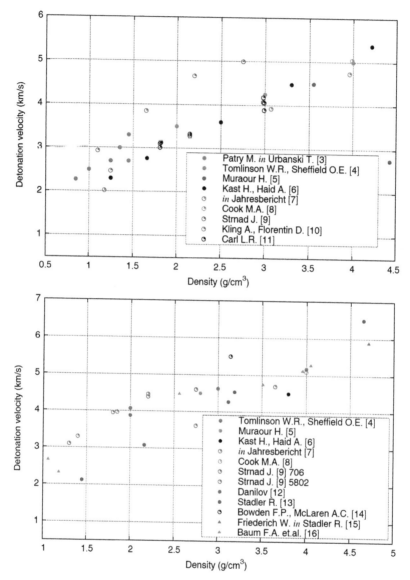

Fig. 2.1 Influence of density on detonation velocity: *top*: mercury fulminate [3–11] and *bottom*: lead azide [4–9, 12–16]. Two samples with specific surface 706 cm^2 g^{-1} and 5,802 cm^2 g^{-1} were listed in [9]

hand MF, DDNP and peroxides can be dead-pressed very easily. The pressure needed for dead-pressing MF is highly dependent on its crystal size [2].

The compaction process reflects in the density, which influences practically all of the other explosive properties. On the following two charts (Fig. 2.1), the relationship between detonation velocity and density for MF and LA is presented.

The values have been compiled from various literature sources and hence obtained under a variety of conditions. Rigorous conclusions based on such data are not possible but they provide a surprisingly good idea of the shape of the relationship.

The values in Fig. 2.1 show that the detonation velocity of both MF and LA increases with increasing density, as expected. In general, one would further expect that it is desirable to press explosives to densities as close to the theoretical maximum density (TMD) as possible. This is, however, not exactly the case for primary explosives in a detonator where it is more important to have good initiation efficiency rather than high detonation velocity.

2.2 Initiating Efficiency

Initiating efficiency, sometimes referred to as initiating power, strength, or priming force, is the ability of a primary explosive to initiate detonation in a secondary explosive adjacent to it. It is usually reported as a minimum amount of primary explosive necessary to cause detonation of the adjacent high explosive with 100% probability.

The initiating efficiency is not a material constant for a particular primary explosive. It depends on many factors including: pressure used for compression, type of ignition, type of confinement, presence of reinforcing cap and its material, type of the secondary explosive, size of the contact surface between primary and secondary explosive, etc. The values of initiation efficiency reported in the literature are therefore difficult to compare due to a variation in these conditions. We have collected initiation efficiencies of some primary explosives with respect to TNT and summarized them in Fig. 2.2.

These values show variations in the minimal amount based on a combination of these factors which are generally not provided in the references. This makes comparison of various results quite a troublesome task. It is important to understand that a single number reported without further specification (as shown in Fig. 2.2) has very low information value. The effects of the most important factors are therefore addressed in the following sections.

2.2.1 Influence of Density and Compacting Pressure

The influence of compacting pressure cannot really be separated from the influence of density as these two parameters are related. Higher compacting pressure leads to material of higher density. This is shown in Fig. 2.3 for MF, LA, and DDNP. The reason why we address both of these factors here is purely practical—the lack of data. Primary charges (for testing of initiating efficiency) are prepared in the form of powder compressed either (a) directly onto an already compressed secondary explosive in a metal cap or (b) into a reinforcing cap which is then pressed onto the secondary charge.

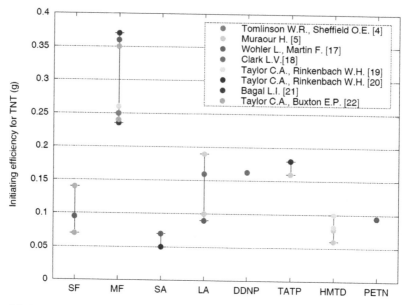

Fig. 2.2 Initiation efficiencies of some primary explosives for TNT (previously unused acronyms: *SF* silver fulminate, *SA* silver azide, *TATP* triacetone triperoxide) [4, 5, 17–22]

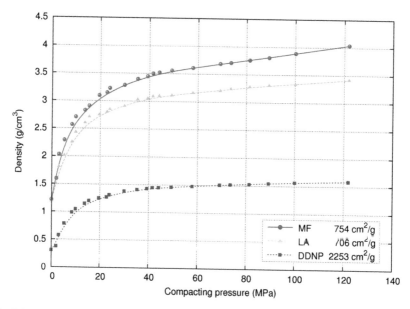

Fig. 2.3 Density of MF, LA, and DDNP as a function of compacting pressure, by kind permission of Dr. Strnad [9]

2.2 Initiating Efficiency

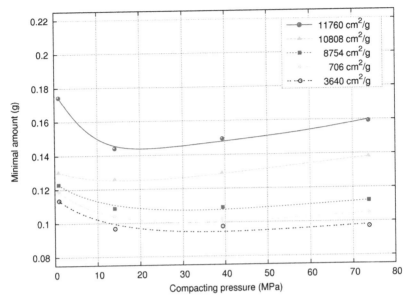

Fig. 2.4 The influence of compacting pressure and specific surface on initiation efficiency of dextrinated LA (acceptor 0.35 g of TNT compressed by 76.5 MPa without reinforcing cap) by kind permission of Dr. Strnad [9]

In both cases it is difficult to determine the exact density before the experiment. The only known (or reported) parameter is therefore compacting pressure.

Very rare results have been obtained by Strnad [9] who used MF, LA, and DDNP of various specific surfaces, compressed them with defined pressures, and experimentally measured the resulting densities (Fig. 2.3). This allowed him to study the influence of the resulting density on the initiation efficiency. More on this issue will be addressed in detail later in this chapter but, to summarize, it can be stated that, in the case of MF, LA, and DDNP, an increase in density first leads to a decrease in the minimal necessary amount of explosive—and what happens subsequently is material specific. Some substances (e.g., DDNP, MF) start to lose their ability to initiate secondary explosives, which is reflected in higher amounts needed for successful initiation, while other substances work in the same way, no matter how hard they are pressed (pure LA).

One important and, unfortunately, not so frequently considered parameter is specific surface. Fine powder of the same primary explosive will behave differently from coarse crystals. Figures 2.4, 2.5, 2.6, 2.7, 2.8, and 2.9 demonstrate that very fine powders show worse initiating efficiency compared to their coarser form. Nevertheless, it can be seen that gently compressing any primary explosive of any specific surface leads to an increase of initiating efficiency.

The influence of compacting pressure on the mean minimal amount is shown in Fig. 2.4 for LA, Fig. 2.5 for DDNP, and Fig. 2.6 for MF. All three exhibit a decrease in the minimal amount by slight compression (9.8 MPa). Further increase of pressure

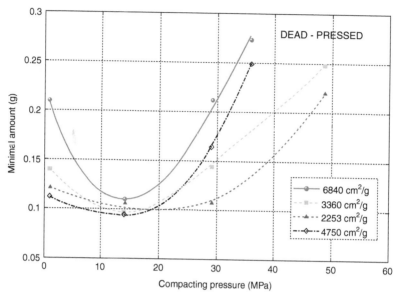

Fig. 2.5 The influence of compacting pressure and specific surface on initiation efficiency of DDNP (acceptor 0.35 g of TNT compressed by 76.5 MPa with reinforcing cap), by kind permission of Dr. Strnad [9]

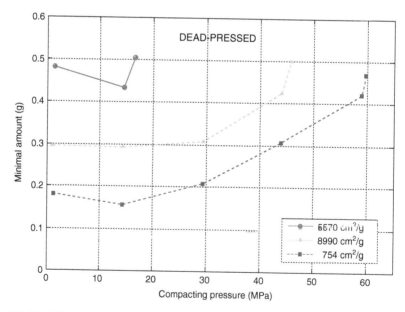

Fig. 2.6 The influence of compacting pressure and specific surface on initiation efficiency of MF (acceptor 0.35 g of TNT compressed by 76.5 MPa with reinforcing cap), by kind permission of Dr. Strnad [9]

2.2 Initiating Efficiency

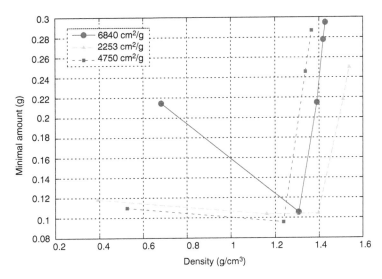

Fig. 2.7 Initiation efficiency of DDNP as a function of density and specific surface (acceptor 0.35 g of TNT compressed by 76.5 MPa with reinforcing cap), by kind permission of Dr. Strnad [9]

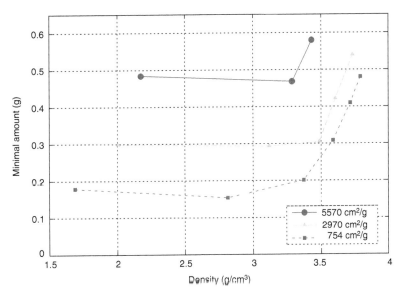

Fig. 2.8 Initiation efficiency of MF as a function of density and specific surface (acceptor 0.35 g of TNT compressed by 76.5 MPa with reinforcing cap) by kind permission of Dr. Strnad [9]

has a negligible effect on LA but significant effects on both MF and DDNP. The decrease of the initiation efficiency for these two depends on their specific surface. Dead pressing occurs earlier for material with higher specific surface [9].

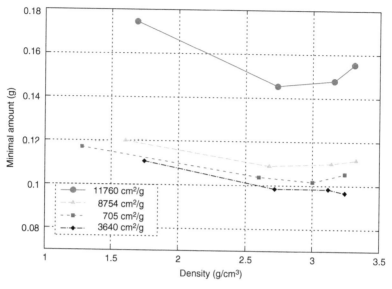

Fig. 2.9 Initiation efficiency of LA as a function of density and specific surface (acceptor 0.35 g of TNT compressed by 76.5 MPa without reinforcing cap) by kind permission of Dr. Strnad [9]

The compacting pressure in Figs. 2.4, 2.5, and 2.6 can be converted to densities of the compacted material and the relationships then plotted as initiation efficiency as a function of density (Fig. 2.7, 2.8, and 2.9). The highest initiation efficiency is obtained for MF at 3.2 g cm^{-3} (72% of TMD) and for DDNP at 1.2–1.3 g cm^{-3} (73.6–80.2 TMD). Using higher pressures leads to an increase in the minimal necessary amount of the explosive, and at density 3.6 g cm^{-3} MF and at 1.3–1.4 g cm^{-3} DDNP became dead-pressed. The density at which dead pressing takes place is lower for material with higher specific surface.

LA shows minimal necessary amount at density 2.7 g cm^{-3} (58% of TMD). Further increase in the compacting pressure does not have any significant effect on the initiation efficiency. The only exception is LA with a very large specific surface (~10,000 cm^2 g^{-1}) which has high minimal amounts that, after exceeding optimal density, increase yet further (Fig. 2.9).

The graphs presented above clearly show that, in the case of primary explosives, optimal rather than high density should be used. All the more so as unnecessarily high pressures used for obtaining material of higher density are also more susceptible to detonations during compression.

The compacting pressure is an important parameter not only for the primary but also for the secondary explosive as it influences the densities of both. The values tabulated in Table 2.1 refer to the amount of primary explosive causing detonation of PETN in 10 out of 10 trials [23]. It can be clearly seen that the uncompressed PETN requires much lower amounts of practically all primary explosives than compressed PETN. The compression of the secondary charge may lead to such a

2.2 Initiating Efficiency

Table 2.1 Influence of processing pressure on the minimal amount of primary explosive needed to detonate PETN in 10 out of 10 trials (copper detonator cap with internal diameter 6.2 mm without reinforcing cap) [23]

Pressure on PETN (MPa)	0	196			
Pressure on initiator (MPa)	0	0	49	98	147
Primary explosive	Minimum initiating charge (g)				
Tetrazene	0.16	0.25	Dead pressed		
Mercury fulminate (white)	0.30	0.34	Dead pressed		
Lead styphnate	0.55	Without detonation with 1 g			
Lead azide	0.015	0.1	0.01	0.01	0.01
Silver azide	0.005	0.110	0.005	0.005	0.005

Table 2.2 Influence of the amount of unpressed lead azide on the probability of detonation of a PETN charge pressed by 0 or 196 MPa [23]

PETN pressed by (MPa)	Weight of LA (g)	Result (+ detonation, − failure)
0	0.01	+ + + + −
0	0.015	+ + + + + + + + +
0	0.02	+ + + + +
0	0.03	+ + + + +
196	0.03	− − − − −
196	0.05	+ + + − −
196	0.09	+ + + − −
196	0.10	+ + + + + + + + +

desensitization of the secondary charge that it is impossible to initiate it by primary explosives with lower initiation efficiency.

The amounts in Table 2.1 are the minimal amounts causing initiation in 10 out of 10 trials. This does not necessarily mean that lower amounts of primary explosives do not initiate the secondary charge. In fact they do, but the probability of failure is higher. This behavior is demonstrated by results for LA in Table 2.2.

2.2.2 Influence of Specific Surface

The influence of the specific surface of the primary explosives on their initiation efficiency has already been included in the graphs above as reflected by their compaction behavior. It was further studied in a standard detonator number 3 cap with 0.35 g of TNT compressed by 76.5 MPa as a secondary charge [9]. LA was compressed by 13.8 MPa without any reinforcing cap while MF and DDNP were compressed by the same pressure with a reinforcing cap. The results are shown in Fig. 2.10. Both DDNP and LA show a performance relatively independent of specific surface with the best performance around 4,000 $cm^2\ g^{-1}$. The behavior of MF is however quite different, as its minimal necessary amount continually increases with increasing specific surface (decreasing crystal size).

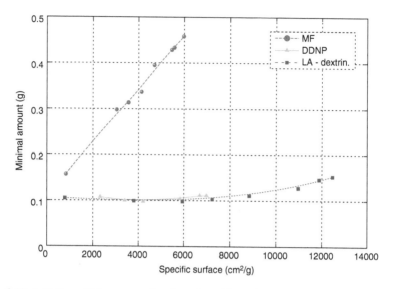

Fig. 2.10 Initiation efficiency as a function of specific surface, LA without reinforcing cap, DDNP and MF with reinforcing cap, compacting pressure 13.8 MPa, acceptor: 0.35 g of TNT compressed by 76.5 MPa, by kind permission of Dr. Strnad [9]

2.2.3 Influence of the Charge Diameter

The detonation velocity is not influenced only by its density but, just as in secondary explosives, also by the diameter of the charge. The fact that LA detonates practically immediately without a predetonation zone does not mean that it always detonates with the same detonation velocity irrespective of the charge size, as can be seen from Fig. 2.11. In these experiments, the LA was measured in layers of varying thickness [14].

2.2.4 Influence of Confinement

The initiating efficiency represented by the minimal initiating charge is further influenced by the overall design of the detonator, especially by the material of the detonator cap and the material of the reinforcing cap. This influence is more significant in the case of primary explosives with a long predetonation zone (DDNP, MF) but has only limited significance in the case of immediately detonating substances (LA). In the cases where it plays a role (DDNP, MF), a larger amount of primary explosive is required with aluminum detonator caps than is required when copper is used.

The influence of the material of the reinforcing cap is shown in Fig. 2.12. The initiation efficiency is in this case represented as a minimum amount of primary

2.2 Initiating Efficiency

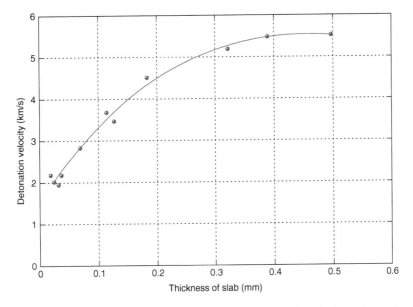

Fig. 2.11 Detonation velocity as a function of thickness of lead azide sheet (mean density 3.14 g cm^{-3}) [14]

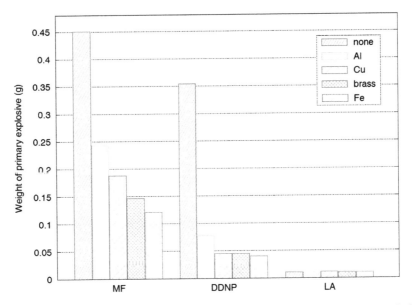

Fig. 2.12 Influence of reinforcing cap material on the minimal amount of primary explosive necessary to detonate PETN with 50% probability [24]

explosive necessary for 50% initiation [24]. It can be seen that tougher confinement significantly decreases the weight of the primary explosive required in the case of MF, to a lesser extent in the case of DDNP, and has practically no effect on LA. This is

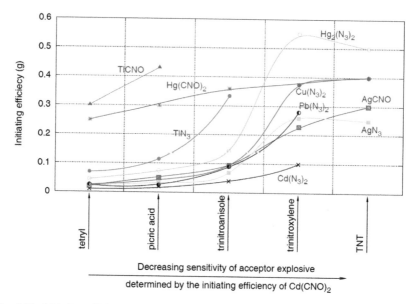

Fig. 2.13 Initiating efficiency of various primary explosives (*lines* are included just to help understanding the chart) [17]

caused by the fact that LA detonates practically instantaneously after ignition while both MF and DDNP detonate via DDT process. A careful optimization of the detonator design with respect to the thickness of the reinforcing cap and the size of the hole can further decrease the minimal necessary amount of some types of primary explosives.

2.2.5 Influence of Secondary Charge Type

For initiation of various secondary explosives different minimum amounts of the same primary explosive are needed. Using a secondary explosive less sensitive to detonation results in the need for an increased amount of the same primary explosive. An amount that works well for tetryl (relatively sensitive substance) is completely insufficient for initiation of TNT (relatively insensitive substance). It is interesting to note that MF, with its long predetonation zone, does not show a steep increase in the necessary minimum amount when going from sensitive to insensitive secondary acceptors (Fig. 2.13). The order of various primary explosives with respect to their initiating efficiency varies with varying type of secondary explosive considered, as can be seen from Fig. 2.13. Substances such as those in Fig. 2.13 are not used in detonators today. They were replaced by PETN, RDX, and for applications requiring high thermal stability by HNS.

2.3 Sensitivity

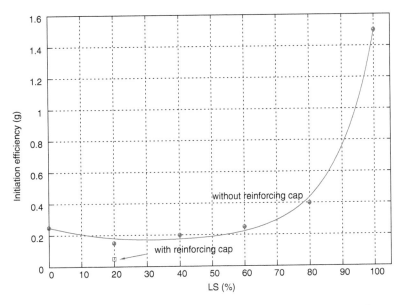

Fig. 2.14 Influence of amount of LS in LS/LA mixture on the initiating charge (values are for detonators with tetryl as secondary charge) [25]

2.2.6 Mixtures

The initiating efficiency of mixtures of primary explosives does not necessarily have to be between the values typical of its components. The classical mixture of LA and LS in which LS serves as a substance sensitive to flame is a typical example. Initiating efficiency of an LS and LA mixture is highly dependent on the ratio of the two substances and the highest values are obtained for mixtures with ratios around LS/LA 20/80 (Fig. 2.14). It is interesting to note that the initiation efficiency of the mixture is higher than that of pure LA up to a 60/40 ratio [25].

2.3 Sensitivity

The usefulness of energetic materials is in their ability to explode when desired. The energy of the stimulus that starts the explosion may range from a simple touch of a feather (nitrogen triiodide) to the impact of a shock wave (in NONEL detonators). Sensitivity of an energetic material can therefore be seen as an amount of energy that the material needs to absorb to attain a certain probability of developing an explosive reaction.

A distinction is sometimes made between the term *sensitiveness* and the term *sensitivity*. The first is related to accidental initiation and the determination of probabilities of initiation by various unwanted stimuli, while the second is related to the reliability of the function. With this approach, impact and friction tests are sensitiveness tests while the gap test is a sensitivity test [26]. In many sources, these two concepts are, however, both referred to using the term sensitivity, and for the sake of simplicity we decided to use this majority approach.

From the perspective of sensitivity, the most sensitive energetic materials are primary explosives, less sensitive are secondary explosives, and very insensitive are tertiary explosives. Rigorous limits between these groups do not exist and new explosives are therefore related to the existing ones through a series of comparative experiments. Some authors define primary explosives as substances being more sensitive than PETN.

The problem of sensitivity is further complicated by the fact that it is influenced by many factors. The most important are the type of initiation, the experimental conditions, the state of the sample tested (crystallography, shape, size), and the method of evaluating the results.

The presence of other materials in primary explosives (additives) also influences the resulting sensitivity values. In some cases, hard particles (e.g., glass dust) are added to increase the sensitivity of a primary explosive which would otherwise be too insensitive for the desired method of initiation. A typical example is addition of glass dust to the LA which increases the sensitivity of the mixture to a level desired for application in stab and friction detonators. The opposite effect is observed after addition of waxes or oils which lubricate the resulting mixture. This desensitizing effect is often utilized when it would be too risky to transport the substance in its pure form.

The tests used for determinating sensitivity of explosives have developed from the historical ones to those that we use today. The progress in development was, however, mainly on a national level and many different tests measuring generally the same thing evolved. The results as absolute values are therefore highly dependent on the country, or even the laboratory, carrying out the tests. Some testing methodologies became standardized (STANAG, MIL-STD, GOST, ADR, BAM) and provide to a certain degree the possibility of comparing results—absolute values—of various researchers. However, the most reliable are still relative results which compare newly referred substances to some well-defined standard. This problem with reported values is not just typical of primary explosives but relates to sensitivity testing of energetic materials in general.

One of the additional complications when trying to compare sensitivity data from various sources is the unspecified methodology of the test. Not only the testing instruments differ but in many cases it is not clear what method was used for the acquisition and evaluation of the results. The most typical ones include 50% or 100% probability of initiation, minimal initiation energy as one positive out of 6 or 10 trials at the same level, at our institute the recently implemented probit analysis [27], etc. Detailed specification of the variety of test methodologies is outside the scope of this work and may be found in the literature [28, 29].

From an application perspective those materials that are easily initiated by relatively small amounts of energy from a nonexplosive event (impact, spark, stab, friction, flame, etc.) are used as the first members of an initiating sequence.

The outcome of their reaction–flame or shock wave–then initiates less sensitive substances which require more energy and are not as easily initiated by nonexplosive stimuli. This leads to an initiation series in which the most sensitive substances initiate the less sensitive ones which then initiate even less sensitive ones, etc. This sequence is called the initiation train or, more specifically, the detonation or ignition train, depending on the desired output. The reason for a sequence of initiation is purely practical. The sensitive substances such as primary explosives are very vulnerable to accidental initiation and do not have the desired performance properties. They are therefore used only in small quantities enclosed in some type of initiating device that prevents as much as possible an unwanted initiation. They may further be stored and transported apart from the main secondary explosives to ensure that an accidental explosion would not initiate the main charges. The secondary explosives, on the other hand, are designed to fulfill specific performance parameters and are used in much larger quantities than primary explosives. Their sensitivity is much lower and they need to be initiated by primary explosives. The even less sensitive tertiary explosives must be initiated by a charge of secondary explosive (so-called booster) that amplifies the output effect of the detonator.

Let us look at a typical example of an explosion train, for example, in a surface mining blasting application. How does it work practically? A hole is first drilled into a rock; some booster with a detonator is inserted and filled with some explosive of low sensitivity—for example, an emulsion explosive. Let us further assume that the detonator is electric. What happens when the electric impulse is discharged into the wires leading to the detonator? First the bridgewire heats up and ignites the pyrotechnic mixture of the fusehead. Flame from the fusehead ignites the delay composition if it is present, flame from the delay composition ignites the primary explosive which undergoes deflagration to detonation transition, and the outgoing detonation wave initiates the adjacent secondary explosive inside the detonator which amplifies the shock wave. As the detonator is placed inside the booster (charge made from secondary explosive) the shock wave, in combination with the kinetic energy of the fragments of the metal cap, initiates it. The detonating booster initiates the detonation reaction of the main explosive (the above-mentioned emulsion).

2.3.1 Impact Sensitivity

Impact sensitivity is probably the most common sensitivity test and, just like other tests of explosive properties, it gives very different results depending on the methodology used and the testing apparatus. Figure 2.15 shows data obtained by various authors for the same substances. Although the idea behind this test is very simple—hitting an explosive by a falling object—ball or hammer—the results obtained show considerable scatter. It can be clearly seen that the values of impact energy cannot easily be compared without exact specification of the test conditions.

An excellent summary of a large number of impact and friction tests of LA and to a lesser extent of some other common explosives has been published by Avrami

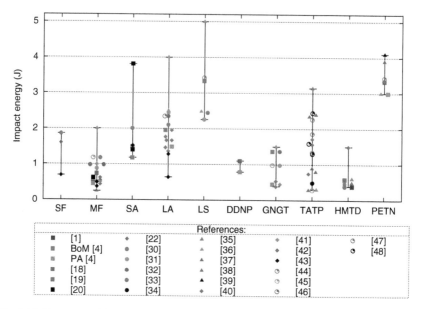

Fig. 2.15 Impact sensitivity of individual primary explosives [1, 4, 18–20, 22, 32–50] (*BoM* Bureau of mines, *PA* Picatinny Arsenal)

and Hutchinson [49]. The importance of methodology, additives, impurities, methods of evaluation, temperature, and many other issues is discussed in great detail especially for lead and copper azides, and will not be repeated here.

Despite the above-mentioned problems, most common primary explosives have been compared and the order of their impact sensitivity has been evaluated by various authors. The sensitivity of LA and SA is lower than that of MF and comparable to that of PETN. The sensitivity of DDNP is mentioned as lower than for MF [4, 18, 50]. 1-Amino-1-(tetrazol-5-yldiazenyl)guanidin (GNGT, tetrazene) is sometimes reported as slightly more sensitive than MF [41] but slightly less sensitive than MF by [33]. SF, often mentioned as a very sensitive substance, has an impact sensitivity comparable to that of LA. Its high sensitivity to friction is sometimes misleadingly attributed to impact. TATP is often reported as extremely sensitive but, as indicated by the figures in Fig. 2.15, the results are relatively evenly spread from about 0.2 to over 3 J. Of the usual primary explosives, LS shows the lowest sensitivity to impact.

Impact sensitivity significantly depends on many aspects. Let us look at some of these properties starting with crystal size of the material under test. Colloidal silver azide prepared from concentrated solutions exhibits significantly lower sensitivity (0.5 kg from 77.7 cm) than coarser crystals prepared from diluted solutions, which required less than half the energy (0.5 kg from 28.5 cm). MF measured under the same conditions for comparison required 12.7 cm with the 0.5 kg hammer [20]. It is interesting to note that the impact sensitivity of SA (in fine powdery form), which is considered very sensitive, is lower than that of MF. Similar investigations have

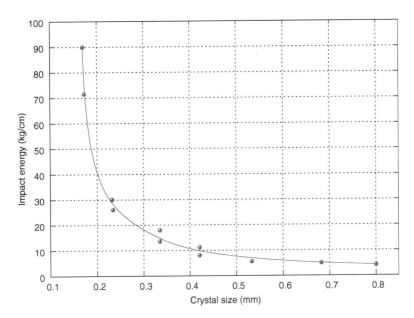

Fig. 2.16 Influence of crystal size on the impact sensitivity of lead azide (50% probability) [51]

been carried out with lead azide and the results are shown in Fig. 2.16 [51]. It can be clearly seen that the fine particles are again much less sensitive.

Impact sensitivity is most often reported as 50% probability which gives a good comparative value but it does not say anything about the steepness of the dependency. An example of results covering the whole range from 10 to 100% probability of initiation for cuprous azide is shown in Fig. 2.17. Results of this type are not very common because it is a very time-consuming process to obtain them. Each point in Fig. 2.17 represents probability calculated out of 15 trials, at various heights, totaling 300 shots. It can be clearly seen that the finer particles are less sensitive and that the drop height more than doubles when going from 10 to 100% probability. In this particular case, the probability curves exhibit roughly the same slope, and the order of sensitivity is the same for all percentages. It will be shown later, in the part on friction sensitivity, that the probability curves may even cross each other. One substance may then appear more sensitive when looking at 10% probability of initiation and less sensitive when evaluating 50% probability. It is therefore desirable to obtain the entire sensitivity curve. Methods such as the probit analysis significantly reduce the number of trials necessary. The steepness of the probability curves depends on the particular explosive [52].

A good comparison of impact sensitivity for various explosives is presented in Fig. 2.18. It shows probability of initiation at specific drop heights expressed as a number of positive trials out of 5.

Impact sensitivity is further influenced to a very large extent by the thickness of the layer of explosive tested. The sensitivity of the azides of silver, lead, and mercury increases with the layer thickness, that of cadmium is almost constant,

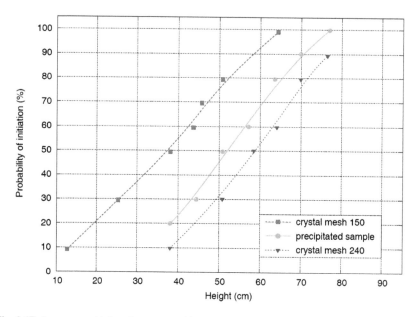

Fig. 2.17 Impact sensitivity of cuprous azide [52]

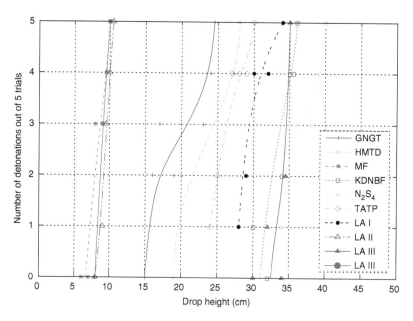

Fig. 2.18 Impact sensitivity of various primary explosives determined as number of positive trials out of five at specific drop heights (KDNBF-potassium salt of 4,6-dinitrobenzofuroxan, for other abbreviations see previous text) [40]

those of copper, manganese, zinc, and thallium decreased in sensitivity with increasing thickness, and those of nickel, cobalt, calcium, barium, and strontium increased to a maximum value with 0.02 g and then the sensitivity decreased [53].

Increasing the temperature significantly increases the impact sensitivity of LS. It is therefore crucial to carefully control temperature and time of drying during manufacture to prevent an unnecessary increase in manipulation risk.

2.3.2 Friction Sensitivity

Despite decades of friction testing, the phenomenon is still relatively poorly understood. The quantitative interpretation of the results is problematic due to numerous factors affecting the mechanism of "friction initiation." In practice, the test is done by placing the tested explosive between two inert surfaces, applying a defined load on the sample and then sliding one or both of the surfaces in a direction normal to the direction of the applied force. Ceramic plates are often the material of choice.

Only a small number of investigations comparing friction sensitivity of primary explosives have been published to date. The most common approach is to synthesize a new compound and compare it to one or two standards, commonly LA, MF, PETN or, more recently, also LS. Comparison of sensitivity of various substances is therefore difficult and mostly based on results gathered from various sources. One of the oldest comparative works was done by Wallbaum [31] who found the following order of decreasing sensitivity of primary explosives: SA > LA > LS > GNGT ~ MF. This is practically identical with the order reported by Meyer [36] for at least one initiation out of 6 trials. The order is consistent with our data of 50% probability of initiation (from probit analysis) [54] which is shown in Fig. 2.19. Sensitivity of organic peroxides (TATP and HMTD) is, however, reported to be very high by Meyer. Measurements of Matyáš [37] indicate that TATP and HMTD are slightly more sensitive to friction than MF. According to our recent results both peroxides, as well as DADP, are less friction sensitive than LA. The results shown in Fig. 2.19 were obtained at our institute under the same conditions and by the same operator.

Some primary explosives are reported to have extreme sensitivity. SF and SA are two such substances. Extreme sensitivity of SF is reported in [33], very high sensitivity (approximately 2–3 times higher than that of LA depending on the testing surface) is reported for SA [55]. Such statements must be carefully considered and evaluation based on solid data. Extreme sensitivity of SA is, for example, commonly found in older sources and could be the result of the method of preparation. In the early days, SA was prepared by direct precipitation of aqueous solutions of sodium azide and the silver salt and such a method of preparation could have led to a more sensitive product. Today's industrial SA (product of BAE Systems) is reported to have sensitivity lower than that of LA (determined by emery friction test) [56].

Comparing friction sensitivity of LA and MF based on published results can be quite complicated. Some authors report LA to be more sensitive [23, 33, 36] than MF while others refer to it being less sensitive [45, 57]. The reason for such

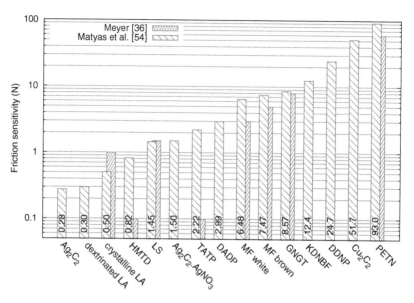

Fig. 2.19 Friction sensitivity of primary explosives—comparison of our data (50% probability of initiation from probit analysis) [54] and data from Meyer (min. one positive out of six trials) [36]

different results could be caused by many factors, e.g., different forms of LA. We have measured dextrinated and crystalline α-LA and both white and brown MF under the same conditions and found MF to be much less sensitive than LA, as can be seen from Fig. 2.19.

A large batch-to-batch variation in surface roughness of porcelain plates used in a BAM testing machine is a well-known problem. The relationship of the plate surface roughness and the resulting friction sensitivity has been studied by Roux [55]. He recommends using sandpaper with a well-defined roughness to obtain better reproducibility and lower the measurement cost. The obtained order of sensitivity of classical primary explosives is the same as the one obtained under standard conditions which are mentioned above. Temperature has also been reported to play an important role in the case of LS whereas in the case of LA, SA, and GNGT it did not [31].

Sensitivity to friction is highly dependent on the method of measurement. The most important factors influencing the final results include the material of the plate surface and the peg, speed of the peg sliding, and humidity. These factors are very difficult if not impossible to compare making results of various authors hardly ever directly comparable.

Figure 2.20 demonstrates another issue regarding sensitivity of primary explosives. Each point on the graph represents the probability of initiation from 15 trials at particular friction force. The curves were obtained by probit analysis [27] for two samples—pure crystalline LA and dextrinated LA. The x-axis shows the used friction force and the y-axis the probability of initiation. Interestingly, the

2.3 Sensitivity

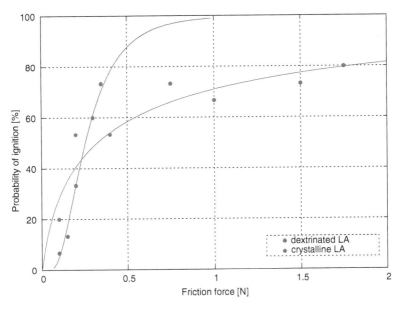

Fig. 2.20 Friction sensitivity of dextrinated and crystalline α-LA [58]

sensitivity curves for crystalline LA cross with those for dextrinated LA showing it to be more sensitive at higher friction forces and less sensitive at lower ones [58].

2.3.3 Sensitivity to Electrostatic Discharge

With sensitivity to electrostatic discharge the situation is even more complicated than in the case of impact and friction sensitivities. The main problem is the variety of testing instruments and testing modes of the discharge circuit (oscillating vs. damped mode) [59]. Due to this variability, it is practically impossible to compare values obtained by different authors. It should always be seen in relation to other substances measured under the exact same conditions. We have therefore included results of measurements of electrostatic discharge sensitivity of some primary explosives measured at our institute, where we can be sure that they have been obtained under the same conditions, and summarized them in Fig. 2.21 (Majzlik J and Strnad J, unpublished work). We are aware of the fact that some published results (e.g., [60]) may be as much as an order of magnitude different from ours due to the variation in methodologies but the order of the individual substances should remain the same. The complete methodology with the description of the measuring device and conditions may be found in [59, 61, 62].

The electrostatic discharge sensitivity of LS is significantly higher than such sensitivity in other primary explosives. This creates problems for the technology used in its production and processing.

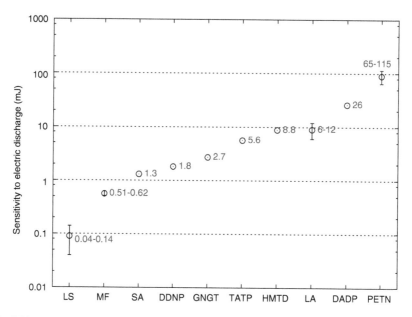

Fig. 2.21 Electrostatic sensitivity of some primary explosives (Majzlik J and Strnad J, unpublished work)

2.3.4 Sensitivity to Flame

Primary explosives differ in the way they respond when subjected to flame and, based on this type of response, may be divided into two groups. The explosives in the first group burn when initiated by flame and may, but do not have to, undergo transition to detonation. The detonation then propagates further with stable detonation velocity if such transition occurs. The typical substances in this group are MF, HMTD, TATP, and DDNP. This group is sometimes called a "mercury fulminate group."

The second group, the so-called lead azide group, does not exhibit a predetonation zone under normal conditions. Initiation by flame results in practically instantaneous detonation. The typical members of this group include, besides lead azide, also silver fulminate and silver azide. Explosives of both groups—MF group as well as LA group—detonate when initiated by shock wave [24].

The sensitivity of primary explosives to flame varies based on their chemical composition and manufacturing process. The pressure by which the material is prepared again plays an important role. Of the classical primary explosives, the most sensitive are LS and GNGT and the least sensitive is LA. The data comparing flame sensitivity of primary explosives are relatively rare. The most compact comparison of flame sensitivity of primary explosives is given by Bagal who used a specially designed pendulum for investigating their ignition behavior (Fig. 2.22) [21]. This pendulum test is, however, not widely used and sensitivity to flame is more often determined by the "ease of ignition" (Bickford fuse) test

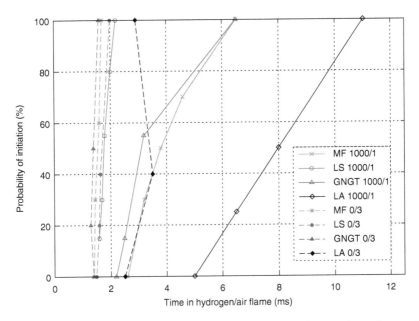

Fig. 2.22 Sensitivity of primary explosives to flame [21]—the first number refers to the compacting pressure (MPa) and the second to the orifice diameter (mm)

where observations are made of the ease of ignition and type of response after the action of a flame from a safety fuse. Such results, although they are easier to achieve, do not provide a quantitative measure of the sensitivity.

It can be clearly seen from Fig. 2.22 that the sensitivity of LA to flame is lower compared to other primary explosives. This is the reason why it is in some applications mixed with other primary explosives with high flame sensitivity, such as LS.

References

1. Urbański, T.: Chemie a technologie výbušin. SNTL, Praha (1959)
2. Krauz, C.: Technologie výbušin. Vědecko-technické nakladatelství, Praha (1950)
3. Patry, M.: Combustion et detonation, Paris 1933 In: Urbański, T. Chemistry and Technology of Explosives, vol. 3. PWN - Polish Scientific Publisher, Warszawa (1967)
4. Tomlinson, W.R., Sheffield, O.E.: Engineering Design handbook, Explosive Series, Properties of Explosives of Military Interest, Report AMCP 706-177. U.S. Army Material Command, Washington, DC (1971)
5. Muraour, H.: Sur la théorie des réactions explosives. Cas particulier des explosifs d'amourçage. Bull. Soc. Chim. Fr. **51**, 1152–1166 (1932)
6. Kast, H., Haid, A.: Über die sprengtechnischen Eigenschaften der wichtigsten Initialsprengstoffe. Zeitschrift für das angewandte Chemie **38**, 43–52 (1925)

7. Jahresbericht VIII der chemisch-technischen Reichsanstalt. 8, 122 In: Urbański, T. Chemistry and Technology of Explosives, vol. 3. PWN - Polish Scientific Publisher, Warszawa (1967)
8. Cook, M.A.: An equation of state for gases at extremely high pressures and temperatures from the hydrodynamic theory of detonation. J. Chem. Phys. 15, 518–524 (1947)
9. Strnad, J.: Iniciační vlastnosti nejpoužívanějších třaskavin a vývoj nových metodik jejich měření. Dissertation thesis, Vysoká škola chemicko-technologická, Pardubice (1972)
10. Kling, A., Florentin, D.: Action des basses températures sur les explosifs. Memorial des poudres 17, 145–153 (1913)
11. Carl, L.R.: The rate of detonation of mercury fulminate and its mixtures with potassium chlorate. Army Ordnance 6, 302–304 (1926)
12. Danilov, J.N., Ilyushin, M.A., Tselinskii, I.V.: Promyshlennye vzryvchatye veshchestva; chast I. Iniciiruyushchie vzryvshchatye veshchestva. Sankt-Peterburgskii gosudarstvennyi tekhnologicheskii institut, Sankt-Peterburg (2001)
13. Stadler, R.: Analytische und sprengstofftechnische Untersuchungen an Azetylensilber. Zeitschrift für das gesamte Schiess- und Sprengstoffwesen 33, 302–305 (1938)
14. Bowden, F.P., McLaren, A.C.: Conditions of explosion of azides: effect of size on detonation velocity. Nature 175, 631–632 (1955)
15. Friederich, W.: Überhöhte Detonationsgeschwindigkeiten. Zeitschrift für das gesamte Schiess- und Sprengstoffwesen 31, 253 (1936)
16. Baum, F.A., Stanjukovich, J.K., Sechter, B.I.: In: Fyzika vzryva, p. 290. Nauka, Moskva (1959)
17. Wöhler, L., Martin, F.: Die Initialwirkung von Aziden und Fulminaten. Zeitschrift für das Gesamte Schiess- und Sprengstoffwesen 30, 18–21 (1917)
18. Clark, L.V.: Diazodinitrophenol, a detonating explosive. J. Ind. Eng. Chem. 25, 663–669 (1933)
19. Taylor, C.A., Rinkenbach, W.H.: H.M.T.D. - a new detonating explosive. Army Ordnance 5, 463–466 (1924)
20. Taylor, C.A., Rinkenbach, W.H.: Silver azide: an initiator of detonation. Army Ordnance 5, 824–825 (1925)
21. Bagal, L.I.: Khimiya i tekhnologiya iniciiruyushchikh vzryvchatykh veshchestv. Mashinostroenie, Moskva (1975)
22. Taylor, C.A., Buxton, E.P.: Silver fulminate, an initiator of detonation. Army Ordnance 6, 118–119 (1925)
23. Wallbaum, R.: Sprengtechnische Eigenschaften und Lagerbeständigkeit der wichtigsten Initialsprengstoffe. Zeitschrift für das Gesamte Schiess- und Sprengstoffwesen 34, 197–201 (1939)
24. Strnad, J.: Primary explosives and pyrotechnics - lecture notes. Katedra teorie a technologie výbušin, Univerzita Pardubice (1999)
25. Grant, R.L., Tiffany, J.E.: Factors affecting initiating efficiency of detonators. J. Ind. Eng. Chem. 37, 661–666 (1945)
26. Zukas, J.A., Walters, W.P.: Explosive Effects and Applications. Springer, New York (1998)
27. Šelešovský, J., Pachman, J.: Probit analysis in evaluation of explosive's sensitivity. Cent. Eur. J. Energ. Mater. 7, 269–277 (2010)
28. Krupka, M.: Testing of Energetic Materials. Univerzita Pardubice, Pardubice (2003)
29. Sućeska, M.: Test Methods for Explosives. Springer, New York, NY (1995)
30. Taylor, A.C., Rinkenbach, W.H.: Sensitivities of detonating compounds to frictional impact, impact, and heat. J. Franklin Inst. 204, 369–376 (1927)
31. Wallbaum, R.: Sprengtechnische Eigenschaften und Lagerbeständigkeit der wichtigsten Initialsprengstoffe. Zeitschrift für das Gesamte Schiess- und Sprengstoffwesen 34, 161–163 (1939)
32. Berthman, A.: Die Werkstofffrage bei der Herstellung der Explosivstoffe und Zündstoffe. Chemische Apparatur 27, 243–245 (1940)

References

33. Fedoroff, B.T., Sheffield, O.E., Kaye, S.M.: Encyclopedia of Explosives and Related Items. Picatinny Arsenal, Dover, NJ (1960–1983)
34. Phillips, A.J.: Technical report no 1202, Report. Picatinny Arsenal, Dover, NJ (1942)
35. Khmelnitskii, L.I.: Spravochnik po vzryvchatym veshchestvam. Voennaya ordena Lenina i ordena Suvorova Artilleriiskaya inzhenernaya akademiya imeni F. E. Dzerzhinskogo, Moskva (1962)
36. Meyer, R., Köhler, J., Homburg, A.: Explosives. Wiley-VCH, Weinheim (2002)
37. Matyáš, R.:Výzkum vlastností vybraných organických peroxidů. Dissertation, Univerzita Pardubice, Pardubice, Česká Republika (2005)
38. Mavrodi, G.E.: Improvements in or relating to explosives of the organic peroxide class. GB Patent 620,498, 1949
39. Šelešovský, J.: Hodnocení stability a životnosti vojenských výbušin. Diploma thesis, Univerzita Pardubice, Pardubice, Česká Republika (2002)
40. Metz, L.: Die Prüfung von Zündhütchen (Initialsprengstoffen) aus Schlagempfindlichkeit und Flammenwirkung. Zeitschrift für das gesamte Schiess- und Sprengstoffwesen **23**, 305–308 (1928)
41. Rinkenbach, W.H., Burton, O.E.: Explosive characteristics of tetracene. Army Ordnance **12**, 120–123 (1931)
42. Matyáš, R.: Influence of oil on sensitivity and thermal stability of triacetone triperoxide and hexamethylenetriperoxide diamine. In: Proceedings of 8th Seminar on New Trends in Research of Energetic Materials, pp. 674–679, Pardubice, Czech Republic (2005)
43. Marshall, A.: Explosives. Butler and Tanner, London (1917)
44. Ek, S., Menning D.: Purification and sensitivity of triacetone triperoxide (TATP). In: Proceedings of 10th Seminar on New Trends in Research of Energetic Materials, pp. 570–574. Pardubice, Czech Republic (2007)
45. Špičák, S., Šimeček, J.: Chemie a technologie třaskavin. Vojenská technická akademie Antonína Zápotockého, Brno (1957)
46. Orlova, E.Y.: Khimiya i tekhnologiya brizantnykh vzryvchatykh vescestv. Khimiya, Leningrad (1981)
47. Hiskey, M.A., Huynh, M.V.: Primary explosives, US Patent 2006/0030715A1, 2006
48. Yeager, K.: Trace Chemical Sensing of Explosives. Wiley, Hoboken, NJ (2007)
49. Avrami, A., Hutchinson, R.: Sensitivity to Impact and Friction In: Fair, H.D., Walker, R.F. (eds.) Energetic materials 2.- Technology of the Inorganic Azides, vol. 2, pp. 111–162. Plenum, New York, NY (1977)
50. Davis, T.L.: The Chemistry of Powder and Explosives. Wiley, New York, NY (1943)
51. Bowden, F.P., Singh, K.: Size effects in the initiation and growth of explosion. Nature **172**, 378–380 (1953)
52. Singh, K.: Sensitivity of cuprous azide towards heat and impact. Trans. Faraday Soc. **55**, 124–129 (1959)
53. Wöhler, L., Martin, F.: Azides; sensitiveness of. J. Soc. Chem. Ind. **36**, 570–571 (1917)
54. Matyáš, R., Šelešovský, J., Musil, T.: Sensitivity to friction for primary explosives. J. Hazard. Mater. **213–214**, 236–241 (2012)
55. Roux, J.J.P.A.: The dependence of friction sensitivity of primary explosives upon rubbing surface roughness. Propell. Explos. Pyrotech. **15**, 243–247 (1990)
56. Millar, R.W.: Lead free initiator materials for small electro explosive devices for medium caliber munitions; Final report 04 June 2003, report QinetiQ/FST/CR032702/1.1, QinetiQ, Farnborough, UK, 2003
57. Military explosives. Report TM-9-1300-214, Headquarters, Department of the Army, 1984
58. Šelešovský, J., Matyáš, R., Musil T.: Using of the probit analysis for sensitivity tests - sensitivity curve and reliability. In: Proceediongs of 14th Seminar on New Trends in Research of Energetic Materials, pp. 964–968. Univerzita Pardubice, Pardubice, Czech Republic (2011)

59. Strnad, J., Majzlík, J.: Technical Description of Apparatus ESZ KTTV, Report. Institute of energetic materials, University of Pardubice, Pardubice (2001)
60. Talawar, M.B., Agrawal, A.P., Anniyappan, M., Wani, D.S., Bansode, M.K., Gore, G.M.: Primary explosives: Electrostatic discharge initiation, additive effect and its relation to thermal and explosive characteristics. J. Hazard. Mater. **137**, 1074–1078 (2006)
61. Strnad, J., Majzlík, J.: Determination of electrostatic spark sensitivity of energetic materials. In: Proceedings of 4th Seminar on New Trends in Research of Energetic Materials, pp. 303–307. University of Pardubice, Pardubice, Czech Republic (2001)
62. Strnad, J., Majzlík, J.: Sample of energetic material as a consumer of electric impulse power during the electrostatic discharge examination. In: Proceedings of 37th International Annual Conference of ICT, pp. 58.1–58.11. Karlsruhe (2006)

Chapter 3
Fulminates

3.1 Introduction

The history of salts of fulminic acid goes back to the seventeenth century when silver and mercury fulminate were discovered by the alchemists. The fulminates—salts of fulminic acid—must not be confused with "fulminating" compounds—such as fulminating silver, gold, or platinum which are most likely nitrides. Fulminating metals are also primary explosives but are not used due to their high sensitivity. They are formed by precipitating the corresponding metal solutions with ammonia.

3.1.1 *Fulminic Acid*

$$H{-}C{\equiv}\overset{+}{N}{-}O^{-} \longleftrightarrow H{-}\overset{-}{C}{=}\overset{+}{N}{=}O$$

Fulminic acid is a gaseous, very toxic, and unstable substance with a hydrogen cyanide-like odor but much more aggressive [1, 2]. Its melting point is −10 °C with decomposition [3]. The structure of fulminic acid is demonstrated above with the left-hand structure being dominant.

In its pure form, fulminic acid polymerizes resulting in an indeterminate product containing molecules with molecular weight about 1,500 g mol^{-1} [3]. In solution, fulminic acid spontaneously polymerizes by a complex chain of chemical reactions resulting in the di- and trimer. The first step of the polymerization is, according to Danilov et al. [3], most probably a dimerization resulting in furoxane formation followed by ring rupture and further reaction yielding the cyclic trimer called "metafulminuric acid" (I) as a main product [3–6]. This "metafulminuric acid" is not an explosive substance [7]. Other minor identified by-products of this spontaneous exothermic polymerization are the tetramer called "α-isocyanilic acid" (II) and another trimer (III). Both trimer (I) and tetramer (II) form via unstable

R. Matyáš and J. Pachman, *Primary Explosives*,
DOI 10.1007/978-3-642-28436-6_3, © Springer-Verlag Berlin Heidelberg 2013

intermediates as shown in a reaction scheme below [4, 6]. Further detailed information on the mechanism of polymerization may be found in [2].

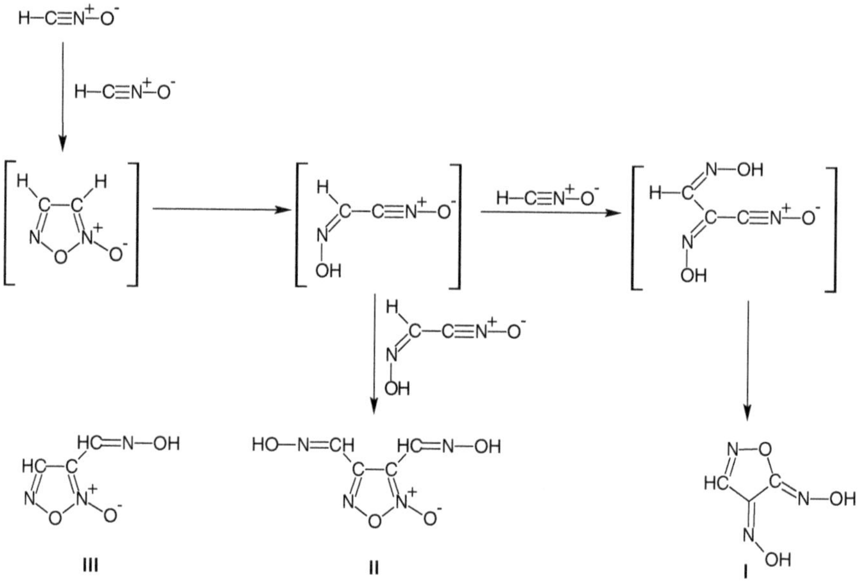

The discovery of the chemical structure of fulminic acid is a long and interesting story that is described in detail in separate scientific papers [8–11]. Fulminic acid was originally regarded as a two-carbon compound ($C_2H_2N_2O_2$) owing to its origin from ethanol [10, 12, 13]. At the turn of nineteenth and twentieth century, Nef proposed fulminic acid as the monomeric oxime of carbon monoxide H−O−N=C [14]. Although other structures of fulminic acid were developed during the twentieth century (the currently accepted structure of formonitrile oxide among others), Nef's formula was used in explosives-related literature throughout virtually the whole of the twentieth century, and even for many years after it was refuted. For example, the structure can be found in Bagal's (1975) [15] or Urbański's (1984) [16] monographs. The currently accepted structure of fulminic acid (in the form of formonitrile oxide H−C≡N$^+$−O$^-$) first appeared in technical literature as early as 1899 [10]. This structure was also proposed by Pauling and Hendrick on the basis of calculated potential energies of all substances formed by combination of H, C, N, and O atoms [17]. However, the structure was only fully accepted after it was unequivocally confirmed by experiments using IR and microwave spectroscopy in the 1960s [8, 10, 18].

The structure of fulminic acid was originally believed to be linear. Based on the IR spectra and theoretical calculations, it was, however, proved that fulminic acid does not have linear, but rather an unusual quasi-linear molecule with angle between H−C−N of precisely 165.13° [8, 10, 11]. The isomers of fulminic acid have attracted the attention of theoreticians and are even the topic of some recent papers [8, 19]. A comprehensive review of fulminic acid and its salts was published relatively recently by Wolfgang Beck [8].

3.1 Introduction

The structure of metallic salts of fulminic acid is usually presented by a Pauling-like structure with metal bonded to carbon [3, 20–22]. This seems to be a reasonable assumption even though some authors published studies supporting bonding between metal and oxygen [23] or even the possibility of the existence of both of these forms [24]. The issue still causes discussion and a report of a theoretical study [22] and X-ray analysis [21], supporting the bonding between metal and carbon, was recently published. The character of the metal to carbon bond of alkaline and thallium fulminate is ionic, whereas fulminates of silver and mercury are covalent [20, 25].

3.1.2 Mercury Fulminate

$$^-O-N^+\equiv C-Hg-C\equiv N^+-O^-$$

Mercury fulminate is one of the oldest known primary explosives probably having been discovered by alchemists. The name of its discoverer is uncertain; however, the discovery is most often ascribed to two alchemists—Cornelius Drebbel, a Dutchman living at the beginning of seventeenth century as well as to the Swedish-German alchemist Johannes Kunkel von Löwenstern living in second half of seventeenth century. The references confirming knowledge of mercury fulminate by these two alchemists can be found in their books [10, 11, 21, 26–32]. Both of them probably obtained this substance by treating mercury with nitric acid and alcohol. Neither of them could, however, find a use for this explosive compound, and mercury fulminate was forgotten until Edward Howard rediscovered it at the turn of the eighteenth century, examined its properties and published his results in "Philosophical Transactions of the Royal Society" [10, 33]. The discovery of mercury fulminate is therefore in some older scientific resources attributed incorrectly to Howard [34–38]. The first patent describing the use of mercury fulminate in initiators is from 1807 and belongs to a Scottish clergyman, Alexander Forsyth [29, 34].

3.1.2.1 Physical and Chemical Properties

The mercury fulminate (MF) formula is $Hg(C\equiv N^+-O^-)_2$ with a covalent bond between the mercury and carbon atoms [20]. Its crystal density is reported to be 4.42–4.43 g cm^{-3} [29, 30, 39, 40], but recent results of X ray analysis updated it to 4,467 g cm^{-3} [21]. Bulk density depends on crystal size and shape—it is reported to be between 1.35 and 1.55 g cm^{-3} [38]. The heat of formation of MF is reported as being between -268 and -273 kJ mol^{-1} [29, 41, 42]. The structure of the MF molecule and its crystal was published recently by Beck et al. [21]. Pure and ordinarily prepared mercury fulminate is, for all practical purposes, not hygroscopic, but its hygroscopicity rapidly increases in presence of impurities (e.g., mercury oxalate, calomel, mercuric chloride), which are generally present in the industrial

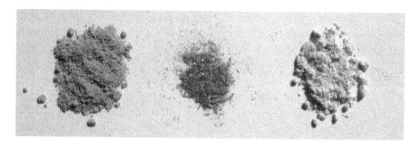

Fig. 3.1 The three common forms of mercury fulminate from *left to right—gray, brown*, and *white*

product [42, 43]. The presence of other substances can increase hygroscopicity as well. For example, a mixture of MF with potassium chlorate (formerly used in primers or blasting caps) is significantly more hygroscopic than each of the chemicals itself. Kast suggested an explanation of this phenomenon by mutual reaction of MF with potassium chlorate producing some hygroscopic substance [15, 43].

The taste of MF is described variously as a sweetish metallic taste by some authors [38] or salty by Walke [12]. MF is, however, a very toxic substance [38].

MF is known to exist in various forms depending on the way of preparation. Many authors mention two forms—white and gray [15, 28]. In reality there are three forms—white, brown, and gray—obtained directly from the reaction mixture (see Fig. 3.1) and other forms obtained by recrystallization from various solvents (Pachman and Matyáš unpublished work).

The most common types of MF and the relations among them are schematically summarized in Fig. 3.2. Some authors misleadingly refer to the brown form as being gray [15, 28]. The white and brown forms are both desired products and are purer than the gray form which is produced by improperly carrying out the reaction. The brown form of MF is, contrary to general expectation, a little bit purer than the white form [16, 44].

The brown product with a well-developed crystal structure (Fig. 3.3) is formed when the reaction is carried out without any additives. Temperatures as well as concentrations and amounts of reacting substances must be optimized.

The same procedure leads to the gray product if nitric acid contains more than 0.5 % of sulfuric acid or if it is too strong (density over 1.4 g cm^{-3}) or too weak (density below 1.38 g cm^{-3}). The gray product is also obtained if the alcohol temperature at the beginning of the reaction is too low or if its initial concentration is lower than 95 %. Such product contains metallic mercury [42]. A preparation route has, however, been published using 90 % ethanol [30].

The white form of MF is less pure than the brown form but purer than the gray form. It is produced in presence of cupric chloride (or some other substances such as KCl, ZnCl$_2$, BaCl$_2$, or Cu$_2$Cl$_2$). In practice copper and hydrochloric acid are added to the dissolving tank containing mercury and nitric acid. Addition of hydrochloric acid itself results in formation of white MF contaminated by oxalic acid. The product is, however, not as clean-white looking and the crystals are not as uniform in shape and size as in the case of combination of copper and hydrochloric

3.1 Introduction

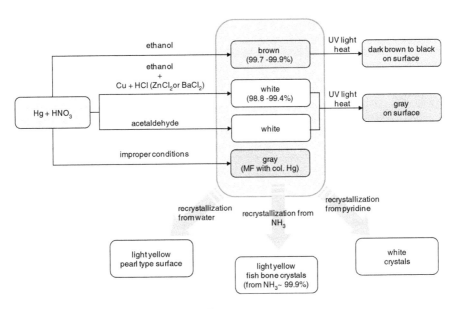

Fig. 3.2 Preparation of various types of MF

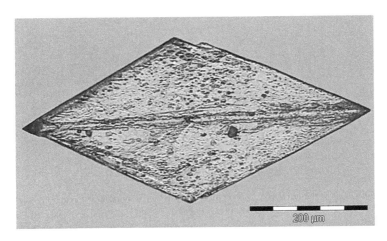

Fig. 3.3 Brown MF crystal

acid [35, 38, 42, 45]. In presence of Cu(NO$_3$)$_2$ the brown form appears indicating that not only copper but also chloride ions play a role in the formation of the white product (Pachman and Matyáš unpublished work). According to Solonina (cited in [38]) the purest white MF forms from addition of cuprous chloride.

Various theories trying to explain the nature of brown versus white MF have been proposed. Probably, the most often cited one is a presence of colloidal mercury in MF crystals [2, 15, 29, 41, 46]. This reasoning has been, however, rejected by other scientists who have examined both color modifications and did not

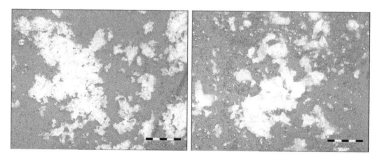

Fig. 3.4 Ground brown MF (left) compared to white MF in "as prepared"

find metallic mercury in either form of MF (for example by Solonia cited by Urbański). It has been proposed that the reason behind the colored mercury fulminate is related to the presence of by-products [47] possibly resinous polymers of fulminic acid [16, 48]. It has been further reported that grinding of brown MF in a mortar under water produces practically white product [35]. The result of grinding well-developed brown crystals (from Fig. 3.3) under water is presented in Fig. 3.4 and it is further compared to the white form (Pachman and Matyáš unpublished work). The resulting product is a white powder.

Our own unpublished investigation shows that brown and white MF does not contain metallic mercury visible by optical microscopy which is in agreement with Krauz [35]. The gray form, however, does contain it. The presence of metallic mercury in gray MF is clearly apparent from Fig. 3.5.

It is possible that the brown product with metallic mercury is obtained if reaction conditions are not maintained within prescribed limits. Such product is, however, not gray but brown with clearly visible mercury drops (Fig. 3.6).

It can be stated that the phenomenon of three basic colors of mercury fulminates—white, brown, and gray—is even today not completely explained.

MF is a photosensitive substance and decomposes under UV light. Originally nicely developed crystals become dark, break into pieces, and elementary mercury, carbon monoxide, and nitrogen are formed [49]:

$$Hg(CNO)_2 \longrightarrow Hg + 2\,CO + N_2$$

Figure 3.7 shows the effect of UV light on crystals of MF. The formation of mercury inside the crystals is shown in Fig. 3.8.

The action of UV radiation on MF has also been published by Bartlett et al. They observed that cracks on MF crystals after exposition on UV radiation can be removed by treatment with moist ammonia vapor (but not with water vapor alone) [50].

3.1 Introduction

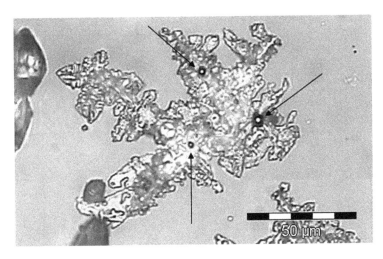

Fig. 3.5 Gray MF with visible metallic mercury

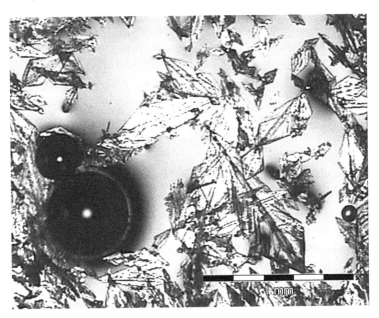

Fig. 3.6 Brown MF with visible metallic mercury

3.1.2.2 Solubility

MF is reported to crystallize in various forms depending on type of solvent used. The resulting form is an anhydride using ethanol or a complex salt using pyridine [28, 29, 35]. Many scientific books that deal with explosives reported that MF crystallizes as a hemihydrate from water [1, 28, 29, 35, 42]. However, this suggestion was disproved

Fig. 3.7 Influence of UV light on MF (*left* before, *center* after 10 min, *right* after 6 h under UV lamp—12 cm distance, 254 nm—15 W, 198 nm—15 W)

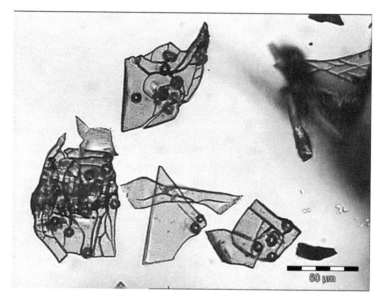

Fig. 3.8 Formation of metallic mercury inside MF crystals after irradiation by UV

by Kast and Selle [15, 51] and by Wöhler and Berthmann [46]. Solubility of MF in water is presented in Table 3.1 [28, 52]. It should, however, be noted that MF is not stable in aqueous solution. The decomposition is not significant at room temperature but becomes observable at higher temperatures (in which case mercury oxide, hydroxylamine, ammonia, and carbon dioxide form) [42, 53]. Despite this instability, purification of MF by dissolving in boiling water followed by decanting supernatant liquid and cooling is recommended by Walke. MF crystallizes in yellowish-white silky crystals [12]. The method of heating of MF with a large volume of water under pressure has been recommended for chemical decomposition of unwanted MF [38].

MF is soluble in many other solvents such as ammonia, pyridine, ethanolamines, and in solutions of inorganic salts (cyanides, iodides), and slightly soluble in ethanol and acetone. It is, however, not stable in these solutions (even in aqueous solutions) and decomposes due to its high reactivity. The rate of decomposition depends on the type of solvent, its concentration, and particularly on temperature. The rate of decomposition can be rapid at high temperature (within several

3.1 Introduction

Table 3.1 Solubility of mercury fulminate in water [28, 52]

Temperature (°C)	Solubility in 100 ml water (g)
12	0.071
49	0.174
100	0.77

minutes). MF is very slightly soluble in hot ether and insoluble in chloroform, glycerol, or benzene [15].

Aqueous ammonia is sometimes recommended for recrystallization of MF in which it is very soluble. MF of very high purity can be obtained by dissolving it in aqueous ammonia and, after filtration, the solution is neutralized with acetic or nitric acid obtaining the white form of MF [2, 28, 39, 42, 54]. This procedure is reported to increase the purity of MF from ~98.5 % to ~99.6 % [54]. Crystallization from aqueous ammonia without neutralization is reported to give white crystals. However, according to our own experience this yields large light yellow fishbone-like crystals (Fig. 3.9). A partial decomposition of MF during recrystallization does not affect the quality of the final product (solution darkens within several days at laboratory temperature, metallic mercury slowly forms and a gaseous decomposition product is liberated [42, 54]). The rate of decomposition in aqueous ammonia rapidly increases at higher temperature (60–65 °C) and mercury oxide, urea, guanidine, and other compounds are formed [1, 15].

Very good solubility is reported for ethanol and acetone saturated with ammonia. The system ammonia/ethanol/water is often recommended as a suitable solvent for preparation of nice well-formed pyramidal crystals (1–2 mm long) for crystallographic and X-ray studies [21, 25, 55, 56].

Recrystallization from pyridine (solubility—1 g MF in 6.9 ml pyridine [15]) is reported to produce complex salt crystals of unreported color [35, 39]. Wöhler and Weber reported the formation of lustrous mica flakes obtained upon pouring the pyridine solution of MF into ether [39]. The stability of MF in pyridine is reported higher than in ammonium hydroxide but it decomposes during long standing or boiling in the same way as it does in other solvents [15]. According to our investigation both brown as well as white MFs dissolve in hot pyridine giving a yellow to brown solution from which MF precipitates on slow cooling in the form of large thin white rhomboids (Fig. 3.10, left). In the case of fast cooling and stirring, these rhomboids break into shapes shown in Fig. 3.10, right. The precipitation of MF back from its solution is possible by pouring it into water or into diluted acid. A yellowish oil forms first which then turns into white crystals of MF (by pouring into diluted acid). Recrystallization from pyridine is recommended by Thorpe as the best method for purification of MF [38].

Mercury fulminate is also soluble in aqueous alkaline cyanides (e.g., KCN). The solubility depends on the concentration of cyanide solution (up to one-to-one ratio of MF to cyanide can be obtained) because the soluble double-salt $Hg(CNO)_2 \cdot KCN$ forms [15]. The MF in its white form can be precipitated back from this solution by addition of diluted nitric acid [3, 15, 57]. However, the purification of MF by recrystallization from cyanide solution is not effective, as the purity increases only slightly, e.g., from 98.39 to 98.60%. Boiling a cyanide

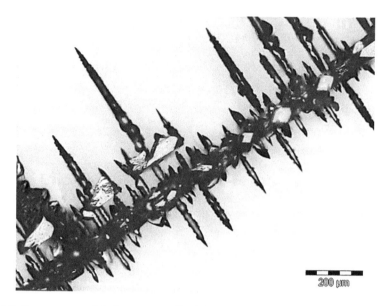

Fig. 3.9 Brown mercury fulminate recrystallized from ammonium hydroxide

Fig. 3.10 Brown mercury fulminate recrystallized from pyridine—*left*—fast cooling and agitation, *right*—slow cooling

solution causes MF slowly to decompose. It is also soluble in alcoholic and acetone solutions of alkaline cyanides [15].

The ethanolamines (mono, di, and tri) have been recommended as a solvent for MF by Majrich (200 g MF/100 g monoethanolamine at 25 °C, 40 g MF/100 g diethanolamine at 30 °C, and 28 g MF/100 g triethanolamine at 30 °C). The MF, however, rapidly decomposes with heat liberation and it is therefore recommended to keep the solution below 30 °C. The MF is precipitated back by pouring the solution into water or by diluting it with acid. The precipitate forms as a white powder [15]. MF is further soluble in many other solvents, e.g., aqueous potassium iodide or potassium thiocyanide.

3.1.2.3 Compatibility with Acids, Bases, and Metals

MF readily decomposes by action of many other chemical agents. It is relatively resistant to the action of dilute acids, e.g., nitric acid. Concentrated nitric acid decomposes it producing carbon monoxide, nitrogen(II) oxide, acetic acid, and mercuric nitrate [1, 3, 35]. The reaction is fast when using fuming nitric acid [15]. Concentrated hydrochloric acid decomposes MF according to the following equation [3]:

$$Hg(CNO)_2 + 2\,HCl \longrightarrow HgCl_2 + 2\,HCNO$$

$$HCNO + HCl \longrightarrow HC\begin{subarray}{l}\diagup Cl\\ \diagdown N-OH\end{subarray}$$

$$HC\begin{subarray}{l}\diagup Cl\\ \diagdown N-OH\end{subarray} + 2\,H_2O \longrightarrow NH_2OH \cdot HCl + HCOOH$$

Free fulminic acid is produced (with typical odor similar to hydrogen cyanide) along with the product of the reaction of fulminic acid with hydrochloric acid (*N*-hydroxyimidoformyl chloride). Concentrated sulfuric acid causes MF to explode, whereas the dilute acid (1:5) decomposes it without explosion [37]. Significant decomposition does not occur in cold dilute sulfuric acid [15]. Furthermore, MF is decomposed even by organic acids (e.g., formic, acetic, or oxalic acids) forming the corresponding mercuric salts [1].

MF also decomposes by the action of inorganic salts such as sulfides, or thiosulfates. MF decomposes to mercuric sulfide by action of aqueous alkaline sulfides. This method of decomposition is used for elimination of MF from waste water [12, 42] or for decomposition of small quantities of solid MF in which case warm ammonium sulfide solution is recommended [38]. The reaction of MF with sulfides or hydrogen sulfide is fast when boiling the mixture, but slow at ambient temperature. Alkaline thiocyanates give the MF double salts (e.g., Hg(CNO)$_2$·KSCN) [15]. The reaction with thiosulfates is very well documented because it is sometimes used for quantitative analysis of MF [10, 15, 29, 35, 58]:

$$Hg(CNO)_2 + 2\,Na_2S_2O_3 + H_2O \longrightarrow HgS_4O_6 + 2\,NaOH + NaCN + NaNCO$$

This reaction can also be used for chemical destruction of MF (on a lab scale). The weight excess 3:1 of sodium thiosulfate in form of 20 % solution is recommended. The decomposition in unsuitable conditions can lead to explosion (e.g., when MF is enclosed in detonators or if carried out at high temperature) [59]. Some poisonous hydrogen cyanide may be produced during the reaction [30].

Concentrated nitric acid is recommended as the decomposition agent for MF and hydrochloric acid as a decomposition agent on a laboratory scale [3]. The concentration of acid has an impact on the rate of decomposition. A 5 % hydrochloric acid mixture decomposes MF very slowly; the 18–20 % acid is faster but still safe and hence more suitable [15].

Most ordinary bases decompose MF even in low concentrations [3] (e.g., in aqueous ammonia solutions within several hours [1]). Decomposition of MF by boiling with aqueous bases is sometimes recommended for destruction of unwanted MF in the laboratory. The yellow form of mercury oxide forms during this decomposition [42].

Nesveda and Švejda proposed reduction of MF (pure or in priming mixtures) in water suspension with metallic magnesium. The decomposition carried out in this manner is suitable for recycling of mercury because the product of the reduction is elementary metal in a form useable for further processing [60].

Dry mercury fulminate does not react with common metals, according to some authors [3, 42] but other authors suggest the contrary (rapidly with aluminum, slowly with copper, zinc, bronze, or brass) [29, 30]. According to Bagal, pure dry MF does not react with metals, but metallic mercury present in the raw product reacts forming amalgams [15]. MF reacts immediately, or rapidly, in presence of low amounts of moisture [3, 42]. The moist MF form reacts with copper giving basic copper fulminate $Cu(CNO)_2 \cdot Cu(OH)_2$ which is less sensitive to impact but more sensitive to friction than MF itself. The presence of moisture is necessary for the formation of basic cupric fulminate. Another side effect of the reaction with copper is precipitation of mercury which forms an amalgam that may weaken a copper cap in which MF is embedded [15, 42]. It is therefore necessary to prevent contact of MF with copper in its applications by protective coating using lacquers or nickeling of the copper surface [15, 29, 34, 35]. Unlike lead azide it rapidly reacts with aluminum (over several hours) forming a large amount of Al_2O_3 and therefore it cannot be used in direct contact with aluminum in its applications. On the other hand, it does not react with nickel even when wet [42].

3.1.2.4 Thermal Stability

MF is significantly less thermally stable than LA. The weight loss of MF is about 7 % (vs. 0.3% for LA) within 45 days at 75 °C [42]. Danilov et al. reported that MF slowly decomposes at 60 °C within 1 month [3]; Urbański reported that decomposition of MF occurs even at 50 °C [1]. The dependence of storage time at 50 °C on the purity of MF and the effect of purification by recrystallization on thermal stability is shown in Fig. 3.11 [30].

MF slowly decomposes at moderate temperatures (as can be seen in Fig. 3.11). The opinions about when MF loses its ignition efficiency differ. The following durations were reported for the time it takes for MF to lose its ignition ability—3 years at 35 °C, 9–10 months at 50 °C, and 10 days at 80 °C [28, 54].

The stability of MF and its mixtures with other substances were measured using a vacuum stability test by Farmer. He found that the presence of moisture accelerates the decomposition of pure MF. The decomposition of dry MF is not accelerated in mixtures with common metals and weak organic acids [61]. MF thermally decomposes to gaseous products (mainly CO_2, some N_2, CO [61–65]) and a brown solid, insoluble and without explosive behavior, forms the remainder of a not very well identified mixture of compounds. It contains HgO (free or in the

3.1 Introduction

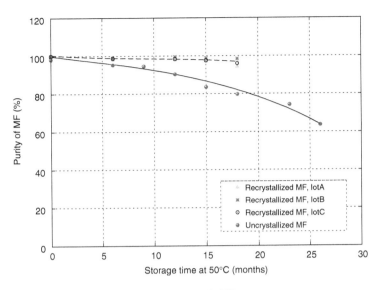

Fig. 3.11 Effect of storage at 50 °C on purity of MF [30]

complex salt HgCO$_3$·2HgO), cyanide complexes, linear chain of C$_n$N$_o$O$_p$, and cluster of Hg$_m$C$_n$N$_o$O$_p$ [63, 66]. These solid decomposition products and impurities accelerate the decomposition of MF [54, 61, 62].

The kinetics of the thermal decomposition of MF has been widely studied by many scientists [50, 61–65, 67]. The decomposition conditions (temperature, pressure, form of MF, weight of MF, etc.) have an impact on the process of decomposition. The decomposition is without explosion at lower temperatures, and the probability of explosion increases with rising temperature. Most reports suggest thermal decomposition up to 85 °C [29] or up to 100 °C [42] without danger of explosion. Explosions are mentioned at temperatures of 100 °C [61, 67, 68] and higher (105–115 °C [64, 65]). The exact temperature above which MF explodes depends on the form of MF and conditions of measurement. Garner [65], and later also Vaughan and Phillips [67], determined the dependence of induction period of explosion on temperature. The white form of MF ignites at a temperature slightly higher than for the brown form [1].

The storage of MF at higher temperatures can be hazardous. Carl reported several spontaneous violent decompositions (probably deflagration) of MF during long-term storage. The form of MF has an impact on the occurrence of this phenomenon. The explosions occurred when MF was pressed as the pressure increases the rate of decomposition [54].

3.1.2.5 Sensitivity

Mercury fulminate is more sensitive to impact than lead azide as was shown and discussed in Chap. 2. Sensitivity to impact rapidly decreases by addition of water. No detonation [35] or only partial detonations [48] were reported for small amounts of MF containing 5 % of water. MF with more than 10 % of water decomposes on impact without detonation [48]. Danilov et al. reported that MF becomes insensitive to mechanical stimuli with water content higher than 30 % [3]. Paraffin, glycerin, or oils could be used as desensitizing agents as well [3, 35, 40, 42]. On the other hand, the presence of hard particles (such as sand, glass), even in low amounts, significantly increases the sensitivity of MF to mechanical stimuli. Long exposure to higher temperatures significantly increases its sensitivity to impact (e.g., by several times when stored for several months even at 50 °C [54]).

Sensitivity to friction is reported lower than for LA by some authors [42, 69] and higher by others [29, 41]. Our own experiments indicate that MF is significantly less sensitive than LA (see Fig. 2.19). The difference between the brown and white modifications of MF is insignificant.

MF is highly sensitive to electrostatic discharge—ESD. The published values show a significant spread, and range from 0.51 to 0.62 mJ (Majzlík and Strnad unpublished work) [70] to 25 mJ [30]. The comparison of sensitivity to ESD with other primary explosives is shown in Fig. 2.21. MF can be initiated by discharge of static electricity generated from the human body [29].

Mercury fulminate is also sensitive to flame (see Fig. 2.22) and can be easily ignited by a safety fuse.

3.1.2.6 Explosive Properties

The explosive decomposition of mercury fulminate is not clearly understood but is often described as below [3, 38, 71, 72]:

$$Hg(CNO)_2 \longrightarrow Hg + 2\,CO + N_2$$

with heat of explosion being 1,540 kJ kg^{-1} (Hg in gas phase), 1,803 kJ kg^{-1} (Hg in liquid phase). On the basis of the analysis of the explosion products in a calorimetric bomb, Kast [43] later corrected his previous work by following a more precise equation for MF decomposition:

$$Hg(CNO)_2 \longrightarrow Hg + 1.9\,CO + 0.05\,CO_2 + 0.05\,C + N_2$$

The heat of explosion calculated from the above equation is 1,543 kJ kg^{-1} (Hg in gas phase) and the volume of liberated gases is 311 dm^3 kg^{-1} [43]. A similar value of 1,660 kJ kg^{-1} was published by Wöhler and Martin who determined heat of explosion in a calorimeter [57].

Dependence of detonation velocity on density is shown in Fig. 2.1. Mercury fulminate belongs to the group of primary explosives with a long predetonation zone. In other words, it means that it takes a long time, and uses significant amounts of charge, before the decomposition reaction accelerates from simple initial impulse to fully developed detonation (slow deflagration to detonation transition with respect to other primary explosives). This reflects in its lower initiating efficiency compared to silver fulminate or lead azide (see Fig. 2.2). The exact values depend heavily on the level of applied pressure, construction, and material of the detonator (see Sect. 2.2). MF shows interesting behavior at lower temperatures. The ignition efficiency of MF decreases with decreasing temperature (tested at +25 °C, −83 °C, and −190 °C) [73] but the detonation velocity does not change [16]. Danilov reported that MF reliably initiates detonation down to −100 °C [3].

Small amounts (several tenths of gram or grams) of unconfined and uncompressed MF only deflagrate with a faint puff when initiated by flame [38]. The burning speed markedly depends on the layer thickness; 0.5-mm-thick film burns with velocity 0.05 m s^{-1} but 2.75 mm thick layer with 8.5 m s^{-1} [74]. At low pressure, MF only burns. This process, as well as the pressure dependencies, has been very carefully investigated, primarily by Russian scientists [75–80]. Large amounts of MF, upon initiation by flame, detonate and can be safely destroyed by burning in a mixture with oil. Ignition of slightly confined MF (even between sheets of papers) leads to explosion [38].

The ability of MF to undergo deflagration to detonation transfer in detonators depends significantly on many factors, primarily on the magnitude of pressure used for pressing and the specific surface of MF. The initiation efficiency reaches its maximum at densities around 3.2 g cm^{-3} after which it steeply decreases to a point where it completely loses its capability to initiate secondary explosives at density 3.6 g cm^{-3} and above (it becomes dead pressed) [81]. The magnitude of applied pressure needed for pressing of MF to this density depends on many factors including the specific surface of MF (Fig. 2.6). The pressure values leading to dead pressed material depend significantly on the construction and material of the detonator. They are generally reported between 15 and 30 MPa [3, 44], although values as high as 60 MPa [36], 68 MPa [54], or even 172 MPa [29] are sometimes mentioned.

The brisance of MF in a sand test is reported as 37.3–48 % TNT [29] or 49 % TNT (vs. 39.6 % for LA) [30]. The power of MF measured by Trauzl test is 51 % TNT (vs. 39 % TNT for LA) [30] or 37–50 % TNT [29]. Both brisance (measured by sand test [29]) and power (measured by lead block test [71, 82]) of MF exceed those for LA.

The ignition temperatures of MF by heating at heating rate 5 and 20 °C are summarized in Table 3.2. The temperature at which explosion takes place within 5 s is 190 °C [54], about 200 °C [40], 205 °C [71], or 215 °C [83].

3.1.2.7 Preparation

As mentioned earlier, MF can be prepared in brown, gray, or white form. The gray form is spoiled material, its formation unintentional, and it can be avoided by maintaining optimized reaction conditions. The brown crystals form by dissolving mercury in nitric acid and pouring the solution into ethanol. The crystals form in solution after a short time. This method of preparation of MF is probably the only one used in industrial production.

Table 3.2 Dependency of ignition temperature of MF on heating

Heating rate (°C min^{-1})	Ignition temperature (°C)	Reference
5	160–165	[35]
5	160–170	[42]
5	180–210	[3]
20	166–175	[71]

Acetaldehyde, or substances which are convertible into acetaldehyde (paraldehyde), can be used instead of ethanol but, unlike in the case of ethanol, yield the white form. The reaction is more vigorous giving higher yields [28, 35, 84, 85]. If substances containing one (methanol, formaldehyde), 3, or 4 carbon atoms are used, instead of ethanol or acetaldehyde, MF is not formed, according to Martin [85]. Kibler, however, contradicts this and reported formation of the high purity white form from propyl alcohol [84]. The Hg^{2+} and Ag^+ cations probably catalyze some of the reaction steps [35]. Without these cations [35], or without the presence of nitrous oxides, fulminic acid is not formed [2, 86]. Due to oxidation processes in the reaction mixture only the higher oxidation state of mercury can originate and therefore mercurous fulminate is not formed in this way [57]. Metallic mercury can be substituted for mercury compounds (e.g., mercury nitrate, basic mercury sulfate) but the yield of MF is lower [9] and the reaction mixture can stand a long time at higher temperature (even several days) before the reaction starts to proceed [54]. Wieland proposed the following mechanism for MF formation [87]:

$$CH_3CH_2OH + HNO_3 \longrightarrow CH_3CH=O + HNO_2 + H_2O \quad (3.1)$$

$$CH_3CH=O + HNO_2 \longrightarrow O=N-CH_2-CH=O + H_2O \quad (3.2)$$

$$O=N-CH_2-CH=O \longrightarrow HON=CH-CH=O \quad (3.3)$$

$$HON=CH-CH=O + HNO_3 \longrightarrow HON=CH-COOH \quad (3.4)$$

$$HON=CH-COOH + HNO_3 \longrightarrow \underset{NO_2}{HON=C-COOH} + H_2O \quad (3.5)$$

$$\underset{NO_2}{HON=C-COOH} \longrightarrow \underset{NO_2}{HON=CH} + CO_2 \quad (3.6)$$

$$\underset{NO_2}{HON=CH} \longrightarrow O^-\!-\!N^{\pm}\!\equiv\!CH + HNO_2 \quad (3.7)$$

$$2\, O^-\!-\!N^{\pm}\!\equiv\!CH + Hg(NO_3)_2 \longrightarrow Hg(CNO)_2 + 2\, HNO_3 \quad (3.8)$$

Ethanol is first oxidized by nitric acid to acetaldehyde (3.1). The nitrosation of acetaldehyde proceeds in the second step to yield 2-nitrosoacetaldehyde (3.2),

3.1 Introduction

which then spontaneously isomerizes to the oxime (3.3). The aldehyde group is then oxidized by nitric acid to the glyoxylic acid oxime (3.4). The glyoxylic acid oxime is nitrated with nitric acid in the following step when nitroglyoxilic acid oxime is formed (3.5). This α-nitrocarboxylic acid easily thermally decarboxylates at 80 °C yielding nitroformaldehyde oxime (3.6) which further yields fulminic acid (3.7). The mercury fulminate forms in the last step of reaction (3.8) when fulminic acid reacts with mercury nitrate. This Wieland reaction mechanism is still considered correct. It is supported by the fact that ethanol can be replacedby the intermediates of the reaction (acetaldehyde, glyoxylic acid oxime, nitroformaldehyde oxime) [2]. However, the reaction mechanism quoted here has been disputed by Dansi et al. (cited in [16]) who claims (unlike source [2]) that MF cannot be prepared from pure glyoxylic acid oxime [3].

The side products of the reaction include ethylnitrate, ethylnitrite, oxalic acid, nitrogen oxides, and carbon dioxide. The raw MF ordinarily contains metallic mercury, mercuric oxalate, calomel, and mother liquor enclosed in crystals [15]. Traces of free mercury can be removed from MF by evaporation by placing it in vacuum desiccator for a few days [38]. The only industrially used method of preparation of MF is the one described above. The technology is described in great detail in the literature [3, 15, 35, 42, 48].

An alternative method of laboratory preparation of mercury fulminate is based on decomposition of the mercury salt of nitromethane published by Nef [14]. The mercuric salt is prepared in the first step of a reaction when mercuric chloride reacts with the sodium salt of nitromethane. The mercuric salt of nitromethane decomposes in a second step by boiling with dilute hydrochloric acid to produce MF [14]:

$$2\ CH_3NO_2 \xrightarrow{+\ NaOH} 2\ H_2C{=}N^+(O^-)(O^-)\ Na^+ \xrightarrow{+\ HgCl_2} H_2C{=}N^+(O^-)(O{-}Hg{-}O)N{=}CH_2$$

$$H_2C{=}N^+(O^-)(O{-}Hg{-}O)N{=}CH_2 \xrightarrow[\Delta T]{HCl} Hg(CNO)_2 + 2\ H_2O + Hg(O{-}N{=}CH{-}O{-}Hg{-}OH)(O{-}N{=}CH{-}O{-}Hg{-}OH)$$

Unfortunately, the yield of MF is too low (only about 5 %) because the majority of the mercuric salt of nitromethane is converted into a basic mercury salt of formhydroxamic acid (also an explosive). This mercury salt cannot be converted into MF [2]. The nitromethane itself can also be converted into fulminic acid by nitrosation with nitrous acid to form nitroformaldehyde oxime. It further decomposes (by heating in water or nitric acid) to fulminic acid which is trapped with mercury nitrate as mercury fulminate [2].

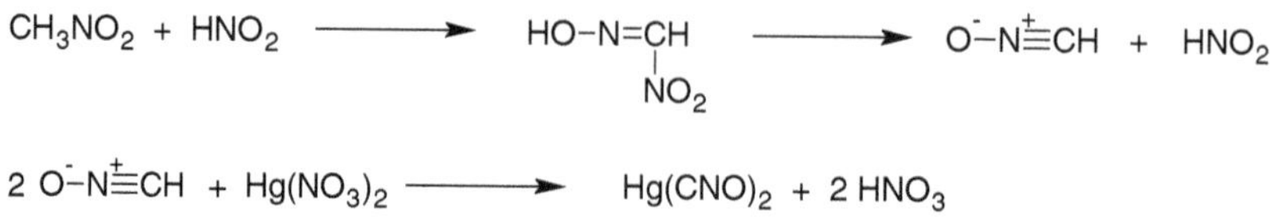

$$CH_3NO_2 + HNO_2 \longrightarrow \underset{NO_2}{HO-N=CH} \longrightarrow O^--N^{\pm}\equiv CH + HNO_2$$

$$2\ O^--N^{\pm}\equiv CH + Hg(NO_3)_2 \longrightarrow Hg(CNO)_2 + 2\ HNO_3$$

Mercury fulminate also forms from malonic acid, mercury, and nitric acid. The first intermediate (hydroxyiminomalonic acid) forms by nitrosation of the starting malonic acid which then decarboxylates to glyoxylic acid oxime. The following reaction is analogous to the Wieland mechanism presented above [2, 16, 35, 88]:

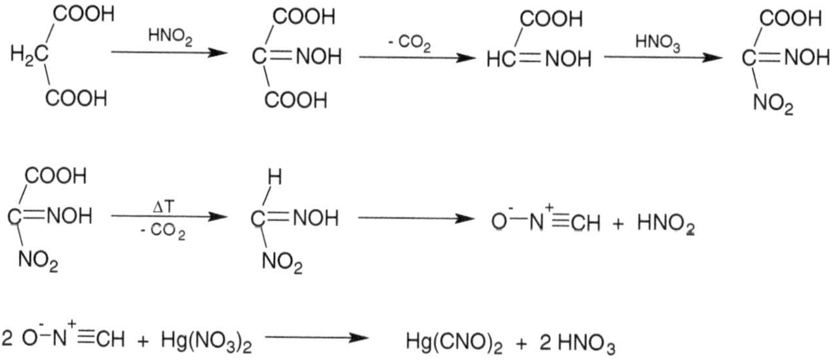

$$\underset{COOH}{\overset{COOH}{H_2C}} \xrightarrow{HNO_2} \underset{COOH}{\overset{COOH}{C=NOH}} \xrightarrow{-CO_2} \underset{}{\overset{COOH}{HC=NOH}} \xrightarrow{HNO_3} \underset{NO_2}{\overset{COOH}{C=NOH}}$$

$$\underset{NO_2}{\overset{COOH}{C=NOH}} \xrightarrow[-CO_2]{\Delta T} \underset{NO_2}{\overset{H}{C=NOH}} \longrightarrow O^--N^+\equiv CH + HNO_2$$

$$2\ O^--N^+\equiv CH + Hg(NO_3)_2 \longrightarrow Hg(CNO)_2 + 2\ HNO_3$$

According to Nef, metallic fulminates can also be synthesized by the reaction of derivates of formaldehyde oxime with the relevant nitrate [15].

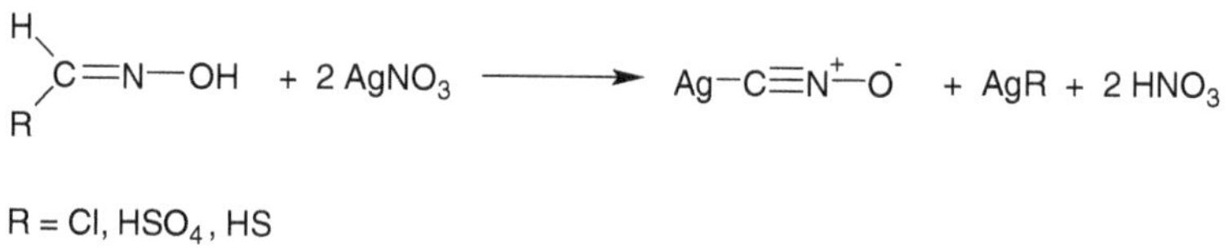

$$\underset{R}{\overset{H}{>}}C=N-OH + 2\ AgNO_3 \longrightarrow Ag-C\equiv N^+-O^- + AgR + 2\ HNO_3$$

R = Cl, HSO$_4$, HS

Other metallic fulminates cannot be synthesized directly by reaction of metal with nitric acid and sequentially with ethanol in the same way as mercury or silver fulminates (and probably also complex Na[Au(CNO)$_2$]). They are therefore mostly prepared by reaction of mercuric fulminate with the relevant amalgam.

3.1.2.8 Storage of MF

MF is highly sensitive when dry but can be safely stored under water at normal temperature for a long time and even transported [29, 40]. Decomposition of MF is very slow under water at normal temperatures. Ingraham reported that the purity of MF decreases from 99.6 to 98.3 % within 5 years with 6 % decrease of brisance. Sea water is not suitable for storage because MF decomposes faster than in fresh water (brisance decreases by 10 % after 2 years and 85 % after 3 years) [53]. At lower temperatures, when the danger of water freezing comes into consideration, it is possible to use a water–methanol mixture instead of water [28, 29].

3.1.2.9 Uses

In 1807, clergyman Alexander Forsyth patented the use of MF and shortly after it found a very broad use in military applications. The first massive use of MF was as a component in percussion priming mixtures. In the early percussion caps, MF was used alone simply agglomerated by means of wax or an aqueous solution of gum, and later in a mixture with other substances. However, the era of percussion priming started in 1786 (before MF discovery) with the French chemist Berthollet who discovered the possibility to initiate mixtures of potassium chlorate with combustible substances by percussion. Purely pyrotechnic compositions such as mixture of potassium chlorate and antimony sulfide were not very suitable for use due to their low conversion rate and insufficient flame ignition temperature. The next development therefore led to the incorporation of a primary explosive into the priming mixture. The presence of a primary explosive such as MF helped to overcome the problems of purely pyrotechnic mixtures. The compositions containing MF, potassium chlorate, antimony sulfide, glass powder, and gum Arabic were used exclusively for over 100 years until the beginning of the twentieth century [38, 42, 89].

The production of mercury fulminate significantly increased from 1867 after Alfred Nobel introduced his "Fulminate Blasting Caps." These caps were effectively the first detonators and were capable of reliably initiating detonation of nitroglycerine. Prior to this invention, initiating the detonation of nitroglycerine was done by fuses loaded with black powder with a very uncertain outcome [35]. Massive development of applications of MF in all types of initiating devices followed Nobel's invention. MF was practically the only primary explosive used at that time and its application ranged from primers of small and artillery ammunition to fuses and detonators. In 1887, Hess used MF, desensitized by addition of 20 % of paraffin, to produce a detonating cord with detonation velocity of 5,200–5,300 m s^{-1}. The use of MF in detonators was first modified by mixing it with black powder, then with potassium nitrate, and later with potassium chlorate. This last mixture was widely used and by 1910 had practically replaced pure MF in detonators in USA. The usual amount of potassium chlorate in the mixture ranged from 10 to 20 % [29, 35, 85, 90]. Such a mixture was cheaper than pure MF, but more sensitive to moisture [28].

The explosion of MF/KClO$_3$ 77.7/22.3 mixture can be described by the following equation:

$$3\ HgC_2N_2O_2 + 2\ KClO_3 \longrightarrow 3\ Hg + 2\ KCl + 6\ CO_2 + 3\ N_2$$

with volume of liberated gases 183 dm^3 kg^{-1}, less than with MF alone [38]. MF/KClO$_3$ mixtures are slightly more brisant than pure MF as shown in Fig. 3.12

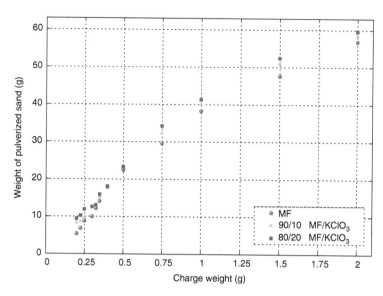

Fig. 3.12 Effect of amount of KClO$_3$ on brisance determined by sand test bomb [28]

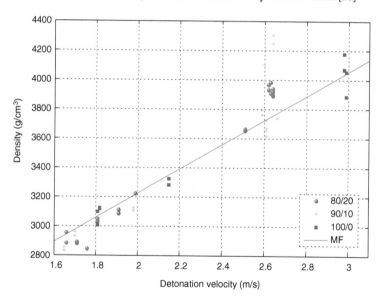

Fig. 3.13 Effect of density and the admixture of potassium chlorate on rate of detonation of MF [91]

by results of sand test [28, 29]. The detonation velocity of pure MF is higher than that of the mixtures at low densities but lower at higher densities (Fig. 3.13) [91]. The initiating efficiency of MF/KClO$_3$ exceeds that of pure MF (0.35 g alone MF vs. 0.30 g for MF/KClO$_3$ 90/10 mixture and 0.275 g for MF/KClO$_3$ 80/20 mixture for tetryl) [38]. Handling and loading of MF/KClO$_3$ mixture was slightly safer than

3.1 Introduction

pure MF according to Davis [28] while Bethelot (cited in [38]) reported that potassium chlorate renders the mixture very sensitive.

The idea of using a dual charge consisting of a small amount of primary explosive pressed on top of a secondary explosive was introduced by Nobel. He was investigating the possibility of lowering the cost of his detonators by replacing at least part of the amount of expensive MF with other explosives. His new detonator, described in French patent 184,129 from 1877, contained picric acid on top of which MF was compressed. The combination of MF and secondary explosive in the same body of the detonator was a great idea as it made the detonators both much cheaper and more powerful. The detonators with picric acid, however, had a major drawback in being sensitive to moisture. The picric acid, in moist environments, immediately reacted with the copper tube of the detonator, giving green cupric picrate. Such detonators were therefore not widely used and practically coexisted side by side with "standard" single component MF ones (sometimes referred to as plain detonators). The era of dominance of MF and its mixtures in detonators ended at the beginning of twentieth century when composite detonators with a partitioned charge (compound or composite caps) started to use explosives other than picric acid as a secondary charge. In 1900, Wöhler and Bielefeld independently proposed to replace picric acid with nitro aromatic substances. Many substances were tested and eventually lead to the incorporation of TNT which was already available at that time and later tetryl. Such detonators used only about 1/3 of the original amount of MF, were much safer to handle, and had higher brisance, as shown in Fig. 3.14 [90]. About the same time, a hollow cap pressed on top of the primary explosive started to be used. This simple measure further increased manipulation safety, decreased the amount of MF required, and increased the reliability of detonators [35, 42].

Although many other primary explosives were discovered, MF was used in a wide variety of applications due to its desirable properties over a long time (high density, flammability, stability, brisance, desired sensitivity, excellent flowability, etc. [35]). Sensitivity of pure MF for applications in detonators (stab or impact initiation) is sufficient for most applications as well as its sensitivity to flame and it is not necessary to alter them by addition of other substance [3]. The versatility of MF is due to its relatively long deflagration to detonation transition under the application conditions. Its capability to deflagrate with high temperature flame was exploited in primer compositions while its capability of steady detonation designated it for the use in detonators and detonating cords. The production of mercury fulminate increased during World War I even though LA was already known. The total production of MF at that time is unknown but in Germany alone production went from 104.8 tons in 1910 to 357 tons in 1917. Development of new primary explosives after World War I slowly displaced MF in priming mixtures and later also in detonators mainly due to its toxicity. As of today, MF is practically completely replaced in detonators and blasting caps by lead azide and in primers by tricinate/tetrazene or, most recently, by heavy metal-free compositions. [29, 34, 89]. It

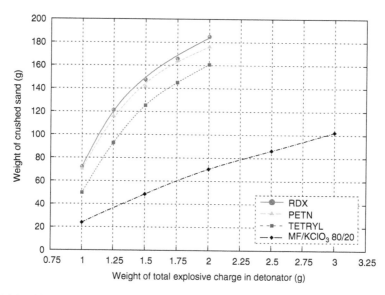

Fig. 3.14 Effect of base charge on initiating efficiency of detonators. The priming charge for all of these detonators was 0.75 g of MF/KClO$_3$ 80/20 [90]

seems probable that MF may still be used in some applications in Russia and China.

3.1.3 Silver Fulminate

The history of silver fulminate is just as long as the history of mercury fulminate. It was probably also discovered by the alchemists Cornelius Drebbel and Johann Kunckel von Löwenstern in the seventeenth century [11, 26] even though some other authors mention Brugnatelli in 1798 [29, 35, 92] or Edward Howard a few years later [15, 35, 92, 93].

3.1.3.1 Physical and Chemical Properties

Silver fulminate (SF) is a white crystalline material with heat of formation 179 kJ mol^{-1} [15]. It crystallizes in the form of small rosettes or star-shaped clusters. Two polymorphic forms have been reported—orthorhombic and trigonal. The crystal shape depends on reaction conditions (temperature, concentration) during silver fulminate preparation [94]. Recrystallization from ammonia leads to needles and multiple growths of leaf-like habits [92]. Long (4–5 mm) needle-shaped crystals were prepared by Singh and investigated by X-ray analysis. It was found that the crystal is orthorhombic [95]. The crystal structure of SF was later investigated in detail by Britton and Dunitz [96, 97] and their results were confirmed by Barrick et al. [98]. Ammonium acetate solution (~20 %) is recommended as the solvent for

3.1 Introduction

Fig. 3.15 Silver fulminate crystals

preparation of SF in the form of thin plate crystals to be used for crystallographic studies [25]. The crystal shape we have obtained at 80 °C starting from AgNO$_3$ and using NaNO$_2$ as a source of NO$_x$ is shown in Fig. 3.15.

Britton and Dunitz observed that SF prepared from an ammoniacal solution after ammonia evaporation is a mixture of needles and multiple growths which tended to a leaf-like habit. The X-ray examination showed that there were two polymorphic forms present, one orthorhombic and the other trigonal, but both forms occurred as needles and as part of the leaf-like clusters. The crystal density determined by X-ray is 4.107 g cm^{-3} for the orthorhombic crystals and 3.796 g cm^{-3} for the trigonal crystals [96, 97]. The density of SF determined formerly by Wöhler and Weber is 4.09 g cm^{-3} [39] or 3.938 g cm^{-3} according to Singh [95]; both determined it by pycnometry.

Silver fulminate is a very poisonous substance. According to Walke it has a strong bitter metallic taste [12].

Just like other silver salts, SF darkens when exposed to light. It is practically insoluble in cold water and can therefore be re-crystallized from hot water with high yields [35]. The solubility of silver fulminate in water at various temperatures is given in Table 3.3 [28, 29]. Solubility of SF in water is considerably higher than silver halogenides.

Silver fulminate is soluble in ammonia and solutions of alkali chlorides forming complex salts NH$_4$[Ag(CNO)$_2$] and K[Ag(CNO)$_2$], respectively. An interesting complex—Ag[Ag(CNO$_2$)]—forms from more concentrated aniline solutions [2].

It is further soluble in alkali cyanides, pyridine, and potassium iodide. Thiosulphate decomposes SF in a similar manner to that of mercury fulminate and may be used for nonexplosive decomposition of this substance. SF is insoluble in nitric acid [28, 35]. Alike MF, silver fulminate reacts with concentrated hydrochloric acid. This

Table 3.3 Solubility of silver fulminate in water [28, 29]

Temperature (°C)	Solubility in 100 ml water (g)
13	0.0075
30	0.018
100	0.25

decomposition is of nonexplosive nature; dry SF decomposes only with hissing noise [99]. Silver fulminate immediately explodes on touch with concentrated sulfuric acid or in contact with chlorine and bromine but not in contact with iodine [28, 35, 100]. The reaction with concentrated sulfuric acid has been used for chemical initiation of explosion [28, 35]. The silver fulminate prepared by precipitation from an aqueous solution of zinc fulminate (prepared by reaction of MF with metallic zinc in water) is more sensitive to chlorine and bromine than commonly prepared SF. It explodes even in presence of 1 % of chlorine [100]. According to Bagal this behavior is probably caused by a small amount of zinc fulminate in SF [15]. Silver fulminate is not hygroscopic and it can remain for years under water at ordinary temperatures without decomposing. It has been reported that SF did not lose its explosive properties even after 37 years under water [93, 101].

3.1.3.2 Sensitivity

Silver fulminate is considered as an extremely dangerous primary explosive to handle due to its high sensitivity, particularly to electric discharge and friction. It is more sensitive than MF with respect to both of these initiation stimuli. The sensitivity to electric discharge is extreme, particularly when it is dry [92]. The sensitivity of silver fulminate depends on its crystal form; the amorphous form is less sensitive to impact than the crystalline form. However, since it is practically impossible to produce purely amorphous material without any crystals the whole mass might be nearly as sensitive as the crystalline form itself [29]. Taylor and Buxton published preparation of SF in form of fine crystals with an impact sensitivity significantly lower than MF (no explosion from 32.7 cm for 1/2 kg hammer vs. only 12.7 cm for MF) approaching values typical for lead azide [93]. Comparison of impact sensitivity of SF with other common primary explosives is shown in Fig. 2.15.

Silver fulminate can explode when moist or even under water [12, 35]. It is known as one of the trickiest primary explosives. A number of accidental initiations have been reported during preparation or handling of SF. Unsuspected explosions have been reported by Liebig who had many years' experience with handling even large amounts—up to 100 g (!) [8]. Collins and Holloway [92] describe an explosion when turning a thermometer in the reaction mixture during SF preparation, when a small wet sample of SF was placed on the glass slide for microscopic examination, and even during filtration under suction. The most probable cause of the last two explosions was an electric discharge. The first was not fully explained but it is believed to be related to the increased rate of nucleation of SF crystals on addition of water (which decreases the solubility of SF in the reaction mixture) [92]. Taylor and Rinkenbach described explosion during filtration and washing of SF after its preparation [94]. Ignition of silver fulminate by exposure to light has been reported by Berchtold and Eggert [102]; Walke reported that extremely sensitive silver fulminate (exploded upon the slightest touch) forms when dried in sunlight after synthesis [12].

3.1.3.3 Explosive Properties

Silver fulminate is a more effective initiating agent than MF although it has nearly the same brisance by sand test [29]. The presence of moisture does not affect its brisance [15]. The initiating efficiency is very high being 2–3 times higher compared to that of MF (0.14 g SF vs. 0.35 g MF without reinforcing cap, 0.07 g SF vs. 0.24 g MF with reinforcing cap both for 0.4 g of TNT in num. 6 detonator capsule [93]).

The only value of detonation velocity the current authors were able to find is 1,700 m s^{-1} reported for 0.5 mm thick unconfined film initiated by hot wire [103].

SF is reported to become dead pressed by loads above 33 MPa in an amount of 120 mg in 1.5 gr type detonator (1 gr = 0.0648 g). For preparation of 100 mg charges it was necessary to press the material in two increments to avoid dead pressing [92].

The 5-s explosion temperature of SF is 170 °C compared to 217 °C for MF and 327 °C for LA [83]. Silver fulminate explodes at 236–241 °C at high heating rates (10–20 °C min^{-1}) and 186–193 °C at slow heating rate (0.2 °C min^{-1}) (Pachman and Matyáš unpublished work). The heat of explosion of SF is 1,970 kJ mol^{-1} (determined in calorimeter [57]).

3.1.3.4 Preparation

Preparation of silver fulminate is similar to the preparation of MF. Silver is dissolved in nitric acid after which the solution is poured into ethanol [14, 29, 33, 35, 93]. The reaction mechanism is the same as in the case of MF and the overall reaction may be simplified by the following equation:

$$Ag + HNO_3 + C_2H_5OH \longrightarrow AgCNO$$

Silver in the above-mentioned reaction may be replaced by silver nitrate. In this case it is dissolved in water, and nitric acid is added followed by ethanol. Temperature is one of the key factors of the reaction. According to Taylor and Rinkenbach it is necessary to heat the reaction mixture to 80–90 °C. When the temperature of reaction mixture is below 30 °C, product does not form even when standing for weeks and only a low yield is obtained after 1 month at 30–34 °C [94]. Addition of small amounts of sodium nitrite to the reaction mixture is often recommended to generate nitrogen oxides necessary for the reaction. These nitrogen oxides form by reaction of nitrite with nitric acid [35, 42, 92, 94, 95]. The reaction conditions affect the size and shape of SF crystals; highly sensitive needle crystals or relatively insensitive fine powder are the two possible forms. The large highly sensitive needle crystals were prepared, e.g., by Liebig [15]. The preparation of fine crystalline and relatively insensitive SF was published by Taylor and Buxton [15, 93].

Silver fulminate also forms when passing nitrogen oxide gas through an ethanol solution of silver nitrate [35, 42]. Other methods of SF preparation are almost analogous to preparation of MF (see Sect. 3.1.2.7).

3.1.3.5 Uses

A combination of the high cost of silver and the extreme sensitiveness of silver fulminate has prevented its wider use in commercial and military applications. Detonators containing silver fulminate (first used by Turpin in 1885 for initiation of picric acid) were apparently used only by the Italian Navy [35, 48]. It was also used in small quantities in fireworks and pyrotechnic toys such as "snaps," or "pullcrackers" for making noise [12, 13, 28, 36, 38, 104, 105].

3.1.4 Other Fulminates

3.1.4.1 Physical and Chemical Properties

In contrast to mercury and silver fulminates, most other metallic fulminates are too sensitive, or physically or chemically unstable. Further, their preparation is too expensive and demanding for practical use. Many fulminates are hygroscopic; stable when dry but decompose in presence of carbon dioxide when moist (cadmium, copper(I), copper(II), thallium) [15, 29, 57, 106].

Alkaline fulminates—sodium and potassium—are soluble in methanol, not soluble in acetone and ethanol, and insoluble in ether and benzene [39, 107]. Sodium fulminate explosively decomposes by action of sulfuric acid in the same way as MF and SF do [15]. The spontaneous explosion is reported even during drying above sulfuric acid [108]. The density of alkali fulminates is similar to alkali azides; sodium fulminate 1.92 g cm^{-3} and potassium fulminate 1.80 g cm^{-3} [39]. Sodium fulminate forms an anhydride or monohydrate depending on preparation procedure. Alkaline fulminates can be stored for a long time in the form of methanol solutions in the dark. These fulminates are hygroscopic; they are not stable in contact with moisture and quickly decompose when wet (white color changes to yellow and brown with loss of explosive properties) [15, 107]. Toxicity of sodium fulminate is about the same as that of sodium cyanide [2].

Thallium fulminate is photosensitive; it turns yellow under light. It very slowly becomes brown and loses explosive properties on exposure to air (unless sealed in air tight container). Thallium fulminate is hygroscopic; it explosively decomposes by action of sulfuric and nitric acid. It slowly decomposes to a nonexplosive substance on long-term heating [15, 57, 109].

3.1 Introduction

Cadmium fulminate forms white crystals that are soluble in alcohol. It is hygroscopic but dry cadmium fulminate is stable (in desiccator above $CaCl_2$) [15].

As briefly mentioned in the previous chapter, fulminates are capable of forming a variety of complexes with metals [8, 39]. The first of the fulminato(metal) complexes (complexes with fulminate ligands) was $K[Ag(CNO)_2]$. This compound was obtained by Liebig from the reaction of SF and potassium chloride [8]. Ammonium fulminate NH_4CNO has not yet been synthesized. A complex salt containing the ammonium cation $NH_4[Ag(CNO)_2]$ can be easily prepared by the reaction of the ammonium salt with SF in methanol [39].

The simple gold (I and III) fulminate has not yet been prepared. Complex $Na[Au(CNO)_2]$, however, exists and is formed by the reaction of sodium fulminate with gold chloride or by direct reaction of $AuCl_3$ with nitric acid and ethanol. The compound is slightly soluble in water, stable at room temperature, and relatively stable against concentrated nitric and hydrochloric acid [39, 110].

Some of the fulminato(metal) complexes are stable only with water of crystallization present and decompose on drying. Examples of such complexes are $Na_4[Fe(CNO)_6]\cdot 18H_2O$ or $Na_4[Fe(CN)_5CNO]\cdot H_2O$. Whether such complexes exhibit explosive properties is unknown to the authors. Anionic (fulminato)metal complexes of alkali and alkaline earth metals were reported by Wöhler as "highly explosive." Explosive properties have also been reported for $Na_2[Hg(CNO)_4]$. The possibility to stabilize fulminate complexes of transition metals by "diluting" energy-rich metal fulminate groups with large cations ($AsPh_4^+$, PPh_4^+, NR_4^+) or with large neutral ligands (PPh_3) has been investigated by Beck [8].

3.1.4.2 Sensitivity

As opposed to alkaline azides which do not have properties of explosives, alkaline fulminates are mostly reported as highly sensitive and explosive substances [8, 107, 108] even though one source mentioned sodium fulminate as not so sensitive (impact sensitivity for NaCNO to be 32 cm with 0.5 kg hammer compared to 7.5–10 cm MF under the same conditions) [27]. Sensitivity of these fulminates is reported as extreme and handling a hazardous operation [8, 107, 108]. Extreme sensitivity is further reported for the rubidium and cesium salts. Alkaline fulminates undergo explosion when initiated by flame, even in small amounts, whereas mercury fulminate only deflagrates. The exact sensitivity data are, however, not reported in this work [107]. Sensitivity of cadmium fulminate to impact is about the same as that of MF, sensitivity of thallium fulminate is higher [15, 57].

3.1.4.3 Explosive Properties

Cupric fulminate and particularly cadmium fulminate are powerful primary explosives. The initiating efficiency of cadmium fulminate is even higher than that of silver fulminate. The initiating temperature and initiation efficiency of some fulminates are summarized in Table 3.4. The ignition efficiency of thallium fulminate is significantly lower than for MF. Wöhler and Martin reported that it is about 20 times lower than LA [57]. In a lead block test, sodium fulminate acts with power of 58 % of MF [112].

The explosive properties of complex fulminates depend on composition. For example, $NH_4[Ag(CNO)_2]$ is explosive but less explosive than SF; $Mg[Ag(CNO)_2]$ detonates violently whereas pyridine complexes only deflagrate or burn [39]. $Na[Au(CNO)_2]$ explodes by action of flame with "sharper" detonation than other complex fulminates [39, 110].

3.1.4.4 Preparation

Fulminates other than those of Hg and Ag (and also complex $Na[Au(CNO)_2]$ from $AuCl_3$ [110]) cannot be prepared directly by reaction of the metal with nitric acid and subsequently with ethanol like MF or SF [15, 33]. Several reasons exist that make direct formation of other fulminates impossible. Wöhler and Martin reported that solubility of most fulminates in the reaction mixture is a reason why only MF and SF can form directly. Most other fulminates are soluble and unstable in the reaction mixture, in which they decompose [57]. Another reason is that the fulminic acid too readily forms complex salts and therefore simple fulminates cannot be isolated [35]. A catalytic effect of noble metals (Hg, Ag) on some of the reaction steps of fulminate formation was reported by Krauz. He mentioned the analogy of the catalytic effect of mercuric ions on hydrocarbons that are oxidized to the relevant nitrophenols in dilute nitric acid [35].

The preparation of alkaline fulminates and fulminates of alkaline earth metals was not successful for a long time. The reaction of MF or SF with relevant salts was not successful due to the formation of double salts. The first successful preparation of sodium fulminate was published by Carstanjen and Ehrenberg only in 1882; about 80 years after the discovery of MF [15, 113].

The majority of fulminates (except those of SF and MF) are prepared by reaction of relevant amalgam of less noble metal with mercury (or silver) fulminate in methanol or ethanol and precipitated by diethyl ether [29, 39, 57, 106, 107, 114].

$$Hg(CNO)_2 + CdHg_x \xrightarrow{MeOH} Cd(CNO)_2 + (x+1)\,Hg$$

Alkaline fulminates form the anhydride while fulminates of alkaline earths (Ca, Sr, Ba) contain one molecule of alcohol (from alcoholic reaction environment). The alcohol cannot be removed from the molecule without its decomposition [15, 39]. The sodium fulminate can also be prepared in an aqueous environment but product forms as the monohydrate [8, 15, 113]. The fulminate turns yellow during evaporation of water [15, 113].

The preparation in an aqueous environment is not recommended for the majority of fulminates due to low stability of the reaction mixture. The hydrolytic dissociation results in free fulminic acid which tends to polymerize [15, 57]. Preparation of

3.1 Introduction

Table 3.4 The explosive parameters of metallic fulminates

	Heat of explosion [57] (kJ g^{-1})	Initiating temperature (°C)		Initiation efficiency [111]	
		Explosion^a [83]	Explosion at heating [107]	Tetryl (g)	TNT (g)
NaCNO	–	215	210–220	–	–
KCNO	–	225	200	–	–
RbCNO	–	–	195	–	–
CsCNO	–	–	220–225	–	–
Cd(CNO)$_2$	1.97	215	–	0.008	0.11
Cu(CNO)$_2$	2.13	205	–	0.025	0.15
TlCNO	0.933	120	–	0.300	–

^{a}Explosion within 5 s, weight 0.02 g

the sodium salt in aqueous environment is also not convenient for another reason—the product formed is not pure and is highly sensitive. A pure and more stable sodium salt forms using ethanol, or preferably methanol, as a reaction environment instead of water [2]. The use of a protective atmosphere of inert gas is recommended during preparation and vacuum drying for moisture sensitive fulminates (e.g., cadmium fulminate) [15, 57].

Significant purification of alkaline fulminates after preparation is recommended especially in case of rubidium and cesium ones (dissolving in methanol and re-precipitated by addition of diethylether). The importance of purification is to avoid the formation of double salts with MF that are more sensitive than the original fulminates themselves (mentioned double salts spontaneously detonate at 45 °C) [107].

The other way of preparation of fulminates is via the sodium salt. For example, thallium fulminate forms as a fine powder by reaction of sodium fulminate with thallium acetate in methanol [25].

$$CH_3COOTl + NaCNO \xrightarrow{CH_3OH} TlCNO + CH_3COONa$$

Some metallic fulminates also form directly by reaction of powdered metal with MF in a water suspension while boiling (e.g., Zn, Al, Cu) [15, 48, 105]. The preparation of thallium fulminate without use of an amalgam starting from MF and thallium is a good example, described by Hawley [109]:

$$2 Tl + Hg(CNO)_2 \xrightarrow{H_2O} 2 TlCNO + Hg$$

The (fulminato)metal complexes are prepared from aqueous sodium fulminate solutions and transition metal salts followed by addition of tetraphenylarsonium chloride [8]. The aqueous solutions of sodium fulminate can be directly prepared from MF and sodium amalgam in water [8].

$$2\,NaHg_x + Hg(CNO)_2 \xrightarrow{H_2O} 2\,NaCNO + (2x+1)\,Hg$$

$$2\,NaCNO + AuCl \longrightarrow Na[Au(CNO)_2] + NaCl$$

3.1.4.5 Uses

Metallic fulminates (except mercury and silver) have never been practically used as explosives due to the difficulties with their preparation and generally low physical and chemical stability. However, the sodium salt can be used for many applications in organic chemistry. This salt is more suitable for this application than the more easily accessible mercury or silver salts due to their tendency to form complexes in solutions. For example, the sodium salt is useable in the preparation of free fulminic acid by acidification of its solution with dilute sulfuric acid [2].

References

1. Urbański, T.: Chemistry and Technology of Explosives. PWN — Polish Scientific Publisher, Warszawa (1967)
2. Grundmann, C., Grunanger, P.: The Nitrile Oxides, Versatile Tools of Theoretical and Preparative Chemistry. Springer, Berlin (1971)
3. Danilov, J.N., Ilyushin, M.A., Tselinsky, I.V.: Promyshlennye vzryvchatye veshchestva; chast I. Iniciiruyushchie vzryvshchatye veshchestva. Sankt-Peterburgskii gosudarstvennyi tekhnologicheskii institut, Sankt-Peterburg (2001)
4. Sarlo, F., Guarna, A., Brandi, A., Goti, A.: The chemistry of fulminic acid. Tetrahedron **41**, 5181–5185 (1985)
5. Wieland, H., Baumann, A.: Zur Kenntnis der polymeren Knallsäuren. Justus Liebigs Annalen der Chemie **392**, 196–213 (1912)
6. Wieland, H., Baumann, A., Reisenegger, C., Scherer, W., Thiele, J., Will, J., Haussmann, H., Frank, W.: Die Polymerisation der Knallsäure. Isocyanilsäure und Erythro-cyanilsäure. Justus Liebigs Annalen der Chemie **444**, 7–40 (1925)
7. Ulpiani, C.: Per la costituzione degli acidi fulminurici. Sull'acido metafulminurico. Gazzetta Chimica Italiana **46**, 1–48 (1916)
8. Beck, W.: The first chemical achievements and publications by Justus von Liebig (1803–1873) on metal fulminates and some further developments in metal fulminates and related areas of chemistry. Eur. J. Inorg. Chem. 4275–4288 (2003)
9. Kurzer, F.: The life and work of Edward Charles Howard FRS. Ann. Sci. **56**, 113–141 (1999)
10. Kurzer, F.: Fulminic acid in the history of organic chemistry. J. Chem. Educ. **77**, 851–856 (2000)
11. Winnewisser, M., Winnewisser, B.P.: Proč trvalo objasňování struktury třaskavé kyseliny 174 let. Chemické listy **70**, 785–807 (1976)
12. Walke, W.: Lectures on Explosives. Wiley, New York (1897)
13. Wisser, J.P.: Explosive Materials – The Phenomena and Theories of Explosion. D. Van Nostrand, New York (1907)
14. Nef, J.U.: Ueber die Constitution der Salze der Nitroparaffine. Justus Liebigs Annalen der Chemie **280**, 263–342 (1894)
15. Bagal, L.I.: Khimiya i tekhnologiya iniciiruyushchikh vzryvchatykh veshchestv. Mashinostroenie, Moskva (1975)
16. Urbański, T.: Chemistry and Technology of Explosives. Pergamon, Oxford (1984)

17. Pauling, L., Hendricks, S.B.: The prediction of the relative stabilities of isosteric ions and molecules. J. Am. Chem. Soc. **48**, 641–651 (1926)
18. Beck, W., Feldl, K.: Die Struktur der Knallsäure HCNO. Angew. Chem. 746 (1966)
19. Shapley, W.A., Bacskay, G.B.: A gaussian-2 quantum chemical study of CHNO: Isomerization and molecular dissociation reactions. J. Phys. Chem. A **103**, 6624–6631 (1999)
20. Sarlo, F., Brandi, A., Fabrizi, L., Guarna, A., Niccolai, N.: Multinuclear magnetic resonance study of organomercury fulminates. The structure of mercury fulminate in solution. Org. Magn. Reson. **22**, 372–375 (1984)
21. Beck, W., Evers, J., Göbel, M., Oehlinger, G., Klapötke, T.M.: The crystal and molecular structure of mercury fulminate (knallquecksilber) [1]. Zeitschrift für anorganische und allgemeine Chemie **633**, 1417–1422 (2007)
22. Beck, W., Klapötke, T.M.: Mercury fulminate: ONC–Hg–CNO or CNO–Hg–ONC – a DFT study. J. Mol. Struct. THEOCHEM **848**, 94–97 (2008)
23. Türker, L., Erkoç, S.: Density functional theory calculations for mercury fulminate. J. Mol. Struct. THEOCHEM **712**, 139–142 (2004)
24. Bottaro, J.: Ideas to Expand Thinking About New Energetic Materials, pp. 473–501. World Scientific Publishing, New Jersey (2005)
25. Iqbal, Z., Yoffe, A.D.: Electronic structure and stability of the inorganic fulminates. Proc. R. Soc. Lond. A Math. Phys. Sci. **302**, 35–49 (1967)
26. Nielsen, A.H.T.: Knaldkviksølvets og knaldsølvets opdagelseshistorie. Dansk kemi **12**, 17–21 (1996)
27. Marshall, A.: Explosives. Butler and Tanner, London (1917)
28. Davis, T.L.: The Chemistry of Powder and Explosives. Wiley, New York (1943)
29. Fedoroff, B.T., Sheffield, O.E., Kaye, S.M.: Encyclopedia of Explosives and Related Items. Picatinny Arsenal, New Jersey (1960–1983)
30. Tomlinson, W.R., Sheffield, O.E.: Engineering Design Handbook, Explosive Series of Properties Explosives of Military Interest, report AMCP 706-177 (1971)
31. Davis, T.L.: Kunckel and the early history of phosphorus. J. Chem. Educ. **4**, 1105–1113 (1927)
32. Davis, T.L.: Kunckel's discovery of fulminate. Army Ordnance **7**, 62–63 (1926)
33. Howard, E.: On a new fulminating mercury. Philos. Trans. R. Soc. Lond. **90**, 204–238 (1800)
34. Brede, U., Hagel, R., Redecker, K.H., Weuter, W.: Primer compositions in the course of time: from black powder and SINOXID to SINTOX compositions and SINCO booster. Propellants Explosives Pyrotechnics **21**, 113–117 (1996)
35. Krauz, C.: Technologie výbušin. Vědecko-technické nakladatelství, Praha (1950)
36. Blechta, F.: Dnešní stav otázky náhražek třaskavé rtuti. Chemický obzor **3**, 330–336 (1928)
37. MacDonald, G.W.: The discovery of fulminate of mercury (1800). Arms Explosives **19**, 24–25 (1911)
38. Thorpe, E.: A Dictionary of Applied Chemistry, pp. 79–84. Longmans, Green, London (1928)
39. Wöhler, L., Weber, A.: Neue Salze der Knallsäure. Berichte der deutschen chemischen Gesellschaft **62**, 2742–2758 (1929)
40. Zhukov, B.P.: Energeticheskie kondesirovannye sistemy. Izdat, Yanus-K, Moskva (2000)
41. Meyer, R., Köhler, J., Homburg, A.: Explosives. Wiley-VCH Verlag GmbH, Weinheim (2002)
42. Špičák, S., Šimeček, J.: Chemie a technologie třaskavin. Vojenská technická akademie Antonína Zápotockého, Brno (1957)
43. Kast, H.: Weitere Untersuchungen über die Eigenschaften der Initialsprengstoffe. Zeitschrift für das gesamte Schiess- und Sprengstoffwesen **21**, 188–192 (1926)
44. Strnad, J.: Iniciační vlastnosti nejpoužívanějších třaskavin a vývoj nových metodik jejich měření. Institute of Chemical Technology, Dissertation thesis, Pardubice (1972)
45. Batrnett, B.: Explosives. D. van Nostrand, New York (1919)
46. Wöhler, L., Berthmann, A.: Über Verunreinigungen und einige Eigenschaften des technischen Knallquecksilbers. Zeitschrift für das angewandte Chemie **43**, 59–63 (1930)

47. Langhans, A.: Graues und weißes Knallquecksilber. Zeitschrift für das gesamte Schiess- und Sprengstoffwesen **15**, 23–24 (1920)
48. Urbański, T.: Chemie a technologie výbušin. SNTL, Praha (1959)
49. Borocco, A.: Contribution à l'étude de l'action des rayons ultraviolets sur le fulminate de mercure. Comptes rendues de l'Académie des sciences **207**, 166–168 (1938)
50. Bartlett, B.E., Tompkins, F.C., Young, D.A.: The decomposition of mercury fulminate. J. Chem. Soc. 3323–3330 (1956)
51. Kast, H., Selle, H.: Über den angeblichen Krystallwasser-Gehalt des Knallquecksilber. Berichte der deutschen chemischen Gesellschaft **59**, 1958–1962 (1926)
52. Holleman, A.F.: Notices sur les fulminates. Recueil des Travaux Chimiques des Pays-Bas **15**, 159–160 (1896)
53. Ingraham, W.T.: The deterioration of mercury fulminate when stored under water. Army Ordnance **10**, 201–202 (1929)
54. Carl, L.R.: Mercury fulminate, its autocatalytic reactions and their relation to detonation. J. Franklin Inst. **240**, 149–169 (1945)
55. Miles, F.D.: The formation and characteristic of crystals of lead azide and of some other initiating explosives. J. Chem. Soc. 2532–2542 (1931)
56. Singh, K.: Initiation of explosion in crystals of mercury fulminate. Trans. Faraday Soc. **52**, 1623–1625 (1956)
57. Wöhler, L., Martin, F.: Über neue Fulminate und Azide. Berichte der deutschen chemischen Gesellschaft **50**, 586–596 (1917)
58. Khmelnitskii, L.I.: Spravochnik po vzryvchatym veshchestvam. Voennaya ordena Lenina i ordena Suvorova Artilleriiskaya inzhenernaya akademiya imeni F. E. Dzerzhinskogo, Moskva (1962)
59. Dupré, F.H., Dupré, P.V.: Notes on the reactions of fulminate of mercury with sodium thiosulfate. Analyst **46**, 42–49 (1921)
60. Nesveda, J., Švejda, M.: Způsob získávání kovové rtuti z třaskavé rtuti a zařízení k provádění tohoto způsobu. CS Patent 267,047, 1990
61. Farmer, R.C.: The velocity of decomposition of high explosives in a vacuum. III. Mercuric fulminate. J. Chem. Soc. Trans. **121**, 174–187 (1922)
62. Vaughan, J., Phillips, L.: Thermal decomposition of explosives in the solid phase. III. The kinetics of thermal decomposition of mercury fulminate in a vacuum. J. Chem. Soc. 2741–2745 (1949)
63. Boddington, T., Iqbal, Z.: Decomposition of inorganic fulminates. Trans. Faraday Soc. **65**, 509–518 (1969)
64. Narayana, P.Y.: The thermal decomposition of mercuric fulminate. Curr. Sci. **13**, 313–315 (1944)
65. Garner, W.E., Hailes, H.R.: Thermal decomposition and detonation of mercury fulminate. Proc. R. Soc. Lond. A Math. Phys. Eng. Sci. **139**, 576–595 (1933)
66. Zhulanova, V.P., Bannov, S.I., Pugachev, V.M., Ryabykh, S.M.: Products of the radiation-chemical and thermal decomposition of mercury fulminate. High Energy Chem. **35**, 26–32 (2001)
67. Vaughan, J., Phillips, L.: Thermal decomposition of explosives in the solid phase. II. The delayed explosion of mercury fulminate. J. Chem. Soc. 2736–2740 (1949)
68. Miszczak, M., Milewski, E., Kostrow, R., Goryca, W., Terenowski, H.: An analysis of the test and assessment methods on physicochemical stability of primary explosives. In: Proceedings of 10th Seminar on New Trends in Research of Energetic Materials, pp. 805–811, Pardubice, Czech Republic, (2007)
69. Military Explosives. Report TM-9-1300-214, Headquarters, Department of the army, (1984)
70. Matyáš, R.: Investigation of properties of selected organic peroxides. University of Pardubice, dissertation, Pardubice, Czech Republic (2005)
71. Kast, H., Haid, A.: Über die sprengtechnischen Eigenschaften der wichtigsten Initialsprengstoffe. Zeitschrift für das angewandte Chemie **38**, 43–52 (1925)
72. Muraour, H.: Sur la theorie des reactions explosives. Cas particulier des explosifs d'amourcage. Mémories présentés a la Société chimique **51**, 1152–1166 (1932)

References

73. Kling, A., Florentin, D.: Action des basses températures sur les explosifs. Mémorial des poudres 17, 145–153 (1913)
74. Evans, B.L., Yoffe, A.D.: The burning and explosion of single crystals. Proc. R. Soc. Lond. A238, 325–333 (1956)
75. Belyaev, A.F., Belyaeva, A.E.: Issledovanie goreniya gremuchei rtuti. Zhurnal fizicheskoi khimii 20, 1381–1389 (1946)
76. Belyaev, A.F., Belyaeva, A.E.: O normalnoi skorosti i kharaktere "goreniya" nekotorykh initsiirujushchikh vzryvchatykh veshchestv. Doklady akademii nauk SSSR 52, 507–509 (1946)
77. Belyaev, A.F., Belyaeva, A.E.: Zavisimost skorosti gorjeniya initsiirujushchikh vzryvchatykh veshchestv ot davleniya. Doklady akademii nauk SSSR 56, 491–494 (1947)
78. Belyaev, A.F., Belyaeva, A.E.: Über das Brennen von Knallquecksilber bei einem Unterdruck. Doklady akademii nauk SSSR 33, 41–44 (1941)
79. Urbański, T., Stanuch, J.: Decomposition of initiating explosives under reduced pressure. Archiwum Procesów Spalania 4, 5–12 (1973)
80. Andreev, K.K.: O zavisimosti skorosti goreniya vtorichnykh i iniciiruyushchikh vzryvchatykh veshchestv ot davleniya. Doklady akademii nauk SSSR 51, 29–32 (1946)
81. Strnad, J.: Primary Explosives and Pyrotechnics—Lecture Notes. Institute of Energetic Materials, University of Pardubice, Pardubice (1999)
82. Wallbaum, R.: Sprengtechnische Eigenschaften und Lagerbeständigke der wichtigsten Initialsprengstoffe. Zeitschrift für das gesamte Schiess- und Sprengstoffwesen 34, 197–201 (1939)
83. Wöhler, L., Martin, F.: Azides; Sensitiveness of. J. Soc. Chem. Ind. 36, 570–571 (1917)
84. Kibler, A.L.: Synthesis of mercury fulminate from propyl alcohol. Orig. Commun. 8th Int. Congr. Appl. Chem. 25, 239–243 (1913)
85. Martin, G., Barbour, W.: Manual of Chemical Technology; III. Industrial Nitrogen Compounds and Explosives. Crosby Lockwood and son, London (1915)
86. Hodgkinson, W.R.: Fulminate of mercury. J. Soc. Chem. Ind. 37, 190t (1918)
87. Wieland, H.: Eine neue Knallsäuresynthese. Über den Verlauf der Knallsäurebildung aus Alkohol und Salpetersäure. Berichte der deutschen chemischen Gesellschaft 40, 418–422 (1907)
88. Adam, D., Karaghiosoff, K., Klapötke, T.M., Hioll, G., Kaiser, M.: Triazidotrinitro benzene: 1,3,5-$(N_3)_3$-2,4,6-$(NO_2)_3C_6$. Propellants Explosives Pyrotechnics 27, 7–11 (2002)
89. Hagel, R., Redecker, K.: Sintox – a new, non-toxic primer composition by Dynamit Nobel AG. Propellants Explosives Pyrotechnics 11, 184–187 (1986)
90. Grant, R.L., Tiffany, J.E.: Factors affecting initiating efficiency of detonators. Ind. Eng. Chem. 37, 661–666 (1945)
91. Carl, L.R.: The rate of detonation of mercury fulminate and its mixtures with potassium chlorate. Army Ordnance 6, 302–304 (1926)
92. Collins, P.H., Holloway, K.J.: A reappraisal of silver fulminate as a detonant. Propellants Explosives 3, 159–162 (1978)
93. Taylor, C.A., Buxton, E.P.: Silver fulminate, an initiator of detonation. Army Ordonance 6, 118–119 (1925)
94. Taylor, C.A., Rinkenbach, W.H.: A sensitive form of silver fulminate. Army Ordonance 6, 448 (1926)
95. Singh, K.: Crystal structure of silver fulminate. Acta Crystallogr. 12, 1053 (1959)
96. Britton, D.: A redetermination of the trigonal silver fulminate structure. Acta Crystallogr. C 47, 2646–2647 (1991)
97. Britton, D., Dunitz, J.D.: The crystal structure of silver fulminate. Acta Crystallogr. 19, 662–668 (1965)
98. Barrick, J.C., Canfield, D., Giessen, B.C.: A redetermination of the silver fulminate structure. Acta Crystallogr. B 35, 464–465 (1979)
99. Divers, E., Kawakita, M.: On the decomposition of silver fulminate by hydrochloric acid. J. Chem. Soc. Trans. 47, 69–77 (1885)

100. Davy, E.: An account of a new fulminating silver, and its application as a test for chlorine. Trans. R. Irish Acad. **17**, 265–274 (1837)
101. Peter, A.M.: On the stability of silver fulminate under water. J. Am. Chem. Soc. **38**, 486 (1916)
102. Berchtold, J., Eggert, J.: Über die Zündfähigkeit von Sprengstoffen dur Stahlung hoher Intensität. Naturwissenschaften **40**, 55–56 (1953)
103. Bowden, F.P., Williams, R.J.E.: Initiation and propagation of explosion in azides and fulminates. Proc. R. Soc. Lond. A Math. Phys. Sci. **A208**, 176–188 (1951)
104. Stettbacher, A.: Explosive toys. Chem. Abstr. **14**, 17792 (1920)
105. Cundill, J.P., Thomson, J.H.: A Dictionary of Explosives. Eyre and Spottiswoode, London (1895)
106. Hanus, M.: Méně známé třaskaviny. Synthesia a.s, VÚPCH, Pardubice (1996)
107. Hackspill, L., Schumacher, W.: Contribution à l'étude des fulminates alcalins. Rocznik Akademji Nauk Technicznych w Warszawie **3**, 84–89 (1936)
108. Ehrenberg, A.: Ueber Natriumfulminat. Journal für praktische Chemie **32**, 230–234 (1885)
109. Hawley, L.F.: Contributions to the chemistry of thallium. I. J. Am. Chem. Soc. **29**, 300–304 (1907)
110. Bos, W., Bour, J.J., Steggerda, J.J., Pignolet, L.H.: Reaction of gold(I) compounds with carbon monoxide to form gold clusters and fulminates. Inorg. Chem. **24**, 4298–4301 (1985)
111. Wöhler, L., Martin, F.: Die Initialwirkung von Aziden und Fulminaten. Zeitschrift für das gesamte Schiess- und Sprengstoffwesen **12**, 18–21 (1917)
112. Wöhler L., Matter, O.: Beitrag zur Wirkung der Initialzünding von Sprengstoffen. Zeitschrift für das gesamte Schiess- und Sprengstoffwesen 244–247 (1907)
113. Carstanjen, E., Ehrenberg, A.: Ueber Knallquecksilber. Journal für praktische Chemie **25**, 232–248 (1882)
114. Wöhler, L.: Neue Salze der Knallsäure und Stickstoffwasserstoffsäure und die explosiven Eigenschaften von Fulminaten und Aziden. Zeitschrift für das angewandte Chemie **27**, 335–336 (1914)

Chapter 4
Azides

Azides are substances containing the N_3^- group. They exist as inorganic salts, organic compounds, organo-metals, or complexes. For the purpose of this book, we have decided to include the inorganic and organic compounds. Some organic substances that contain the azido group are included in other chapters (e.g., tetrazoles, other substances). We have also decided to separate out complex compounds containing the azido group and place them into a separate chapter with other complexes.

4.1 Azoimide

$$HN_3$$

Azoimide (or hydrazoic acid) was discovered by Curtius in 1890 and in pure form was first prepared by Curtius and Radenhauser in 1891 [1, 2]. It is a colorless liquid with a strong odor which boils without decomposition at 37 °C and freezes at −80 °C [3–5]. Azoimide, even in gaseous form, is highly toxic with a very strong unpleasant odor [4]. The pure liquid form is dangerous to handle as it explodes with a blue-colored flash with the slightest mechanical or thermal shock. One of its discoverers was wounded by the explosion when he removed a test tube with azoimide from the cooling mixture [2, 6]; explosion can occur even by friction of an azoimide bubble (at boiling point of azoimide) against a sharp glass edge [4]. An aqueous solution containing as little as 17% azoimide can explode [7]. Pure liquid azoimide may be stored but, after months of storage, it has a tendency to undergo spontaneous explosion [5]. It is dangerous in gaseous form in which it may also spontaneously explode. The probability of explosion is higher at higher pressures. Azoimide explosively decomposes to nitrogen and hydrogen:

$$2\,HN_3 \longrightarrow H_2 + 3\,N_2 + 600.5\text{ kJ}$$

Heat of formation of azoimide is 300.25 kJ mol^{-1} [3] and density is 1.13 g cm^{-3} (from 0 to 20 °C) [5]. The detonation velocity is 8,100 m s^{-1} according to Urbański [8]

R. Matyáš and J. Pachman, *Primary Explosives*,
DOI 10.1007/978-3-642-28436-6_4, © Springer-Verlag Berlin Heidelberg 2013

or 7,000–7,500 m s^{-1} according to Danilov [3]. On contact with flames, it explodes with a blue flash. Azoimide is weakly acidic and may be liberated from its salts using acids [4, 6]. Aqueous solutions react with common metals such as iron, copper, or aluminum even when quite diluted (7% m/m). More concentrated solutions react also with silver or gold. A mixture with hydrochloric acid dissolves gold and platinum. Platinum black decomposes azoimide to ammonia and nitrogen. Inorganic salts of azoimide have chemical properties similar to those of halides [4].

4.2 Lead Azide

$$Pb(N_3)_2$$

The discovery of lead azide (LA) is attributed to Curtius who first prepared this substance and characterized its explosive properties in 1891 [2]. In 1893, some experiments with lead, silver, and mercury were carried out in Spandau in Prussia. However, an unexpected explosion occurred during testing of azides with fatal results, which caused termination of further experiments. They were not re-started until 1907 when Wöhler drew attention to azides once again as he saw it as a possible substitute for expensive MF [5].

The real era of LA started in 1908 after Hyronimus patented its use as "a primer for mines and fire-arms consisting of charge of trinitride of lead" [9]. In 1911, Wöhler proposed the use of LA as a substitute for MF in military applications. Problems with industrial application of LA, mostly related to its high sensitivity, were overcome in the 1920s and, despite the previous tragic accidents and numerous doubts about its practical utilization, lead azide gradually superseded mercury fulminate as a primary explosive [8]. The manufacture of LA probably began in Germany around 1914 following the extensive investigations by Wöhler. Science itself was, however, not persuasive enough to convince the armies to use LA. The development of the course of World War I played a key role in the incorporation of LA into munitions. The Italian advance upon Gorizia brought the Austrian mercury mines within their range. This created a problem as the mercury was used for preparation of MF. Central powers had no choice but to replace MF with LA where possible. Since 1920, the use of LA has developed considerably and other nations followed the Germans [4, 10, 11]. Significant production of LA is reported from Russia since 1929 [3].

4.2.1 Physical and Chemical Properties

Lead azide forms white or yellowish crystals. It is known to form four allotropic modifications: α, β, γ, and δ (older literature refers only to the first two of them). The orthorhombic α-form with crystal density reported from 4.68 to 4.716 g cm^{-3} [12–14] is the main product of precipitation, with traces of other forms present, and is the only form acceptable for technical applications [15]. A variety of crystal

4.2 Lead Azide

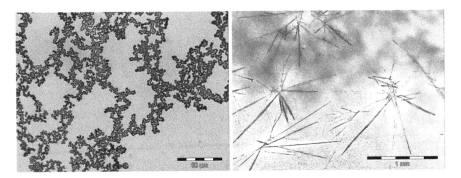

Fig. 4.1 Crystals of lead azide: left α-form, right β-form

structure modifiers was tested for modification of the crystal form of precipitated LA. The most commonly noted ones include dextrin, carboxymethyl cellulose, and PVA. The presence of dextrin promotes formation of the α-form [16] while the presence of organic dyes (eosin, erythrosin, or neutral red) at precipitation time enhances the formation of the β-form [15]. The crystals of β lead azide are formed in the shape of long needles (Fig. 4.1).

The dry monoclinic β-form has a crystal density reported from 4.87 to 4.93 g cm^{-3} [12, 13] and is stable. In older literature, the β-form is referred to as extremely sensitive to mechanical stimuli with the formation of long needles that may explode simply by the breaking of a single crystal [11]. Crystals 1 mm in length were reported liable to explode spontaneously because of the internal stresses within them [17]. This extreme sensitivity of the β-form was, however, later disproved and shown to be a common myth (more in part on sensitivity of LA).

The other two allotropic forms can be obtained from pure reagents by maintaining the pH in the range 3.5–7.0 for monoclinic γ and 3.5–5.5 for triclinic δ. They usually precipitate from the solution simultaneously [8]. The γ-form is also created in presence of polyvinyl alcohol [18]. A method of exclusive preparation of a particular polymorph (γ, δ) was not known to Fair who recommended hand selection under the microscope as the only possible way, since the crystals differ in shape and size [15]. γ LA may be prepared reproducibly by using PVA (degree of polymerization does not play a role) which is free from unhydrolyzed polyvinyl acetate [19].

The α-form is the most stable of all four versions. For example, the enthalpy of formation of the γ-form is higher by 1.25 kJ mol^{-1} than the α-form [3]. Thermal decomposition of both the α-form and β-form has been studied and various results were reported. At temperatures around 160 °C, the β-form transforms irreversibly to the α-form [20].

LA is practically insoluble in cold water, ammonia, and in most common organic solvents. It decomposes in boiling water liberating azoimide [6, 8]. LA is soluble in solutions of sodium or ammonium acetate (see Sect. 2.6), acetic acid, and ethanolamine (Table 4.1).

Lead azide is photosensitive and its crystals turn yellow and later gray in daylight. Lead and nitrogen form during photodecomposition. Decomposition, accompanied by

Table 4.1 Solubility of lead azide [16, 21]

Temperature (°C)	Solubility in 100 ml solvent (g)	
	Water	Conc. aqueous sodium acetate
18	0.023	1.54
70	0.090	–
80	–	2.02

the change of color, takes place only in the outer layer and does not propagate throughout the crystal. The decomposition of the surface layer is not reflected noticeably in any deterioration of explosive properties of the bulk material [3, 5, 15, 22]. Lead azide can be ignited by high-intensity light (argon flash) [23]. LA decomposes even when illuminated under water in which case basic lead azide forms.[1] Wöhler and Krupko described this reaction by the following sequence of equations [24]:

$$Pb(N_3)_2 + 2 H_2O \longrightarrow Pb(OH)_2 + 2 HN_3$$

$$Pb(N_3)_2 \xrightarrow{h\nu} Pb + 3 N_2$$

$$Pb + HN_3 + 2 H_2O \longrightarrow Pb(OH)_2 + N_2 + NH_3$$

$$Pb(OH)_2 + Pb(N_3)_2 \longrightarrow PbO \cdot Pb(N_3)_2 + H_2O$$

Thermal stability of dry LA in absence of moisture is quite good and degradation at 60 °C after 12 months or at 80 °C after 1 month is scarcely detectable. The presence of water plays the most important role in decomposition of this substance and it has been reported that heating periods at 250 °C in vacuum, and hence excluding moisture, had to be increased 80 times to produce the same reduction of detonation velocity compared to the same trials under moist air [25, 26]. Gray and Waddington published heat of formation of LA 484 kJ mol^{-1} [27]; however, according to Yoffe its decomposition is reported to release 443 kJ mol^{-1} [28].

$$PbN_6 \longrightarrow Pb + 3 N_2 + 443 \text{ kJ}$$

Metallic lead, formed during thermal decomposition of LA, may react with oxygen from the air and form lead oxides [29]. LA is a weak photoconductor [7].

4.2.2 Chemical Reactivity

Lead azide decomposes by action of acids liberating azoimide and the relevant lead salt. It easily reacts even with weak acids (such as acetic acid, carbonic acid,

---

[1] Basic forms of lead azide are described by various formulas most often as PbO·Pb(N$_3$)$_2$ or Pb(OH)N$_3$.

4.2 Lead Azide

atmospheric CO_2, or even lead styphnate) liberating azoimide. An example of such decomposition is its reaction with nitric acid which decomposes LA into azoimide and lead nitrate [3].

$$Pb(N_3)_2 + 2\,HNO_3 \longrightarrow Pb(NO_3)_2 + 2\,HN_3$$

The too narrow blank in comparison with others liberation of poisonous azoimide limits the use of acids themselves for decomposition of LA. This problem can be overcome simply by addition of sodium nitrite which eliminates azoimide, since the reaction changes in the following way [3, 5]:

$$2\,Pb(N_3)_2 + 6\,HNO_3 + 2\,NaNO_2 \longrightarrow 2\,Pb(NO_3)_2 + 2\,NaNO_3$$
$$+ N_2O + 6\,N_2 + 3\,H_2O$$

It is therefore recommended to add sodium nitrite solution before using acid for destruction of unwanted LA (or also sodium azide) residues in the laboratory or even in industrial applications [3, 5, 21]. Urbański recommends the use of 8 % solution of sodium nitrite and 15 % nitric acid for LA [30], whereas 92 % sulfuric acid is recommended for sodium azide [5]. Many other reactions have been proposed for the decomposition of LA, including reaction with sodium polysulfide [21] or dissolving LA in ammonium acetate and adding sodium or potassium bichromate until no more lead chromate precipitates [5].

Lead azide is stable in air under normal conditions when dry. However, it slowly decomposes in presence of moist air containing carbon dioxide. Detailed analysis of the reactions of LA with water and carbon dioxide has been presented by Lamnevik [31]. According to this author, basic lead azide forms and gaseous azoimide is liberated by reaction of LA with moisture:

$$Pb(N_3)_2 + H_2O \rightleftharpoons Pb(OH)N_3 + HN_3$$

If the partial pressure of carbon dioxide is lower than 1.2 kPa azoimide, basic lead azide forms in a similar way as with water (see above). At higher carbon dioxide partial pressures, dibasic lead carbonate and also azoimide form according to the following equation:

$$3\,Pb(N_3)_2 + 4\,H_2O + 2\,CO_2 \rightleftharpoons Pb_3(CO_3)_2(OH)_2 + 6\,HN_3$$

Both reactions are in equilibrium and reach that state at very low partial pressure of azoimide. For example, the equilibrium pressure of azoimide in the first reaction is 5.3 Pa (at 25 °C). To give an idea, this pressure can be reached by the reaction of 0.6 mg of LA with 0.04 mg water in space with volume of 1 l. Therefore, the normal moisture content of LA is more than sufficient to establish equilibrium conditions. The deterioration of LA due to these reactions is therefore not significant as only a very small amount reacts before reaching equilibrium. When the system with lead azide is not well sealed azoimide diffuses into atmosphere and equilibrium conditions are not obtained. The consequence is gradual conversion of LA to basic lead azide or basic lead carbonate. This is one of the reasons why detonators with LA must be hermetically sealed. The other problem related to degradation of LA due to azoimide consumption, and hence impossibility of reaching equilibrium, is the reaction of LA with some components of the system mainly with some metals like copper and its alloys or zinc commonly used in construction of detonators. The reaction of azoimide with metals, especially copper, is without question the biggest drawback of LA. Various copper azides that form as a result of this reaction exhibit differing degrees of sensitivity, some being more sensitive than lead azide itself. It is therefore necessary to prevent lead azide coming into contact with copper in its applications. Aluminum, stainless steel, tin, and lead do not react with lead azide even in presence of water. The preferred material for applications containing LA is therefore aluminum [5, 8, 32–34]. Lead azide is also incompatible with many polymeric substances [35].

It was originally assumed that the deterioration of LA in a detonator, because of its decomposition by carbon dioxide and the formation of basic lead azide or basic lead carbonate, would decrease its initiation efficiency and hence decrease overall ability of the detonator to perform with the desired strength. Danilov et al., however, published that the lead carbonate that forms during the reaction of LA with carbon dioxide creates a surface layer which protects LA from further decomposition [3]. The mechanism of deterioration of LA depends upon hydrolysis conditions. According to Blay and Rapley, if the hydrolysis is not accelerated by abnormal conditions, the deterioration does not proceed beyond an acceptable level in service detonators [36]. The same thinking was reported in 1975 by Lamnevik who did not notice any loss of function in an LA detonator due to LA degradation [3, 31].

Bases only decompose the surface layer, which turns to lead oxide and the relevant azide. The lead oxide protects LA from further attack by the base [3]:

$$Pb(N_3)_2 + 2\,NaOH \longrightarrow PbO + 2\,NaN_3 + H_2O$$

Other possible reactions are described for decomposing LA by hydroxides. The two reactions below are such examples. The first takes place with a sodium hydroxide solution concentration of 10 % while the second takes place with the concentration at 20 % [8]. Some authors even suggest the formation of basic salts $Pb(N_3)_2 \cdot xPbO$ [15, 34].

$$Pb(N_3)_2 + 2\,NaOH\ (10\ wt.\ \%) \longrightarrow Pb(OH)_2 + 2\,NaN_3$$

$$Pb(N_3)_2 + 4\,NaOH\ (20\ wt.\ \%) \longrightarrow Na_2PbO_2 + 2\,NaN_3$$

Lead azide easily reacts with many inorganic salts. Its reaction with cerium(IV) sulfate [21, 34] or with ammonium cerium(IV) nitrate [3, 5] is used for quantitative analysis or decomposition of unwanted LA:

4.2 Lead Azide

Table 4.2 Sensitivity of LA polymorphs to impact, friction and electrostatic discharge (EDS) [37]

	Service LA[a]	β-LA[b]	γ-LA
Impact (50 %) (cm)	15.24	15.04	7.35
Friction (50 %) (m s^{-1})	1.36	1.32	1.92
ESD (μJ)	2	1.2	1.1

[a]Service lead azide refers to "British military service LA" whose manufacture is based on nucleation by 2.5 % of lead carbonate (formed from sodium carbonate and lead acetate) [38].
[b]Crystals were broken up by a rubber spatula and therefore were of mixed particle size.

Table 4.3 Sensitivity of LA polymorphs to impact, friction, and electrostatic discharge (ESD) [19]

	Service LA[a]	90 % Service LA 10 % β-LA	90 % Service LA 10 % γ-LA
Impact (50 %) (cm)	15.24	11.91	10.31
Friction (50 %) (m s^{-1})	1.36	1.10	1.09

[a]Service lead azide refers to "British military service LA" whose manufacture is based on nucleation by 2.5 % of lead carbonate (formed from sodium carbonate and lead acetate) [38].

$$Pb(N_3)_2 + 2\,Ce(SO_4)_2 \longrightarrow PbSO_4 + Ce_2(SO_4)_3 + 3\,N_2$$

$$Pb(N_3)_2 + 2\,(NH_4)_2Ce(NO_3)_6 \longrightarrow Pb(NO_3)_2 + 4\,NH_4NO_3 + 2\,Ce(NO_3)_3 + 3\,N_2$$

4.2.3 Sensitivity

The sensitivity of lead azide depends on its crystalline form. Experimental values of the sensitivities of α, β, and γ-forms published by Wyatt are summarized in Table 4.2. The γ modification of LA is more sensitive to impact but less sensitive to friction than the α and β forms. The electrostatic discharge sensitivity is about the same for all three polymorphs according to the author. The difference between 1.2 μJ and 2.0 μJ was claimed to be not too significant. On the other hand, the sensitivity of the γ-form to mechanical stimuli and electrostatic discharge is reported not to be significantly higher than that of the α and β forms according to Taylor and Thomas [18].

Table 4.3 shows that the addition of both β- and γ-LA to the service LA increases its sensitivity to friction and impact. It was further reported that crystal size does not influence sensitivity of α-LA [19].

The α-form of lead azide is less sensitive to impact than MF and approaches PETN (see Fig. 2.15). Sensitivity to friction is reported lower than for MF by some authors [3, 4, 21, 39] and higher by others [5, 40]. This inconsistency in published results is most likely due to a different crystal form of samples tested [34]. According to our own results (see Fig. 2.19) is LA significantly more sensitive to friction than MF and its sensitivity exceeds other commonly used primary explosives (e.g., LS, GNGT, DDNP). The sensitivity of pure LA and dextrinated one is about the same [41, 42]. The sensitivity of LA to mechanical stimuli is generally influenced by several factors: by its method of preparation, by crystal size (see Fig. 2.16), and by the presence of side products with particles softer than LA (e.g., basic lead carbonate decreases its sensitivity) [4, 21]. The presence of water does not decrease its sensitivity as significantly as it does in the case of MF [6]. Other authors even report the sensitivity for wet and dry LA to be the same [17]. The sensitivity to mechanical stimuli may be diminished, just as with MF, by addition of desensitizing substances such as paraffin, glycerin, or oils [21]. Dextrinated lead azide is generally considered less sensitive to mechanical stimuli than pure LA [31]. Mixtures with harder particles (e.g., glass) are significantly more sensitive than LA [4].

An interesting finding was published by Clark et al. They observed that a mixture of LA with tetryl is more sensitive to impact (maximum at about 10 % content of LA) than pure LA even though tetryl crystals are considerably softer than crystals of LA. A reasonable explanation of this phenomenon is a chemical reaction between LA and tetryl. Tetryl decomposes to picric acid, which happens readily on heating or impact. The picric acid so formed further reacts with lead azide to produce a minute amount of highly sensitive lead picrate. However, no accident as a consequence of this reaction was observed [43].

LA has relatively low sensitivity to flame which is one of its biggest drawbacks. This low sensitivity is further lowered by formation of basic lead carbonate on its surface. This layer of carbonate isolates LA and makes the initiation by flame difficult, and as mentioned earlier, the formation of carbonate is favored in a humid and CO_2-rich atmosphere (e.g., in mines). In order to introduce and maintain good ignitability by flame, LA is mixed with some other primary explosive sensitive to flame (particularly lead salts of di- or trinitroresorcine).

LA also shows very low sensitivity to stab initiation. This problem is usually solved by adding harder particles or tetrazene to the mixture.

4.2.4 Explosive Properties

Lead azide explosively decomposes to nitrogen and metallic lead. The volume of gaseous products is only 231 dm^3 kg^{-1} [3]. Dependence of detonation velocity on density is shown in Fig. 2.1. Danilov et al. [3] published equation for determination of detonation velocity of LA in dependence on its density:

$$D_\rho = D_0 + 860(\rho - \rho_0) \text{ m.s}^{-1}, \text{ where } \rho_0 \text{ is 2 g.cm}^{-3}$$

and detonation pressure:

$$P = (58\rho - 99).10^2 \text{ MPa}$$

The detonation velocity, D_0, of LA is 3,880 m s^{-1} at density 2 g cm^{-3} and it is slightly higher than for SA (3,830 m s^{-1} at the same density) [3]. Lead azide easily detonates in thin layers with high detonation velocity. The detonation velocity for

4.2 Lead Azide

Table 4.4 Properties of pure, dextrinated and gelatined (G.A.M) LA [44]

	Pure LA	Dextrinated LA	G.A.M
Bulk density (g cm^{-3})	1.6–1.8	2.1–2.3	1.3–1.5
Particle size (μm)	–	60–100	200–400
Detonation velocity (m s^{-1})[a]	4,550	3,850	3,750
Ignition efficiency to PETN (g)[a]	0.02	0.03	0.04
Ignition efficiency to RDX (g)[a]	0.04	0.06	0.07

[a]Measured in 6 mm copper tube; pressed at the same pressure.

0.45-mm-thick sheet is 5,500 m s^{-1} [26]. The dependence of LA's detonation velocity on thickness is shown in Fig. 2.11. The detonation velocity in slim line diameters (10–200 μm) was recently measured by Jung [22]. He observed accelerative reaction of LA after ignition (depends on sample diameter) before reaching steady detonation velocity. The steady detonation velocity was in the range of 2,670–3,440 m s^{-1} for diameters from 50 to 200 μm.

Initiation efficiency of LA is significantly higher than that of MF. Unlike mercury fulminate, LA belongs to the group of primary explosives with a short predetonation zone, so-called detonants—substances with a fast deflagration to detonation transition with respect to other primary explosives. This is reflected in its high initiating efficiency compared to MF. The method of LA preparation, its composition, density, confinement, physical dimensions—all influence initiation efficiency. Even the presence of a crystal structure modifier during LA preparation has an impact on initiation efficiency. The efficiency of pure LA is superior to LA prepared in presence of a colloid agent. According to Lamnevik, approximately twice the amount of dextrinated LA must be used compared to pure LA to get the same output from a secondary explosive in a detonator [31]. Gelatin was one of the modifying agents tested in the 1960s and at first it was regarded as less favorable than dextrin. Later it was found that electrostatic sensitivity was excellent but cohesion was poor. This was overcome by addition of a small amount of molybdenum disulfide. LA precipitated in presence of gelatin and molybdenum disulfide is known as G.A.M. Although it had excellent electrostatic and friction sensitivity its flowability was not as good as that of dextrinated LA and was influenced by gelatin type [44]. The comparison of initiation efficiency of pure LA with azide prepared in presence of a colloid agent is presented in Table 4.4.

In practice, the quantity of LA used in detonators is generally several times higher than the minimum amount required to cause detonation of the secondary explosive. This ensures satisfactory performance for a number of years [44].

The brisance and initiating efficiency of LA enclosed in standard detonators is not significantly affected by decomposition of LA following the reaction with water and carbon dioxide in the air during long-term storage. In the cases when there appears risk of function loss, the humidity inside the detonator should be checked. Aging of detonators was investigated by Lamnevik who found that detonators which failed after more than 20 years of storage functioned properly after being dried [31].

The power of LA (measured by Trauzl block) is relatively low, 36.7 % TNT [40] or 39 % TNT [45], which is lower than for MF [46–48].

Lead azide is a primary explosive with exceptionally high thermal stability. The ignition temperature of LA is relatively high, 330 °C at a heating rate of

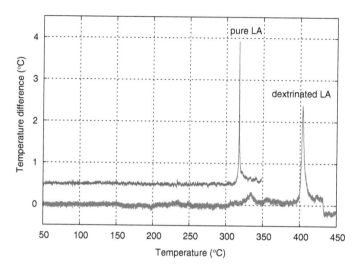

Fig. 4.2 Graph DTA thermogram of pure and dextrinated α-LA showing exceptional thermal stability of the dextrinated product (heating 5 °C min^{-1}, 5 mg samples, static air atmosphere, open test tube)

5 °C min^{-1} (LA type RD 1333) [35]; 357–384 °C at a heating rate of 40 °C min^{-1} [43]. The 5-s explosion temperature of LA is 340 °C [49]. LA has been found not to have changed after being stored for 4 years at 80 °C. The thermal stability threshold is, according to Danilov, 200 °C at which temperature it keeps its explosive properties for 6 h. It is therefore used in thermostable detonators TED-200 with maximum application temperature 200 °C [3]. We have tested various types of lead azides at our laboratory and found that standard industrial dextrinated LA (product of Austin Detonator) is surprisingly even more thermally stable than pure crystalline product. The decomposition of both types (pure and dextrinated) of LA under the same conditions is shown in Fig. 4.2.

LA also does not lose brisance at low temperatures [50] and can be used as an initiating substance at temperatures as low as the boiling point of oxygen (−183 °C) [3].

As opposed to mercury fulminate, pure LA cannot be dead-pressed in normal circumstances (even by a pressure of 200 MPa). However, LA prepared in presence of a crystal structure modifier (dextrin, carboxymethyl cellulose, and others) or desensitized (e.g., paraffin) can be dead-pressed; dextrinated LA cannot be dead-pressed below 118 MPa and paraffined below 78 MPa [3].

4.2.5 Preparation

The first preparation of LA, by Curtius, was based on addition of lead acetate to a solution of sodium or ammonium azide [51]. This has not changed and, even today,

4.2 Lead Azide

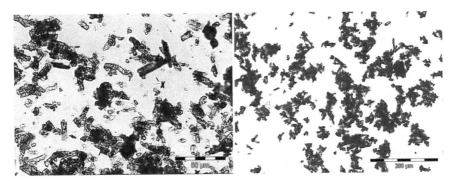

Fig. 4.3 Comparison of lead azide prepared with the colloid agents—dextrin and PVA

preparation of lead azide is based on the reaction of sodium azide and lead nitrate or acetate followed by precipitation of the insoluble lead azide:

$$2\,NaN_3 + Pb(NO_3)_2 \longrightarrow Pb(N_3)_2 + 2\,NaNO_3$$

A summary of published LA preparation methods before the 1930s is presented in Audrieth's review [11]. Preparation and technology of LA production is described in many books that deal with explosives in detail [3, 5, 8, 15, 21, 30, 46].

The pure α-form of LA can be grown from a solution of LA in ammonium acetate or acetic acid [7]. In the early years, the main problem in the manufacture of LA using the above reaction was the formation of a very sensitive product—desired α-LA mixed with other forms. Re-crystallization was industrially unacceptable and this problem was therefore solved by addition of a colloid, fast agitation, and use of diluted solutions. The colloid agent is added to the reaction mixture to prevent growth of large crystals and to support formation of the α-form with suitable shape and size. With the colloid present, crystals of LA form agglomerates (consisting of small crystals Fig. 4.3) with better flowability than the pure crystalline form (compare with Fig. 4.1). Furthermore, the colloid decreases the sensitivity of the LA product. The first colloids used in this application were gelatin and dextrin [5, 44, 52]. Some other tested surface active agents include polyvinyl alcohol and the sodium salt of carboxymethylcellulose [8].

Dextrin does not cover the formed crystals but is rather entrapped inside. The dependency of the amount of dextrin inside the crystals of LA on the concentration of the dextrin in the solution is shown in Fig. 4.4.

It is evident that the absorption of colloid at first increases rapidly with the concentration in the reacting solution, but less and less rapidly as the concentration rises. The curve for 60 °C becomes almost level more rapidly than the others and tends to reach much lower limiting absorption. The data for 20 °C show that when the acidity of the dextrin was not neutralized, the limit of absorption was more readily reached.

At 60 °C the influence of this acidity was found to be very small [16]. When dextrin is added to already developed crystals it adsorbs on the surface. The amount

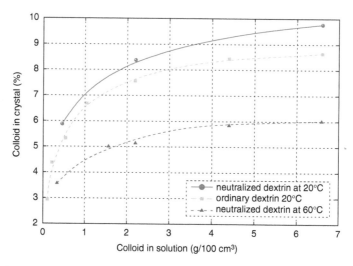

Fig. 4.4 Absorption of dextrin by crystals of LA [16]

adsorbed on the crystal surface in the solution with concentrations in range from 0.2 to 2.17 % was from 0.151 to 0.174 % calculated on the weight of azide [16].

The type of agitator, concentration, pH, and order of adding the reactants all have a significant influence on the size and shape of LA crystals and their composition. Sodium azide and lead nitrate aqueous solutions used for preparation in today's technologies are considerably diluted and the reaction mixture is intensively stirred during reaction. The reason for this is to prevent the formation of large sensitive crystals that form from saturated solutions when using slow diffusion of both reaction agents. Spontaneous explosions of reaction mixtures occur under these conditions (see Sect. 2.6).

In the 1960s, production of a spheroidal form of the dextrinated azide was developed [8, 53]. The technology is not covered here as it is described in great detail in the easily obtainable literature [6, 8, 21, 46].

The order of adding of reactants is an important factor. Even though simultaneous addition exists [8], a solution of sodium azide is most often added to a solution of lead nitrate (in discontinuous technology) [3, 5, 6, 8, 21, 30, 50]. The main reason for this order is to eliminate potential hydrolysis in aqueous sodium azide and the formation of basic lead azide (basic reaction of sodium azide) which would occur if the order of adding were reversed (Valenta Private Communication). The product formed by addition of aqueous sodium azide into a solution of lead salt is easy crumbled into a free flowing powder. If the order of reactant mixing is reversed, a hard and scarcely pulverizable product forms. The process of grinding of such product is dangerous and therefore its formation is undesirable [50].

The α-form (Fig. 4.5) is the only one that is really acceptable in any manufactured product. However, the β, γ, and δ forms are also sometimes created if the manufacturing process is not carried out in a correct and consistent manner. Their presence in the product is not acceptable as it increases the overall sensitivity. In industrial production each batch is checked and if it contains any of these other modifications the whole batch is destroyed [3].

4.2 Lead Azide

Fig. 4.5 Agglomerates of the dextrinated α-form of lead azide prepared by batch process (*left*) and the same prepared without addition of crystal structure modifier (*right*) (by kind permission of Pavel Valenta, Austin Detonator)

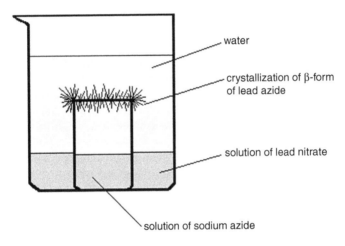

Fig. 4.6 Method for preparing β-form of lead azide [46]

The β-form is not a desirable commercial product as it increases the sensitivity of the final product but, if needed, it may be prepared by so-called diffusion method [46]. In this method, a small beaker filled to about one half with a 10 % solution of sodium azide is placed into a bigger beaker. The space between the smaller beaker and the larger one is filled with 5 % solution of lead nitrate to the level of sodium azide in the smaller beaker. Both beakers are then carefully filled with water up to the level where the smaller beaker is completely immersed under the water surface. In a few hours, crystals form on the edges of the smaller beaker (Fig. 4.6) and after 15–20 h the beaker fills up with nice long crystals of the β-form of lead azide.

It is also possible to fill both beakers with pure water to the level when the smaller one is completely immersed. Then, using a funnel, a few crystals of sodium azide are dropped into the smaller one and a few crystals of lead nitrate into the bigger one. The salts dissolve, diffuse, and needles of the β-form of lead azide appear [46]. Another method uses a big dish into which two small beakers are placed and one is filled with sodium azide and the other with lead nitrate. The big dish is then carefully filled with water in such a way that the two solutions do not

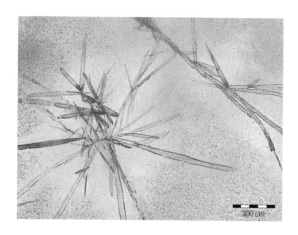

Fig. 4.7 Needles of the β-form of lead azide prepared by the diffusion process captured by optical microscopy

mix. The crystals of the β-form of lead azide form between the beakers [13]. Figure 4.7 shows the needle crystals formed by the diffusion process. Evans reports another way of preparing the β-form of lead azide by a different type of diffusion process, in which azide and nitrate solutions are separated by sodium nitrate. Haze forms in the nitrate layer close to the original azide boundary just before crystals of β-LA begin to grow out from the walls of the container into the solution [7].

Another method for preparing the β-form is based on the reaction of lead nitrate and sodium azide, at room temperature and in the presence of organic dyes. When solutions of lead nitrate (20 %) and of sodium azide were run continuously into water at 20 °C only α-LA was observed, but if even as little as 0.02 % of eosin or other organic dye is present in the reaction mixture, the β-form appears. The crystals are best handled as single specimens as they are sensitive; breaking one will set off the whole batch [16].

The γ and δ forms form together by precipitation from an unstirred diluted aqueous solution of sodium azide and lead nitrate in presence of nitric acid (in molar ratio 2:2:1) at room temperature. The reaction mixture is admixed and swirled once after mixing of the nitric acid and lead nitrate with solution of sodium azide. Both forms appear after 3 days in an unstirred mixture; the crystals need to be collected and separated by hand [15].

The pure γ-form of lead azide may be prepared by simultaneous slow addition of solutions of sodium azide and lead acetate into a stirred solution of PVA containing a small amount of sodium azide at a temperature not exceeding 15 °C. The temperature must be kept low as with temperatures above 25 °C mainly α-LA forms. The γ-LA may also be prepared by action of hydrazoic acid on a lead acetate solution containing 0.2 % of PVA. The final product may be recrystallized from ammonium acetate to give crystals of better quality. The γ-form may also be prepared from α- or β-LA by recrystallization from a 5 % aqueous solution of ammonium acetate containing PVA. The solution is left standing at temperatures below 40 °C. The preparation of the γ-form of LA without the use of additives requires employment of a buffer of hydrazoic acid/sodium azide at pH 3.0–5.0 [19].

4.2 Lead Azide

Table 4.5 Time to explosion for LA dissolved in 5 % aqueous ammonium acetate at 70 °C and allowed to cool down to room temperature without disturbance [18, 54]

	Concentration of LA (%)						
	1.0	0.9	0.8	0.7	0.6	0.5	0.4
Time to explosion (min)	40 50	75 65	80 85	210 220	225 255	7 explosions out of 10 experiments in 18 h	0 explosions out of 4 experiments in 4 days

4.2.6 Spontaneous Explosions During Crystal Growth

Spontaneous explosions of LA occur without apparent reason during the course of its crystallization. This phenomenon is known but not very well understood. These explosions are not exclusive to azides of lead but have been observed also in the case of cadmium, cupric, and mercuric [15]. In the case of lead azide, it has been frequently attributed to the formation of β-lead azide or the growth of large crystals to a point where internal stresses become important [12, 18, 34]. Both of these theories have little or no experimental support [18].

The experimental results show conclusively that the suppression of spontaneous explosion during crystallization by metathesis from solutions of soluble lead salt (usually lead acetate or nitrate) and soluble azide (usually sodium azide) is obtained by (a) rapid and thorough mixing of the solutions and (b) the use of certain additives especially hydrophilic colloids (PVA, dextrin, eosin). Amounts as small as 0.02 % of PVA completely suppressed spontaneous explosions [18].

It has been further shown that explosions still occur when lead nitrate is replaced by lead acetate and sodium azide with calcium azide [12]. The explosions are reported to occur at room temperature as well as at 0 °C and 60 °C [12].

With the reduction of mixing efficiency during crystallization by metathesis the probability of explosion increases. When mixing was reduced until the solutions merely diffused as is usual in the method for preparation of the β-form (!), the incidence of explosions increased greatly—up to a 50 % probability [18].

The most reliable way to produce explosions of LA is recrystallization from ammonium acetate. A reasonable explosion control time is possible by carefully controlling the concentration, temperature, and conditions of cooling. Table 4.5 gives results of a series of experiments with times to explosion from concentrations from 0.6 to 1.0 % LA. The frequency of explosions decreases below 0.6% and falls to nil at 0.4 %. An interesting observation was made that intact crystals of LA could be found after the vessel had been shattered by an explosion [18].

An experiment has been reported where a solution of 0.7 % LA was made in exactly the same way as is reported in Table 4.5 and allowed to cool down from 70 °C. After approximately 190 min (220 min was the expected time of explosion!) a number of large crystals had formed and these were filtered off and separated from the mother liquor. Thirty minutes later the mother liquor, still free of crystals, exploded shattering the vessel. The separated crystals remained intact. This leads to the conclusion that spontaneous explosions in LA crystallization originate from the mother liquor and not from the crystals already formed [18].

The capability of the LA/ammonium acetate solution to propagate explosions was tested. It proved unable to propagate explosions in a variety of conditions [18].

Aqueous ammonium acetate is recommended as the solvent for preparing large crystals of LA (α-form, size 0.5–1 cm) for further studies. The growing solution must have a near neutral pH. This prevents the formation of basic salts of LA (at higher pH) or high concentrations of azoimide (at low pH). Ammonium acetate is a suitable solvent since the hydrolysis products (acetic acid and ammonium hydroxide) have equal dissociation constants and therefore buffer the solution at a neutral pH value. LA is dissolved in a hot solution of ammonium acetate (70 °C) and very slowly cooled (e.g., 0.5 °C per day). LA slowly crystallized from the solution and it was isolated by filtration when the temperature reached 40 °C. If the cooling continues below 35–40 °C, explosions occur! The collection of LA crystals is a critical step. If the growing solution is not immediately washed from the crystals' surfaces it will saturate by evaporation and cause spontaneous explosion. The remaining LA dissolved in the solution must be destroyed, e.g., by ceric solution. The second method of preparing large crystals of LA is based on the reaction of stoichiometric concentrations of ammonium azide and lead acetate to give LA solution in aqueous ammonium acetate. The details are in the literature [55, 56].

4.2.7 Uses

Lead azide is today one of the most widely produced primary explosives. It has almost completely replaced mercury fulminate in some of that product's former applications, primarily in detonators (a summary of published patents up to the 1930s is presented in Audrieth's review [11]) due to its high initiation efficiency and extremely short predetonation zone in normal conditions. High initiating efficiency, high density, low cost, reasonably good stability, ability to withstand high pressures without becoming dead pressed, and long-term experience are the main reasons for its extensive application as a charge for detonating secondary explosive in flash, stab, electric, and other types of detonators.

Lead azide is used in many applications accompanied by other substances that compensate for its drawbacks, particularly its low sensitivity to flame and stab. The most common additive in detonators is lead styphnate which improves the inflammability of resulting mixture. A typical composition of this binary mixture is 30 % LS and 70 % LA. It is sometimes presented that lead styphnate can serve as a protective layer against access of water and carbon dioxide to LA surface [3, 4]. However, lead styphnate increases the level of acidity and accelerates the rate of hydrolysis of LA in presence of moisture [35, 49]. Regardless of this fact a combination of LA/LS is still used in detonators.

Probably, the most commonly known LA/LS mixture for detonators is the ASA composition. It contains 68 % dextrinated LA, 29 % LS, and 3 % aluminum. The small amount of aluminum was added to overcome the tendency of the LA/LS mixture to build up on the punches during pressing. On the other hand, aluminum unfortunately has a sensitizing effect on the composition. The possible problems with this mixture caused by static discharge between the leading wires and the tube of electric detonators have been discussed in detail by Medlock and Leslie [44]. These problems led to a number of accidents in the past [44]. The ASA mixture has been subsequently discontinued.

A single step method of preparing an LA and LS mixture by precipitation of the two salts together in one solution has been reported by Herz [57, 58]. In this method, a solution of sodium azide and a soluble styphnate salt (e.g., magnesium) is allowed to slowly flow into the solution of lead nitrate. This mixture, however, also suffers from the above-mentioned incompatibility [35].

LA is further not very sensitive to stab initiation. In compacted form it requires an initiation energy as high as 1 J. However, addition of only 2 % of GNGT lowers the required initiation energy to 3 mJ [59].

Despite all its strengths and manageable weaknesses, its toxicity, caused by the presence of lead, resulted in the search for nontoxic primary explosives which could fully replace LA.

4.3 Other Substances Derived from Lead Azide

There exists a variety of other substances which are formed from lead, azide, and other ions. Such substances mostly do not have sufficient explosive properties to be considered as initiating explosives but may form in the process of production or aging of normal lead azide.

4.3.1 Basic Lead Azide

Basic lead azides were in the early days believed to have varying content of $(OH)^-$ groups. The monobasic lead azide was sometimes abbreviated by the formula $PbN_3(OH)$ [15, 19] and other times as $Pb(N_3)_2 \cdot PbO$ [60]. Todd and Tasker [61] have carried out X-ray diffraction analysis and infrared analysis of gamma modification of basic LA and concluded that the product is oxyazide containing neither hydroxide nor water of crystallization. Other reported forms include: $3Pb(N_3)_2 \cdot 5PbO$, $Pb(N_3)_2 \cdot 2PbO$ to $Pb(N_3)_2 \cdot 3PbO$, $2Pb(N_3)_2 \cdot 7Pb(OH)_2$, $Pb(N_3)_2 \cdot 4PbO$ to $Pb(N_3)_2 \cdot 9PbO$ [5]. Fourteen types of basic lead azides are reported by Todd and Tasker [61], and twelve by Sinha [62].

Basic LA sometimes appears in service lead azide in trace amounts [19]. This substance may be prepared by recrystallization of α-LA from a 5 % ammonium acetate solution containing free ammonia to a pH of 8.7. The product takes the form of well-formed white crystals on pellets of α-LA [19]. A pure form of basic lead azide—$Pb(N_3)_2 \cdot PbO$ may be prepared by the following methods:

1. Bubbling of carbon dioxide-free air through a boiling suspension of lead azide in water until the calculated amount of azoimide is evolved [24]
2. Heating an aqueous suspension of $Pb(N_3)_2$ and $Pb(OH)_2$ in molar ratio 1:1 in a sealed tube at 140 °C for 12–15 h [24]
3. Heating the desired amounts of $Pb(N_3)_2$ and $Pb(OH)_2$ on a water bath for 20 h [24]

During slow cooling of the tube, the second method tends to give long needle crystals which tend to detonate. Wöhler recommended the third method as the most suitable one (out of the first three) [24].

Reacting stoichiometric quantities of lead nitrate solution with mixed solution of sodium azide/sodium hydroxide (molar ratio 2:1) at 35 °C leads to pure $3Pb(N_3)_2 \cdot 2PbO$ [63]. Varying molar ratio and acidity of the reaction mixture lead to 12 different types of basic LA. It was observed that only one type of basic LA forms, if mixture of sodium azide and sodium hydroxide is added to the solution of lead nitrate. The opposite way of reactants mixing leads to formation of mixture of basic azides. It indicates that in neutral or acidic reaction mixture only one type of compound forms while from basic medium mixture of azides emerge [62].

Basic lead azide—$3Pb(N_3)_2 \cdot 2PbO$—is much less sensitive to mechanical stimuli than LA but it keeps relatively acceptable explosive properties. It explodes violently when heated, with an explosion temperature of 350 °C at heating rate 5 °C min^{-1}. Basic lead azide is not attacked by moisture like LA at 30 °C and 90 % humidity during 3 months. Ignition efficiency is 0.07 g for tetryl, and bulk density 1.177 g cm^{-3} or 1.25 g cm^{-3} if prepared in presence of 0.02 % of carboxymethyl cellulose. Sinha et al. suggested basic lead azide as an LA replacement despite its lower initiating efficiency [63]. Agrawal [64] reported identical explosion temperature and initiation efficiency; however, deviates from the report of Sinha et al. [63] in bulk density by reporting 2.00 ± 0.2 g cm^{-3}.

Substances analogous to basic lead azide having the assumed OH$^-$ group replaced with a halogen have also been reported. One example of such substance is PbN_3Cl which may be prepared from lead nitrate and azide/chloride mixtures. This substance has inferior explosive properties compared to normal lead azide.

4.3.2 Lead (IV) Azide

The compound that would correspond to the overall formula $Pb(N_3)_4$ is not stable. Attempts to prepare it from PbO_2 and hydrazoic acid in an aqueous medium failed, as the resulting red solution or dark red needles quickly decomposed to yield LA and nitrogen [15]. Different lead azides form when using Pb_3O_4 and aqueous HN_3. The yellow to red solution contains compounds with overall formula PbN_9 to almost PbN_{12} based on the reaction conditions. These aqueous solutions are unstable and decompose evolving nitrogen and LA [5].

4.4 Silver Azide

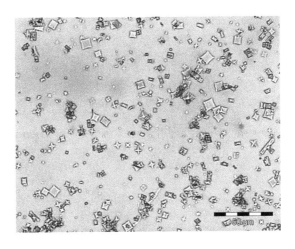

Fig. 4.8 Silver azide prepared by diffusion process

AgN_3

Silver azide (SA) was first prepared by Curtius in 1890 by passing azoimide into a silver nitrate solution [1].

4.4.1 Physical and Chemical Properties

Silver azide forms a white crystalline compound presented in Fig. 4.8. It forms orthorhombic crystals with reported crystal density 4.81 g cm^{-3} [5, 21] or 5.1 g cm^{-3} [3, 5, 14, 40, 45]. Heat of formation of SA is 311 kJ mol^{-1} [27].

Silver azide is nonhygroscopic [8, 65] and its chemical properties are similar to silver chloride; silver azide is practically insoluble in water (0.765 mg/100 g at 25 °C) and common organic solvents; soluble in pyridine and ammonia with which it forms complexes [1, 8, 15, 65–67]. Pure silver azide may be precipitated back from its solution in the form of orthorhombic crystals [5, 8]. Recrystallization of silver azide from its aqueous ammonia solution, by evaporation while standing, yields needles up to 20 mm long and 0.3 mm thick. Thin needle form of SA is reported to be extremely sensitive and with a tendency to explode by breaking even under water [2, 6, 15, 68]. Solonina (cited by Bagal), however, reported that these crystals do not explode when broken or even cut by a steel knife. He further reported an accident during which a bottle with SA needles fell to the floor, all crystals broke, the glass bottle shattered, but no explosion occurred [46].

Unlike lead azide SA does not react with water in presence of carbon dioxide [6, 69] or, more precisely, it yields very low partial pressure of azoimide, too low for formation of copper azides in applications with copper or brass [70].

Silver azide is often reported as being compatible with most usual metals, even though it reacts with the two most common ones—copper and aluminum. Taylor does not mention aluminum and reports that among common metals only copper reacts under moist conditions [71]. Blay and Rapley reported that copper azides form when SA comes into contact with copper in moist conditions [36]. They further reported that SA reacts with aluminum as well. The corrosion of aluminum is quite fast but requires water in liquid phase in direct contact with both SA and aluminum. This is not the situation that would normally be found inside a detonator and, if it were the case, then the presence of liquid water would cause the detonator to fail for other reasons than corrosion. A humid environment itself is not sufficient to cause any significant degree of reaction and the use of SA in aluminum detonators has not presented a problem [36]. The decomposition products of SA are not hazardous substances and mainly contain metallic silver [36].

The incompatibility of SA with sulfur or its compounds (e.g., Sb_2S_3 present in many stab priming mixtures) leading to a silver sulfide [72] is a well known and often mentioned issue [3, 36, 73, 74]. SA slowly decomposes when it is in contact with some plastics and rubbers (possibly due to sulfur content). Silver azide is compatible with common explosives such as HMX, RDX, or the lead salt of dinitroresorcine [36]. According to Bates and Jenkins, it is incompatible with GNGT [74], which is supported by findings of Blay and Rapley, who reported that explosions may occur during storage of SA/GNGT mixtures [36]. SA, unlike silver chloride, dissolves in nitric acid with decomposition and evolution of azoimide [6, 15]. Suspensions of silver azide decompose in boiling water in a similar way to that during photodecomposition [15].

It is more difficult to decompose SA chemically than LA for two reasons. First, it is more resistant to hydrolysis and secondly the reaction with sodium nitrite solution results in soluble silver salt which gives a flocculent yellow precipitate of silver nitrite which is insoluble in the reaction mixture. It is therefore recommended to use nitric acid for acidifying the solution rather than acetic acid as used in the destruction of LA. Nitric acid first reacts with SA yielding silver nitrite which then dissolves in the excess of nitric acid. The rate of nitric acid addition must be carefully controlled. It is important not to add the acid too quickly as that would result in reaction with unreacted SA and liberation of azoimide [71].

Silver azide is photosensitive (but much less than a halogen silver salt [75]); it turns violet at first and finally black, as colloidal silver is formed on exposure to light (and nitrogen is released) [7, 76]. The photolysis takes place only on the crystal surface and the material does not lose its explosive properties [6, 8, 15, 77, 78]. Photodecomposition is faster than in the case of LA and is quite rapid when SA is irradiated by ultraviolet light [22] and, if sufficiently intense, it may cause initiation. The critical light absorption for initiation of a crystal of silver azide corresponds to a total energy input of 0.19 mJ into each square millimeter of the crystal [77]. The initiation of decomposition is photochemical and the growth to explosion is thermal. The mechanism of the decomposition has been suggested by

4.4 Silver Azide

Evans and Yoffe [7, 76, 79]. The complete equation of SA decomposition is given by the following equation:

$$2 AgN_3 \longrightarrow 2 Ag + 3 N_2 + 621 kJ$$

Yoffe published a different value of reaction heat for the above-mentioned equation, 568 kJ [28]. Earlier papers described the photochemical mechanism of initiation as thermal, meaning that light absorbed by the thin surface layer of the crystal is converted to heat, and very quickly (less than 1/50 μs) results in ignition [7, 30, 77].

Silver azide is exceptionally thermally stable exceeding the stability of other common primary explosives. Unlike most of them, silver azide melts when heated. By heating a small amount of silver azide it first begins to turn slightly violet at about 150 °C, the color increasing somewhat in intensity until, at a temperature somewhat above 250 °C, the compound melts giving a blackish looking liquid which begins to liberate gas or vapor at about 254 °C. It decomposes rapidly above melting point to silver and nitrogen gas—so quickly that it looks as if the liquid is boiling [65, 76, 80]. Yoffe reported that the explosion temperature depended on ambient pressure, the mass of SA, and the thermal conductivity of the vessel. SA does not explode even at 400 °C if it is heated in a vacuum but it explodes at 340 °C in a nitrogen atmosphere at 26.7 kPa with a white flash [28].

Temperature of ignition highly depends on heating rate and on the testing methodology. Millar [81] reported it as 390 °C (by STANAG) but onset of decomposition on DSC thermogram is 303 °C (heating rate 10 °C min^{-1}). Taylor and Rinkenbach [82] reported 273 °C (hot iron dish method, which determines the temperature at which material explodes instantaneously or within 1 s); Hitch gave 300–301 °C [80], and Wöhler and Martin 297 °C (explosion within 5 s) [83]. The rate of decomposition of SA depends not only on sample purity, sample size, or heating rate, but on the sample history as well. Particularly whether or not SA has been exposed to light, because metallic silver nuclei are believed to influence the rate of decomposition [81].

Ionizing radiations generally decompose azides without explosion, which however may occur if the intensity is high enough owing to the crystals heating up during decomposition by irradiation [79].

4.4.2 Sensitivity

Sensitivity of silver azide to impact is several times lower than that of mercury fulminate. Comparison with lead azide is reported a little bit higher than LA by some authors [45] (Bureau of Mines) [82], same as LA [45] (Picatinny Arsenal Apparatus), or lower than LA by others [21, 81] (see Fig. 2.15). The sensitivity of SA highly depends on crystal size and shape. Colloidal silver azide prepared from concentrated solutions exhibits significantly lower sensitivity (0.5 kg from 77.7 cm) than coarser crystals prepared from diluted solutions which required less than half the energy (0.5 kg

Table 4.6 The sensitivity of SA to mechanical stimuli [81]

Compound	Impact sensitivity		Friction sensitivity	Sensitivity to ESD
	Ball and Disc test (h_{50}, cm)	Rotter test (F of I)	Emergency paper friction test 50 % prob. (m s^{-1})	Advanced test no. 7 (μJ)
SA[a]	47.4	30	2.6	0.21 ign., 0.118 not
LA[b]	19	29	1.2	2.5

[a] RD 1374, producer BAE system.
[b] RD 1343, precipitated from sodium carboxymethylcellulose/sodium hydroxide.

from 28.5 cm) [65]. Values for impact sensitivity of SA (commercial product) determined by Millar are summarized in Table 4.6 in comparison with values for LA.

Silver azide is generally considered to be highly sensitive to friction and much more than other common primary explosives including LA [21, 84] and, just as with impact, friction sensitivity depends considerably on its crystalline form [65]. The values reported by Millar [81] for two specific kinds of LA (RD 1343) and SA (RD 1374) are presented in Table 4.6. It follows from the results that the sensitivity of at least some commercially produced SA is lower than that of LA. Sensitivity of SA to electrical discharge is higher than that of LA (Table 4.6), while sensitivity to flame is about the same as that of mercury fulminate [6, 21].

4.4.3 Explosive Properties

Silver azide has a very small critical diameter. Needles of SA having a diameter of 25 μm still explode. Rapid combustion (about 7 m s^{-1}) was observed when the diameter was decreased to approx. 10 μm [85]. The values of detonation velocity usually range from about 1,000 m s^{-1} for very thin layers (~0.1 mm thick) to about 5,000 m s^{-1} for layers with dimensions ensuring ideal detonation [7]. Bowden published an average detonation velocity of 1,500 m s^{-1} for 0.5-mm-thick unconfined films (initiated by hot wire) and 1,700 m s^{-1} for confined product, initiated by impact [86]. A detonation velocity of 1,500 m s^{-1} was determined for crystals with diameter 0.3–0.4 mm and length about 1 cm by Deb (cited as unpublished work in [76, 87]). Detonation velocity in capillaries and micro-capillaries (10–200 μm) has been relatively recently measured by Jung [22]. He observed an accelerating reaction of SA after ignition (depends on sample diameter). Steady detonation velocity was in the range from 1,880 to 2,400 m s^{-1} for diameters from 50 to 200 μm. The density of SA was not determined.

Danilov et al. [3] found the detonation velocity of SA to be 3,830 m s^{-1} at density 2 g cm^{-3} and 4,400 m s^{-1} at maximum obtainable density. For the dependence of detonation velocity on density of SA, they proposed the following equation:

$$D_\rho = D_0 + 770(\rho - \rho_0) \text{ m.s}^{-1}, \text{ where } \rho_0 \text{ is 2 g.cm}^{-3}$$

and for detonation pressure:

4.4 Silver Azide

Table 4.7 Results of sand test for SA and MF [65]

Charge weight (g)	Weight of crushed sand (g)	
	SA	MF
0.05	1.4	0
0.1	3.3	0
0.2	6.8	4.2
0.3	10.4	8.9
0.5	18.9	16.0
0.75	30.0	26.1
1	41.1	37.2

$$P = (40\rho - 61) \cdot 10^2 \text{ MPa}$$

The initiating efficiency is very high; it surpasses mercury fulminate (several times) and even LA [48, 71, 88] (see Table 2.1). Taylor and Rinkenbach measured brisance of SA by the sand test [65]. They observed that brisance of SA is not that much greater than that to be expected for MF. Values for both primary explosives are summarized in Table 4.7.

The power of SA measured by the Trauzl test is a relatively low 115 cm^{-3} for 10 g SA (38.3 % TNT, 88 % MF). Volume of gaseous products of detonation is 244 dm^3 kg^{-1} [3, 40, 45]. Silver azide cannot be dead-pressed [3].

4.4.4 Preparation

Preparation of silver azide is based on precipitation of silver azide from a sodium azide solution after addition of a solution of silver nitrate:

$$NaN_3 + AgNO_3 \longrightarrow AgN_3 + NaNO_3$$

The extremely high friction sensitivity of the crystalline form of SA prohibited its usefulness in the early days. Taylor and Rinkenbach found that mixing fairly concentrated solutions of silver nitrate and sodium azide yielded silver azide in a form of colloidal aggregates. This form of SA is less sensitive than crystals [5, 65]. Another method of decreasing the sensitivity of SA is based on the addition of an inert absorbent material to the solution of SA. In this way, individual crystals are separated from each other by the inert material [89]. The presence of some substances (e.g., cyanamide ion) during precipitation of SA can, however, increase the decomposition rate and decrease its explosion temperature [90].

A technologically interesting way of preparation of silver azide is based on mixing a solution of sodium azide with some explosive (MF, tetryl, TNT, picric acid, etc.) before adding silver nitrate. Silver azide then forms a thin film on the surface of the added substance [5].

The initiating efficiency of the mixture of MF with approximately 3–5 % of SA is several times higher than for pure MF and practically the same as for pure SA.

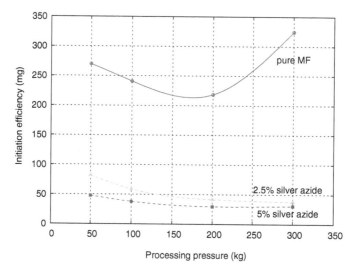

Fig. 4.9 Effect of silver azide addition on initiation efficiency of MF (conditions—tetryl as acceptor charge, detonator number 6, varying processing pressures) [6]

Figure 4.9 [6] also shows that addition of SA prevents dead pressing of such mixture in the pressure range concerned. This mixture, prepared by mixing moist SA with MF, was named Astryl and was used in blasting caps [5, 91, 92].

The mixture described above may raise a question as to why one should not mix MF with LA which is cheaper and accelerates MF in the same way as silver azide does. There are two reasons why such a mixture has been rejected. The first one is the very high sensitivity of LA/MF mixture (higher than SA/MF) and the second reason is its incompatibility with both aluminum and copper [6].

The main historical obstacle for the practical use of pure silver azide (apart from its price) was its unsuitable form—unsuitable free flowing properties for volumetric loading and pressing into detonators. This problem was successfully solved by the development of a process for production of granular silver azide. An example of commercially manufactured spherical shaped crystals of silver azide is presented in Fig. 4.10 [22].

Silver azide prepared by direct reaction of sodium azide and the soluble silver salt forms as a fine powder with a low bulk density. This is caused by its extremely low solubility and a tendency to nucleate profusely which results in extensive nucleations and very small crystals. Several processes have been developed for preparation of a product with a more suitable crystal structure.

The most commonly known historical technique is based on the simultaneous addition of sodium azide and silver nitrate solutions to a vigorously stirred solution of sodium hydroxide. The product obtained, in the form of small granules made of very small crystals, had good sensitivity but low initiating ability. Larger crystals were obtained by substituting sodium hydroxide with ammonium hydroxide, which increased the solubility of SA but created two problems (a) low yield and (b) dangerous waste (ammoniacal silver solution). Simultaneous addition of nitric acid to provide a nearly neutral solution solved the two problems and yielded a

4.4 Silver Azide

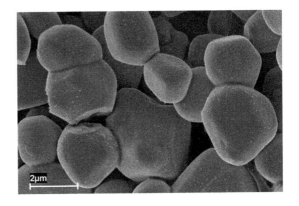

Fig. 4.10 Micrograph of Sandia produced silver azide [22]. Image courtesy of Sandia Corporation

product commonly known as RD 1336 [71]. This product forms as a free-flowing irregular aggregate approximately 0.1 mm in size (individual crystals in aggregates are approximately 0.02 mm in size) and has a bulk density about 1 g cm^{-3} [71, 93, 94]. The possibility of neutralization of ammonia by carbon dioxide (rather than by nitric acid) has also been patented [94].

The RD 1336 process leads to a SA with relatively low bulk density. Technologies leading to a product with not only good flowability but also higher bulk density are based on the modification of the RD 1336 process. They differ from each other only in partial technological steps as all of them are based on increasing the solubility of silver azide in aqueous medium by addition of other complex-forming substances such as ammonia, pyridine, cyanides, thiosulfates, thiourea, etc. [69, 95].

A method of preparing a reduced sensitivity SA introduced in USA is called the "Costain" process after Thomas Costain who improved the original procedure for RD 1336 developed in England in the ERDE laboratories shortly after World War II [96]. In the Costain process, aqueous solutions of silver nitrate and sodium azide are added to the dilute aqueous ammonia (or an aqueous solution of sodium azide is added to an aqueous solution of silver nitrate and ammonia). The reaction mixture is then heated and part of the ammonia is distilled from the solution. When the first silver azide precipitate appears, a small amount of acid (e.g., acetic acid) is added to induce crystal seeding and results in profuse nucleation ("shock crystallization"). The distillation of ammonia then continues and the precipitation of silver azide is total. Costain reported several improvements for his product, first of all bulk density 1.4 g cm^{-3} [96]. Hirlinger and Bichay later reported a further improvement leading to a product with density 1.6 g cm^{-3} [97] (vs. 1.0 g cm^{-3} for original ERDE silver azide). Further, concentration and addition parameters are not as critical as for the ERDE process [96]. Not much has been published about the Costain process but some details have been published in [98].

Three years after publication of the Costain process in the USA, McGuchan published a conference paper summarizing areas of primary explosives research at PERME in UK which included production of improved SA—product RD 1376 [95]. The increased solubility of SA by addition of ammonia was confirmed as a best option (all other substances performed worse). The neutralization of the ammoniacal solution of SA was recognized as a major aspect. The type of acid was found to affect the nucleation process and even weak acids (e.g., acetic acid) were found to cause the nucleation at a rate which was too quick, yielding small crystals. It was recognized that the process needed to be modified by addition of crystal modifiers (e.g., carboxymethylcellulose) and controlled rate of product nucleation. The technical operations such as stirring, temperature regime, rate of reactant feeding, and method used for ammonia evaporation have a significant impact on the properties of the final product [8, 69, 95, 96].

One of the oldest but least known methods for obtaining silver azide (unfortunately in the form of nice white needles) is based on addition of a cold saturated aqueous solution of silver nitrite to a cold saturated solution of hydrazine sulfate [46]. A modification of this method (developed in the KHTOSA LTI department in Lensoveta [3]) uses hydrazine sulfate (or hydrazine), silver nitrate, and sodium nitrite. Careful adjustment of the reaction conditions can result in an optimized method giving yields up to 90 % of the theoretical maximum [46, 99].

$$N_2H_4 \cdot H_2SO_4 + AgNO_2 \longrightarrow AgN_3 + H_2SO_4 + 2H_2O$$

$$N_2H_4 \cdot H_2SO_4 + AgNO_3 + NaNO_2 \longrightarrow AgN_3 + NaNO_3 + H_2SO_4 + 2H_2O$$

Some rather unusual situations in which silver azide forms may be found in the literature. One such case is the formation of SA during electrolysis of solutions of sodium or ammonium azides with a silver anode [100].

4.4.5 Uses

The only remaining problem preventing wider use of SA was the high cost which limited the use of pure silver azide only to special applications. The first occurrence in non-American ammunition is reported from 1945 [45]. Urbański published the use of silver azide in small and strong detonators in USA, Great Britain, Germany, and some other countries [8]. In industrial applications, silver azide was used practically exclusively in the form of a mixture with other substances. The first reported use of silver azide in detonators for underground mining is from 1932 [92]. Even today, SA is used particularly in small size detonators. For these applications it is more suitable than LA because of its high brisance, being superior to LA. It is also used in applications requiring a thermally stable primary explosive [3].

4.5 Copper Azides

Copper forms several azides including cuprous CuN_3, cupric $Cu(N_3)_2$, and a few basic azides. The only copper azide that used to have at least some practical application was cupric azide. Even though the practical application is rather limited, it is important to understand the formation of various copper azides because their formation represents significant risk for munitions containing lead azide.

Fig. 4.11 Amorphous form of cupric azide

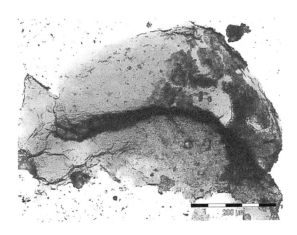

Some of the copper azides are extremely sensitive and may form wherever copper comes into contact with lead azide in presence of weak acids. Even carbonic acid forming from moisture and carbon dioxide has the ability to decompose lead azide liberating azoimide. Copper and its alloys react with liberated azoimide and, depending on conditions, form various copper azides.

This process may lead to an increase of sensitivity of the original ammunition resulting in higher manipulation risks. Furthermore, copper azides form many complex compounds that will not be covered in this monograph.

4.5.1 Physical and Chemical Properties

Cuprous azide CuN_3 is a white crystalline solid easily oxidized by atmospheric oxygen, as a result of which it is usually obtained as a gray-green powder [15]. Heat of formation is 281 kJ mol^{-1} [27]. CuN_3 is photosensitive and changes color to deep red with a violet tinge in sunlight while decomposing into copper and nitrogen [5, 15, 78]. On heating, the surface turns slightly red at 205 °C, then black, then melts to a viscous blob and explodes [76]. It can explode when irradiated with an intense light flash [76]. Photoconductivity of CuN_3 has not been observed [7, 76]. It is practically insoluble in water and most common organic solvents [5, 15, 21] and decomposes in acids and bases at higher temperatures [21].

Cupric azide $Cu(N_3)_2$ forms a dark black-brown crystalline solid with a red tinge (Fig. 4.11). The color depends on the way it is prepared [101].

Cupric azide $Cu(N_3)_2$ is not reported to be light sensitive [15]. It is slightly soluble in water, insoluble in most common solvents except those that form complex compounds [5, 6, 15]. It undergoes hydrolysis in boiling water transforming to CuO [6, 102]. Cupric azide $Cu(N_3)_2$ is decomposed by mineral acids liberating HN_3 [46]. $Cu(N_3)_2$ is slowly converted to basic cupric azide in the presence of water or after a

Table 4.8 Color of various copper azides (unless stated otherwise [15])

	Formula	Density (g cm^{-3})	ΔH_f (kJ mol^{-1})	Color
Cuprous azide	CuN$_3$	3.26 (pycnometer) 3.34 (X-ray) [104]	281 [27, 103]	White
Normal cupric azide	Cu(N$_3$)$_2$	2.2–2.25 [5] 2.58 [105]	587 [27]	Black-brown
Basic cupric azides	Cu(N$_3$)$_2$·Cu(OH)$_2$			Yellow-brown
	Cu(N$_3$)$_2$·2Cu(OH)$_2$			Yellow-green
	Cu(N$_3$)$_2$·3Cu(OH)$_2$			Green
	Cu(N$_3$)$_2$·8CuO			Blue-green

long exposure to the air [46, 101]. Hydrazine and hydroxylamine reduce Cu(N$_3$)$_2$ to white CuN$_3$ [102]:

$$CuN_3 \xrightleftharpoons[N_2H_4]{O_2} Cu(N_3)_2$$

Cupric azide Cu(N$_3$)$_2$ forms several basic salts that are described in Table 4.8 [15]. Basic cupric azides are insoluble in water and undergo hydrolysis above 80 °C. They are soluble in acids and bases [5, 15].

The basic cupric azides form in principle by partial hydrolysis of cupric azide or by partial reaction of azoimide with Cu(OH)$_2$ [15].

$$Cu(N_3)_2 + H_2O \longrightarrow CuN_3(OH) + HN_3$$

$$Cu(OH)_2 + HN_3 \longrightarrow CuN_3(OH) + H_2O$$

It has been stated that copper azide easily forms when copper or brass is in direct contact with lead azide. Ammunition containing lead azide is therefore tested for the presence of copper azide using a 30 % solution of ferric chloride (Swedish copper azide test). A positive test results in the appearance of an intense red color [70]. A solution of lower concentration—only 1 % was reported to be sufficient—was further recommended, as it does not attack the tested metal's surface. It is also possible to use ammonium hexanitrato cerate (NH$_4$)$_2$Ce(NO$_3$)$_6$ which is a much more sensitive reagent capable of detection of the "invisible layer" formed in the early stages of corrosion [106].

4.5.2 Explosive Properties

Cuprous and cupric azides are very sensitive to impact and friction [15, 21, 101, 102]. Sensitivity of cuprous azide CuN$_3$ depends on the crystal size (Fig. 2.17). The larger crystals of CuN$_3$ (~3 mm) may explode even by the touch of a feather [5].

4.5 Copper Azides

Table 4.9 Sensitivity of normal and basic cupric azide to impact in comparison with lead azide and mercury fulminate

	1 kg (cm) [5]	0.5 kg (cm, 50 % prob.) [46]	0.13 kg ball (cm, 50 % prob.) [49]	0.13 kg ball (cm/% prob.) [107]
CuN_3	–	16	10	10/40 12.5/80
$Cu(N_3)_2$—crystals	1	9[a]	–	–
$Cu(N_3)_2$—amorphous	2	–	–	–
$Cu(N_3)_2 \cdot Cu(OH)_2$	7–8	–	10	10/60; 12.5/78
$Cu(N_3)_2 \cdot 2Cu(OH)_2$	–	–	70 no fire	70/0
$Pb(N_3)_2$	4	–	20	20/50
$Hg(CNO)_2$	–	14	–	–

[a]Unspecified crystal form.

Table 4.10 Sensitivity of normal and basic cupric azide to friction in comparison with lead azide [107]

	Load (g)	Probability of explosion (%)
CuN_3	30	50
	40	80
$Cu(N_3)_2 \cdot Cu(OH)_2$	120	75
	150	100
$Cu(N_3)_2 \cdot 2Cu(OH)_2$	60	15
	120	60
$Pb(N_3)_2$	80	50

Moist $Cu(N_3)_2$ cannot be ignited by flame and is rather insensitive to friction and shock [101]. Unfortunately, discrepancies in the literature exist and other authors mention high sensitivity even when wet [5]. $Cu(N_3)_2$ explodes with a green flame and red smoke [101].

The sensitivity of normal and basic cupric azides to impact varies significantly. The impact sensitivity is represented in Table 4.9 showing that cuprous and cupric azides are more sensitive than LA. Monobasic cupric azide is reported to be more sensitive than LA according to Kabik and Urman [49] and Lamnevik [107] while Fedoroff's encyclopedia reports its sensitivity lower than LA [5]. Basic cupric azides with two or more bases are less sensitive to impact than LA [107].

The friction sensitivity of various copper azides is often generalized as "very high." The values measured by Lamnevik [107] (Table 4.10), however, show that such generalization cannot be accepted.

From the above data it can be seen that copper azides are not significantly more sensitive than LA to mechanical stresses with just one exception—cuprous azide. This, however, does not apply to sensitivity to electrostatic discharge where copper azides are shown to be in most cases below the lowest possible limit obtainable with standard apparatus. The estimations of initiation energies by Lamnevik are 1–10 µJ [107], Holloway obtained 0.1–0.2 µJ for both CuN_3 and $Cu(N_3)_2$ [108].

The thermal stability of copper azides is not very high. The 5 s explosion temperature of cuprous azide is around 216 °C compared to 340 °C for LA [49].

Table 4.11 Response of copper azides to open flame [107]

	Behavior
CuN_3—both forms	Detonates
$Cu(N_3)_2$	Detonates
$Cu(N_3)_2 \cdot Cu(OH)_2$	Detonates
$Cu(N_3)_2 \cdot 2Cu(OH)_2$	Flashes like a black powder
$Cu(N_3)_2 \cdot 3Cu(OH)_2$	Burns rapidly
$Cu(N_3)_2 \cdot 3Cu(OH)_2$	Burns rapidly
$Cu_xZn_{1-x}(OH)N_3$	Burns rapidly

Lamnevik reports copper azides as "rather heat sensitive with explosion temperatures at ca. 180 °C" [107]. The response of various copper azides to open flame is summarized in Table 4.11.

4.5.3 Preparation

CuN_3 forms as an intermediate product during the reaction of hydrazoic acid with copper.

$$2\,Cu + 3\,HN_3 \longrightarrow 2\,CuN_3 + NH_3 + N_2$$

Cuprous azide is further oxidized by atmospheric oxygen to $Cu(N_3)_2$ [15, 102] or $Cu(N_3)_2 \cdot Cu(OH)_2$ [107]. The second stage of the reaction is slow and usually takes about a month [107].

Turrentine and Moore reported different reaction; the metallic copper reduces hydrazoic acid and yields cupric azide, ammonium azide, and nitrogen as its reduction products [109].

$$Cu + 4\,HN_3 \longrightarrow Cu(N_3)_2 + NH_4N_3 + N_2$$

The common method for preparing CuN_3 is by reaction of a cuprous salt with sodium azide. Aqueous solution of cupric salt (sulfate) is reduced with potassium sulfide giving cuprous sulfide as precipitate which is then dissolved by addition of acetic acid. In the following step, sodium azide is added to the solution, precipitating CuN_3 [78, 107].

$$CuSO_4 \xrightarrow{K_2S} Cu_2S \xrightarrow{NaN_3} CuN_3$$

Singh published details of the preparation of cuprous azide from cuprous chloride dissolved in saturated aqueous sodium chloride in presence of a small amount of potassium bisulfite and acidified by acetic acid. The cuprous azide was precipitated by addition of aqueous sodium azide [76, 110]. Another way of preparing CuN_3 is based on the reaction of freshly prepared cuprous oxide (from cupric hydroxide and

4.5 Copper Azides

hydrazine) with one drop of hydrazine to which solutions of sodium azide and sulfuric acid are added simultaneously over the period of 30 min [108].

Cupric azide $Cu(N_3)_2$ forms by exposing finely ground CuO to hydrazoic acid. The $Cu(N_3)_2$ forms after several days [15, 101, 102, 108].

$$CuO + 2 HN_3 \longrightarrow Cu(N_3)_2 + H_2O$$

Cupric $Cu(N_3)_2$ azide can be precipitated from a soluble cupric salt and sodium azide in an aqueous medium [108, 111].

$$Cu(NO_3)_2 + 2 NaN_3 \longrightarrow Cu(N_3)_2 + 2 NaNO_3$$

The product of the above reaction is, however, contaminated with a hydrolyzed surface layer (the product often has a yellow tinge). This layer can be removed by adding hydrazoic acid and letting the mixture stand for one day. Another possibility for precipitation of pure cupric azide is using a mixture of hydrazoic acid with sodium azide (pH 4.5–5.5) on cupric perchlorate. The product forms in the form of clusters of $Cu(N_3)_2$ needles [107]. Hydrolysis may be prevented by carrying out the reaction in a nonaqueous medium [15]; the dried product must be stored in a desiccator [107]. $Cu(N_3)_2$ may be prepared by the reaction of sodium azide with cupric perchlorate in acetone [112] or by the reaction of cupric nitrate with lithium azide in ethanol [102].

$$Cu(NO_3)_2 + LiN_3 \xrightarrow{EtOH} Cu(N_3)_2$$

$$Cu(ClO_4)_2 + NaN_3 \xrightarrow{acetone} Cu(N_3)_2$$

Laboratory preparation of basic cupric azides is based on the reaction of a soluble copper salt with sodium azide (or hydrazoic acid) in presence of a hydroxide (sodium, barium). The reaction conditions influence the type of basic cupric azide formed [15, 108].

$$2 NaN_3 + 2 CuSO_4 + 2 NaOH \longrightarrow Cu(N_3)_2 \cdot Cu(OH)_2 + 2 Na_2SO_4$$

$$2 NaN_3 + 3 Cu(CH_3COO)_2 + 4 NaOH \longrightarrow Cu(N_3)_2 \cdot 2Cu(OH)_2 + 6 CH_3COONa$$

$$2 NaN_3 + 4 Cu(CH_3COO)_2 + 6 NaOH \longrightarrow Cu(N_3)_2 \cdot 3Cu(OH)_2 + 8 CH_3COONa$$

Lamnevik published details of the preparation of mono, di, and tribasic basic cupric azides. The monobasic salt was prepared by hydrolysis of cupric azide at 50 °C and dibasic cupric azide by hydrolysis of copper diamine azide at 80 °C. Tribasic salts, together with the monobasic form, are prepared by reaction of copper tetramine perchlorate with sodium azide. The type of resulting product depends on the reaction conditions [107].

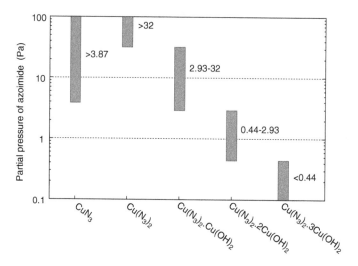

Fig. 4.12 Stability ranges of copper azides (from equilibrium measurements at 25 °C, the existence of CuN_3 is possible in absence of oxidizing agents) [107]

4.5.4 Undesired Formation of Copper Azides

Lead azide, otherwise a very good primary explosive, has one major drawback, apart from its toxicity. It reacts with water and forms volatile azoimide (b.p. 35 °C) which may react with copper forming sensitive copper azides. This process is accelerated by the presence of acidic substances such as carbon dioxide (from the air) or even lead styphnate, which is often used together with LA [3, 5, 8, 40].

The liberated azoimide attacks copper or its alloys (brass) and copper azides form on the metal surface. The concurrent use of LA and copper (or brass) in fuses or blasting caps therefore represents an inherent safety hazard. Although the LA–copper (copper alloys) reaction has been known since 1913, it has been a cause of many incidents decades later [49]. These incidents have been sometimes erroneously reported as spontaneous explosions. The reality, however, is that, in all known cases, they have been linked to some type of movement and therefore should not be considered spontaneous [70].

The chemical processes involved in the formation of copper azides may be summarized by the following set of equations:

$$Pb(N_3)_2 + H_2O \longrightarrow Pb(OH)N_3 + HN_3$$

$$2\,Cu + 3\,HN_3 \longrightarrow 2\,CuN_3 + NH_3 + N_2$$

$$4\,CuN_3 + O_2 + 2\,H_2O \longrightarrow 2\,Cu(N_3)_2 \cdot Cu(OH)_2$$

The azoimide is the first substance formed from LA in presence of moisture. It reacts with copper giving cuprous azide—CuN_3—which appears as a white film on the copper surface. The cuprous azide is then oxidized to yellow-brown

4.5 Copper Azides

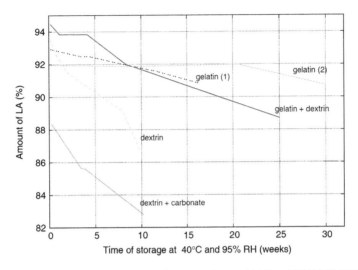

Fig. 4.13 Decomposition of various types of LA over time at 40 °C and 95 % RH [113]

$Cu(N_3)_2 \cdot Cu(OH)_2$ (some older reports incorrectly mention oxidation to $Cu(N_3)_2$). Depending on the conditions in the particular system, corrosion may proceed further to di-, tri-, and octa-basic cupric azides, yellow-green to blue-green in color. The partial pressure of azoimide is the key factor influencing which azide is formed. The pressure ranges and corresponding forms of copper azides are graphically summarized on Fig. 4.12. The presence of light and lack of ventilation accelerate the transformation of the first formed very sensitive form to the less sensitive ones [49].

While corrosion of the copper surface is usually concentrated in the vicinity of the source of the LA, diffusion of the azoimide vapor can occur and result in the formation of the sensitive products at some distance from the LA (for example in moving parts of fuses).

The corrosion rate of copper and other metals is affected by the following general conditions [31]:

- Increasing concentration of moisture
- A high degree of metal surface roughness
- A heavy contamination of the metal surface with hygroscopic materials (e.g., fingerprints)
- A large number of cold-worked areas

The corrosion of copper by azoimide cannot take place without decomposition of the LA. Depletion of LA can therefore be used for indirect studies of copper corrosion. Experiments where copper tubes were held above LA at constant humidity and temperature have shown that different types of LA decompose with different rates as shown in Fig. 4.13 [113].

Materials not corroded by azoimide, or corroded without formation of explosive products, include plastics, glass, stainless steel, aluminum, tin, titanium, silver, gold, zinc, and magnesium. Among these options only aluminum and stainless steel are used in significant amounts in ammunition (for economic and other reasons). Tin, zinc, and silver are used as a thin (ca. 50 µm) protective layer on brass and copper. Tin and silver do not react with azoimide and zinc forms zinc oxide. Both tin and silver must be carefully deposited on the surface in a form with minimal porosity [70].

Nonmetallic protective coatings such as varnishes and lacquers are penetrated by azoimide in a relatively short time if applied in usual thickness. The only reported acceptable protective finish is a high molecular weight silicon resin [70]. A coating by PVA with 1 % 1,2,3-benzotriazole proved to successfully prevent corrosion even with a very thin layer (5 µm) for a duration of some 2 months at 35 °C and 100 % RH [70]. Another option is the addition of a small amount of *o*-aminophenol to an ASA mixture (mixture of LA/LS/Al used in detonators). This prevents copper azide corrosion for 4–5 years [31].

The copper azides resulting from corrosion of surface copper by lead azide has been found to propagate detonation in layers with concentration exceeding 0.40 mg cm^{-2}; in the discussion following Lamnevik's presentation, Dr. Ball stated that layers of 0.0254 mm thickness detonate with a velocity of 1,400 m s^{-1} [70].

The corrosion of copper and formation of copper azides has been said to be the result of the presence of azoimide. Various methods exist that reduce partial pressure of azoimide in ammunition. The first method involves the use of moisture and carbon dioxide absorbents in air-tight elements (e.g., American artillery fuses). The carbon dioxide absorbents unfortunately accelerate decomposition of LA. The second method involves a ventilated design so that the azoimide can easily escape. This method yields, in a worst case scenario, the less dangerous basic copper azides. The drawback of this method is the fact that such a system is open to environmental effects and that the lead azide gradually decays. A third method uses azoimide destroying agents such as zinc or magnesium disks placed just above the LA. Azoimide preferentially reacts with these metals forming nonexplosive products (oxides, nitrogen, ammonia, hydrazine, and hydroxylamine). The passivation of the metal surface may, however, present a significant problem. Use of palladium and platinum has not been successful due to catalyst poisoning and formation of highly explosive products on the surface of these metals. Another way to prevent copper azide formation is to avoid use of lead azide and replace it with an alternative primary explosive such as silver azide or avoid copper by using stainless steel [70].

The incompatibility of LA and copper is the reason why it could not be used directly in detonators with the otherwise very popular copper tubing [70]. This, however, does not mean that LA could not be present inside a copper tube detonator. Many designs have been developed in which LA is encapsulated inside a protective tube made of compatible material and inserted into the copper tube of a standard detonator. The tin, zinc, or silver coating may also be used in some special applications.

4.5.5 Uses

The copper azides—both normal and basic—have a very limited field of application with the exception of cupric $Cu(N_3)_2$ which has been reported as a primary

4.6 Other Metallic Azides

explosive for detonators. The shape and size of crystals of $Cu(N_3)_2$ was modified by addition of polyvinylalcohol or gelatin [114].

4.6 Other Metallic Azides

Other azides do not have significant importance for practical applications. Most of them are moisture sensitive hygroscopic solids and some of them hydrolytically decompose. They also easily form complex compounds. Azides other than those of lead, copper, mercury, silver, palladium, and thallium are soluble in water and their isolation from aqueous solutions by crystallization is problematic on an industrial scale.

4.6.1 Physical and Chemical Properties

Some basic physical and chemical properties of selected azides are summarized in Table 4.12.

Mercurous azide $Hg_2(N_3)_2$ is probably composed of the linear molecules, N–N–N–Hg–Hg–N–N–N [76]. Melting point is about 205–210 °C [76]. Hitch reported that mercurous azide does not melt; it turns yellow, sublimes, and starts to decompose by heating to temperature 215–220 °C [80]. It is very sensitive to light and it easily photodecomposes, liberating nitrogen and turning into a solid brown product (probably nitride) on the surface without loss of brisance. There is no evidence of photoconduction [7, 50] and unlike LA it does not react with carbon dioxide [50].

Mercuric azide $Hg(N_3)_2$ comes in two forms, α and β (Fig. 4.14). The stable α-form has orthorhombic crystals, whereas the β-form consists of long extremely sensitive thin needles [15, 16]. The α-form may be obtained by recrystallization of $Hg(N_3)_2$ from hot water [7]. Miles, however, reported that recrystallization from water results in a mixture of both crystal forms. In every case when the β-form is present the material is extremely sensitive to touch and it may spontaneously explode [12]. The presence of dextrin does not have a significant influence on the form produced [16].

Azides of mercury darken in color and start to decompose when heated above 212 °C. The decomposition process is accompanied by evolution of nitrogen prior to its explosion [80].

Thallous azide TlN_3 is one of the few metallic azides that melt (m.p. 330 °C in vacuum [80] or 334 °C [28, 76]) followed by sublimation (at 340 °C) prior to its explosive decomposition with a green flash (at 430 °C) [76, 80]. Heat of formation published by Gray and Waddington is 234 kJ mol^{-1} [27], heat of decomposition to elements published by Yoffe is -238 kJ mol^{-1} [28]. TlN_3 should be stored in a completely dry, dark place as the damp substance has a tendency to discolor [15]. Pure crystals of TlN_3 show marked photoconduction when irradiated with light [7, 76]. Figure 4.15 shows crystals of thallous azide prepared by diffusion process. This preparation method is based on placing one drop of $TlNO_3$ on glass surface of a watch glass and one drop of NaN_3 next to it but not in contact. The two drops are

Table 4.12 Physical and chemical properties of selected azides [5, 15, 83, 115, 116]

	Density (g cm^{-3})	ΔH_f (kJ mol^{-1})	Color	Properties
Co(N$_3$)$_2$	–	–	Reddish-brown	Forms hydrates, hygroscopic, soluble in water in which it slowly hydrolyzes
Ni(N$_3$)$_2$	–	–	Grainy green	Hygroscopic, soluble in water in which it slowly hydrolyzes
Cd(N$_3$)$_2$	3.24 [3, 5] 5.15 [117]	445 [103]	White	Soluble in water, hygroscopic, tends to hydrolyze
Hg$_2$(N$_3$)$_2$	–	592 [103] 593 [27]	White	Slightly soluble in water, photosensitive (turns yellow and gray under light—colloidal mercury forms), very toxic
Hg(N$_3$)$_2$	–	–	Clear to lemon yellow	Slightly soluble in water, photosensitive (turns yellow under light due to formation of colloidal mercury) but less than Hg$_2$(N$_3$)$_2$
Zn(N$_3$)$_2$	–	–	White	Hygroscopic, soluble in water, has a strong tendency to decompose hydrolytically, forms basic salt with water
TlN$_3$	5.74 [117]	234 [103]	Pale yellow	Slightly soluble in water, highly toxic

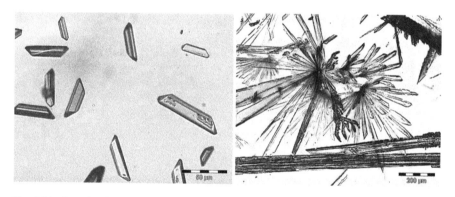

Fig. 4.14 Crystals of mercuric azide: α-form (*left*), β-form (*right*)

then connected, the solutions diffuse into each other, and crystals begin to appear. However, different crystals form on the side of TlNO$_3$ and different on the side of the azide. The left picture on Fig. 4.15 shows what forms on the side of TlNO$_3$ and the right one what forms on the side of sodium azide.

Cadmium azide is not physically stable; it easily hydrolyzes. The melting point of cadmium azide is 291 °C accompanied by decomposition [3].

4.6.2 Explosive Properties

Cadmium azide is a powerful primary explosive being superior to LA [3]. It is, however, extremely sensitive to mechanical stimuli (explodes when rubbed with a

4.6 Other Metallic Azides

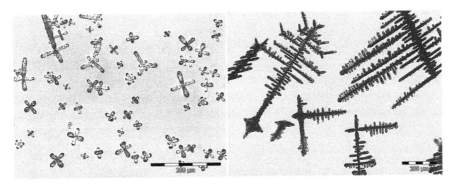

Fig. 4.15 Crystals of thallous azide prepared by diffusion process (*left*—what forms on the TlNO$_3$ side, *right*—what forms on the NaN$_3$ side)

Table 4.13 Impact sensitivity, explosion temperature, and initiating efficiency to tetryl and TNT of azides of mercury, cadmium, and thallium and comparison with lead azide

			Initiating efficiency (g)	
	Impact sensitivity	Ignition temperature (°C)	Tetryl	TNT
Hg$_2$(N$_3$)$_2$	6 cm/0.5 kg [82]	281^a [119]; 270 [80]	0.045 [5]	0.145 [5]
α-Hg(N$_3$)$_2$	6.5 cm/0.6 kg [21]	300 [80]	0.005 [21]	0.03 [21]
Cd(N$_3$)$_2$	Extremely high [15, 118]	291^a [119]; 360^a [3]	0.01 [5]	0.04 [5]
TlN$_3$	–	430^b [80]; 500^c, 530^d [28]	0.07 [5]	0.335 [5]
Pb(N$_3$)$_2$	43 cm/0.5 kg [82]	327^a [119]	0.025 [5]	0.09 [5]

^{a}Temperature at which explosion occurred within 5 s.
^{b}Ignition temperature depends on ambient pressure.
^{c}Instantaneous explosion.
^{d}Explosion within 1 s.

spatula) [118] or even spontaneously [15]. Danilov reported cadmium azide as highly thermostable with an ignition temperature of 360 °C (within 5 s) [3] while Wöhler and Martin reported only 291 °C [119]. The detonation velocity is 3,760 m s^{-1} at 2 g cm^{-3}. Values for other densities can be calculated using the following equation [3]:

$$D_\rho = D_0 + 360(\rho - \rho_0) \text{ m.s}^{-1}, \text{ where } \rho_0 \text{ is 2 g.cm}^{-3}.$$

Detonation pressure may be calculated in a similar way from the following equation:

$$P = (59\rho - 106).10^2 \text{ MPa}$$

The ignition efficiency of cadmium azide is higher than that of LA (see Table 4.13) [3, 5].

The sensitivity of mercurous azide Hg$_2$(N$_3$)$_2$ to impact and friction is significantly higher than that of mercury fulminate [82, 120]. When Hg$_2$(N$_3$)$_2$ is heated over a flame the solid melts, volatilizes, and explodes if the diameter of the molten layer is sufficiently large. A thin layer of $Hg_2(N_3)_2$ only burns with a blue flame. It explodes if the diameter of the fused azide is greater than a few tenths of a millimeter [76]. The ignition temperature is 298 °C with explosion taking place instantaneously or within 1 s [82].

The sensitivity of mercuric azide $Hg(N_3)_2$ depends on its crystal modification and the shape and size of the crystals. α-$Hg(N_3)_2$ is not abnormally sensitive, its sensitivity to impact being lower than for cupric azide $Cu(N_3)_2$ but higher than for mercury fulminate. However, as mentioned earlier, β-$Hg(N_3)_2$ explodes at the slightest touch or even spontaneously [12]. Generally, $Hg(N_3)_2$ is considered to be highly sensitive to impact and friction, more than mercury fulminate [5, 80]. $Hg(N_3)_2$ explodes with a beautiful white-blue flash resembling an electrical discharge when initiated by flame (Pachman and Matyáš Unpublished). The summary of sensitivities of mercury, cadmium, and thallium azides to impact, their explosion temperature, and initiating efficiency is listed in Table 4.13. The heats of explosion of many metallic azides have been measured by Wöhler and Martin [115].

Another disadvantage of mercuric azide $Hg(N_3)_2$ is its tendency to agglomerate into bigger blocks that cause problems during loading. The grinding of such agglomerates is an operation with high risk of explosion [21].

Although impact and friction sensitivity of thallous azide TlN_3 are generally considered to be lower than for LA it has one unusual property that should be brought to general attention. Its sensitivity to friction depends on the thickness of the sample layer and is highest in very thin layers. This phenomenon has been reported to be a cause of several explosions in laboratories [15]. Its initiation efficiency is lower than that of LA [3].

Some other metallic azides, like cadmium azide, are extremely sensitive to mechanical stimuli. Nickel azide is extremely sensitive to friction; it explodes by slight pressure or rubbing between metal and glass. Cobalt azide is even more sensitive than nickel azide. On the other hand, zinc azide detonates only under strong impact [115].

4.6.3 Preparation

Insoluble azides such as AgN_3, $Pb(N_3)_2$, $Hg_2(N_3)_2$, $Hg(N_3)_2$, CuN_3, $Cu(N_3)_2$, TlN_3 are prepared by precipitation from a solution of a soluble metallic salt by addition of sodium azide [1, 15, 121].

$$TlClO_4 + NaN_3 \longrightarrow TlN_3 + NaClO_4$$

For example, mercurous azide is prepared by this method; it is precipitated by combining aqueous solutions of a mercurous salt (nitrate) and sodium azide. The use of tetramethoxysilane is recommended for producing better crystal structure [116].

4.6 Other Metallic Azides

Mercuric azide (α-form) is prepared by mixing a saturated aqueous solution of mercuric chloride with the equivalent amount of sodium azide solution in presence of hydrazoic acid [12].

$$HgCl_2 + 2\,NaN_3 \xrightarrow{HN_3} Hg(N_3)_2 + 2\,NaCl$$

Some of the water-soluble metallic azides may be prepared by reaction of the metal oxide, hydroxide, or carbonate with excess of hydrazoic acid. This method is reported for preparation of cadmium azide [3, 5].

$$CdO + 2\,HN_3 \longrightarrow Cd(N_3)_2 + H_2O$$

The azide is obtained from the aqueous solution by evaporation. The product (depending on the process of evaporation) often forms long needles that are extremely sensitive and often explode by touch [50]. It is also possible to obtain water soluble azides by reaction of barium azide with the relevant soluble metallic sulfate. The precipitate of barium sulfate is filtered off and the relevant azide can be crystallized from the aqueous solution.

$$CdSO_4 + Ba(N_3)_2 \longrightarrow Cd(N_3)_2 + BaSO_4$$

Soluble azides can be purified by recrystallization from hot water. TlN_3 is considered insoluble at normal temperature and therefore can be prepared by above-mentioned precipitation and may be purified just like the soluble azides because it does dissolve in hot water [7].

In many cases (Mg, Mn, Zn, Cr, Co, etc.) metallic azides are not obtainable from aqueous solution as they give insoluble basic salts or mixtures of varying composition on standing or evaporation. In the case of Al, Zr, Th the corresponding hydroxides form [11]. The azides that easily undergo hydrolysis are therefore prepared by using ether or chloroform solutions of azoimide. A typical example is shown below. First the basic zinc azide forms by reaction of metallic zinc with an aqueous solution of azoimide. Then the basic salt is converted to normal zinc azide by additional reaction with an ether solution of azoimide. The end of the conversion can be recognized by the complete dissolution of the suspension in water when insoluble starting metallic salt is used [5, 52, 111, 115].

$$Zn + HN_3 \xrightarrow{H_2O} Zn(N_3)(OH) \xrightarrow[Et_2O]{HN_3} Zn(N_3)_2$$

Azides soluble in methanol can be prepared by shaking the dry metal salt in absolute methanol with sodium azide (e.g., ferric azide from ferric sulfate) [115].

Arsenic triazide forms by treatment of arsenic trichloride on sodium azide in trichlorofluoromethane [122]; antimony triazide by reaction of antimony triiodide with silver azide in acetonitrile [123].

$$AsCl_3 + 3\ NaN_3 \xrightarrow{CFCl_3} As(N_3)_3 + 3\ NaCl$$

$$SbI_3 + 3\ AgN_3 \xrightarrow{CH_3CN} Sb(N_3)_3 + 3\ AgI$$

The preparation of arsenic, antimony, and bismuth triazides in pure crystalline form (suitable for X-ray crystallography) was published by Haiges et al. These azides form by reaction of arsenic or antimony trifluoride in sulfur dioxide with trimethylsilyl azide [124, 125].

$$AsF_3 + 3\ (CH_3)_3SiN_3 \xrightarrow{SO_2} As(N_3)_3 + 3\ (CH_3)_3SiF$$

The same method can be used for preparation of neat pentaazides of arsenic and antimony whose preparation was not successful for a long time [126].

$$AsF_5 + 3\ (CH_3)_3SiN_3 \xrightarrow{SO_2} As(N_3)_5 + 5\ (CH_3)_3SiF$$

4.6.4 Uses

Due to low physical and chemical stability (e.g., $Zn(N_3)_2$) or high sensitivity (e.g., $Hg(N_3)_2$) most metallic azides covered in this chapter have never been practically used despite the high initiating efficiency that most of them have. Cadmium azide (despite its tendency to hydrolyze [35]) is the only one that is reported to have some limited application thanks to its high thermostability. It has been used in hermetically sealed detonators for high temperatures in oil wells as they are able to withstand 2 h at 250 °C or 6 days at 160 °C [3].

Mercuric azide was one of the first metallic azides studied because of its similarity with mercury fulminate. Unexpected and unfortunately fatal accidental explosions occurred during testing of this substance because of the contamination of mercuric azide with mercurous azide (highly sensitive substance). Because of these accidents, further examinations of salts of azoimide were stopped for several years [3].

4.7 Organic Azides

A variety of organic azides exists and has been described in scientific literature. However, only some of them have the characteristics of primary explosives. The sensitivity of these substances to the mechanical stimuli necessary for primary explosives generally increases with content of azido group in the molecule.

4.8 Cyanuric Triazide

For example, 1-azido-2,4,6-trinitrobenzene with one azido group in the molecule has the character of a secondary explosive (its sensitivity is too low for primary explosive) whereas 1,3,5-triazido-2,4,6-trinitrobenzene is a typical primary explosive.

Most organic azides do not possess the properties required for primary explosives and generally can be characterized by low physical and chemical stability (lower alkyl azides, sulfuryl azide explode even spontaneously [8, 11]), high sensitivity to mechanical stimuli (lower alkyl azides, acyl azides e.g., carbonyl diazide explode on contact with a glass rod [30]), low thermal stability (e.g., cyanogen azide, 1,3,5-triazido-2,4,6-trinitrobenzene, dicyanamid azide, esters of azidoacetic acid [8]), or sensitivity to light [127].

The thermal stability of organic azides generally decreases in the following order [3]:

alkyl azides > aryl azides > azidoformates > sulfonyl azides > acyl azides

Organometallic azides do not have the characteristics of explosives (e.g., triethyl lead azide) [3].

There are two basic methods for preparing organic azides [3]:

1. By nucleophilic substitution of the halogens, nitroxy group, etc. by the azido group:

$$RX + N_3^- \longrightarrow RN_3 + X^-$$

R = Alk

X = halogen, OAlk, ONO_2, OTos, RSO_2, AlkC=O

2. Via the diazonium salt by action of nitrous acid with hydrazino-compounds or by reaction of basic azides with the diazonium salt.

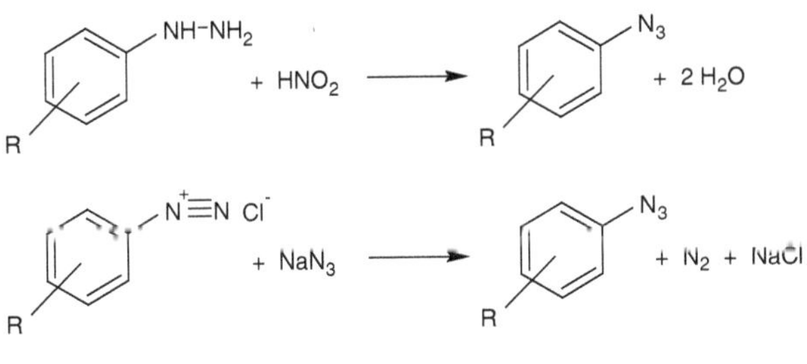

4.8 Cyanuric Triazide

Cyanuric triazide (2,4,6-triazido-1,3,5-triazine; TAT; sometimes called simply "triazide") was probably first prepared by H. Finger in 1907 by the reaction of 2,4,6-trihydrazino-1,3,5-triazine with sodium nitrite in an acidic environment [6, 46, 128].

He, however, did not examine this new compound too closely and did not discover its possible usefulness in priming [46, 128]. It is therefore Erwin Ott who is sometimes cited as the discoverer of cyanuric triazide as he was the first to perform a detailed examination of this compound and he patented it as a primary explosive [129, 130].

4.8.1 Physical and Chemical Properties

Cyanuric triazide forms white crystals with density 1.54 g cm^{-3} [5, 45, 131] (1.5 g cm^{-3} reported in older literature [132, 133]) or 1.71–1.72 g cm^{-3} [3, 46, 134, 135] with melting point 94 °C. This compound is easily soluble in acetone, hot alcohol, chloroform, benzene, pyridine; slightly soluble in cold alcohol; and insoluble in water [46, 129, 131, 132, 136, 137]. Yoffe studied the behavior of cyanuric triazide during heating under various pressures. He reported that cyanuric triazide can be melted without decomposition (even at 112 °C no decomp. over several hours) or even at 198 °C in a vacuum without any observable decomposition. The azide simply condensed on a cold part of the vessel [28]. Other authors, however, reported that decomposition begins when this material is heated above 100 °C and explosion occurs in the range from 150 to 180 °C depending on the heating rate. It may entirely decompose without explosion when heated slowly [129, 131, 132]. Gillan described its thermal decomposition as follows: clear liquid forms during melting (94–95 °C), gas liberating starts at 155 °C, orange to brown solution coloration appears around 170 °C, solidification to an orange-brown solid by 200 °C, and rapid decomposition at 240 °C (capillary examination) [138].

Cyanuric triazide is slightly hygroscopic and slightly volatile [6, 17, 30, 47, 136], but significantly more hygroscopic than MF and LA [47]. Danilov et al., however, reported noticeable volatility even at 30 °C [3]. Muraour published heat of formation 917 kJ mol^{-1} [46, 139], but according to Meyer it is 931 kJ mol^{-1} [40] and according to recent work of Gillan $\Delta H_f = 1,053$ kJ mol^{-1} [138]. It does not react with metals, water, and carbon dioxide [132, 136, 137].

Cyanuric triazide is reduced by action of hydrogen sulfide to melamine—nitrogen is evolved and sulfur is precipitated [130]. It also decomposes in aqueous sodium hydroxide (0.1 M) at 50 °C within several minutes yielding the sodium salt of cyanuric acid and sodium azide [46, 52]. The substance is not irritating to the skin [17].

4.8.2 Explosive Properties

Small crystals of cyanuric triazide are more sensitive to mechanical stimuli than small crystals of MF [47]. It explodes when pressed into a detonator capsule. Large

4.8 Cyanuric Triazide

Table 4.14 Impact sensitivity of cyanuric triazide

Cyanuric triazide		Comparison with LA				
Converted to energy (J)	As reported in literature	Converted to energy (J)	As reported in literature	Value reported	Apparatus	Reference
0.34	7 cm/0.5 kg[a]	2.1	43 cm/0.5 kg	h_{min}	Small impact machine	[82]
0.69	7 cm/1 kg	2.0	10 cm/2 kg[b]	h_{50}	BM apparatus	[45]
0.18	0.9 cm/2 kg	1.6	32 cm/0.5 kg	–	BM apparatus	[5]
–	25.4 cm	–	17.8–28 cm	–	Ball Drop Impact	[140]

[a]Needles of cyanuric triazide from 0.04 to 0.2 mm in diameter and averaging 0.1 mm in length.
[b]Pure LA.

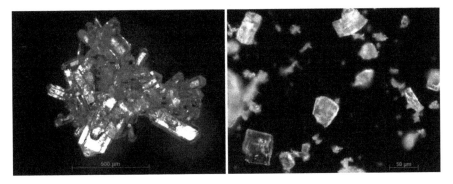

Fig. 4.16 Crystals of raw cyanuric triazide (*left*) and recrystallized sample (*right*) [140]. Reprinted by permission from Neha Mehta

crystals (obtained by recrystallization) can explode when broken by pressure of a rubber [17]. Long needles of cyanuric triazide (obtained by evaporation of organic solutions) can explode even spontaneously [136]. Some of the values of impact sensitivity of cyanuric triazide are summarized in Table 4.14.

The sensitivity of cyanuric triazide to friction and electrostatic discharge has recently been measured by Mehta et al. [140]. They observed that sensitivity to friction of a recrystallized sample is lower than that of a nonrecrystallized one contrary to sensitivity to electrical discharge. The photographs of both samples are presented in Fig. 4.16. The values observed for these two types of cyanuric triazide are compared with lead azide in Tables 4.15 and 4.16 [140].

Taylor and Rinkenbach [82] determined the ignition temperature of cyanuric triazide to be 252 °C (the sample dropped on the surface of the molten alloy with a constant temperature, explosion took place immediately or within 1 s) [82]. The substance explodes when heated to 205–208 °C (heating rate 20 °C min^{-1}) [47] and at 170 °C (heating rate 5 °C min^{-1}) [5]. The experimentally determined values of the detonation velocity for cyanuric triazide are summarized in Table 4.17.

Cyanuric triazide decomposes explosively into molecular nitrogen and cyanogen (C_2N_2) when initiated under vacuum. Ignition in a confined space leads to formation of a black sooty residue (with high yield >91 wt.%) consisting of

Table 4.15 Sensitivity to friction of cyanuric triazide and LA measured as a piston loading (in g) on small BAM friction test [140]

	Cyanuric triazide (g)		LA (g)
	Nonrecrystallized	Recrystallized	
No go at	–	10 (0/10)	10–20
Go at	10[a]	20	30

[a]The lowest level of the apparatus.

Table 4.16 Sensitivity to electric discharge of cyanuric triazide compared with LA (in mJ) [140]

	Cyanuric triazide (mJ)		LA (mJ)
	Nonrecrystallized	Recrystallized	
No go at	31 (0/20)	1.2 (0/20)	1.2
Go at	–	–	1.6

Table 4.17 Detonation velocity of cyanuric triazide

Density (g cm^{-3})	Detonation velocity (m s^{-1})	Note	Reference
1.02	5,500	Unconfined	[40]
1.15	5,440–5,650		[47]
	5,550		[139]
	5,600		[141]

Table 4.18 Initiation efficiency of cyanuric triazide

Initiating efficiency (g)				
TNT	Tetryl	PETN	Comment	Reference
0.1 (with RC)	0.04	–	–	[5, 17]
0.12 (without RC)[a]	–	–	1.4 g cm^{-3}	[139]
0.08 (with RC)[a]				
0.20 (with RC)[a]	–	–	1.5 g cm^{-3}	[139]
–	–	0.005	49 MPa	[6]
0.13	0.06	–	–	[21]
0.07	0.02	–	–	[46]
(0.1 for LA)	(0.02–0.03 for LA)			

[a]TNT compressed to a density 1.35 g cm^{-3} in copper detonator tube.

nanostructured carbon materials [142, 143]. Interesting material—carbon nitrides (e.g., C_3N_4, C_3N_5) form by thermal decomposition (185 °C) of cyanuric triazide in a high-pressure reactor [138].

Cyanuric triazide belongs to the group of explosives having brisance superior to mercury fulminate:

- Brisance by sand test: 32.2 g (67 % TNT and 264 % MF) [5, 45]
- Lead block test: 415 cm^{-3}/10 g (138 % TNT [40]; 140 % TNT [5])

The values of initiating efficiencies for TNT, tetryl, and PETN are summarized in Table 4.18.

After ignition, small crystals of cyanuric triazide burn (unlike silver azide) even though Ficheroulle and Kovache reported them exploding violently by the action of flame [136]. The outcome of initiation by hot wire depends on temperature. At 20–50 °C it only cracks while at 75 °C it explodes [136].

Burning speed of a 3-mm-long crystal with diameter 0.35 mm is about 0.3 m s^{-1} and increases with increasing diameter. Crystals with length 1.8–3 mm and cross-section 0.6–0.7 × 0.6–1.0 mm burn with velocity from 1 to 1.5 m s^{-1}. The burning has an accelerating character and with larger crystals explosions may occur (e.g., burning of crystal 2 × 2 × 1 mm transitions to explosion) [85]. Fogelzang et al. found a linear dependence of combustion rate on pressure in range 0.1–40 MPa [144].

Cyanurazide can be dead-pressed by pressure above about 20 MPa [3, 46, 52].

4.8.3 Preparation

Cyanuric triazide can easily be obtained by the action of cyanuric chloride on sodium azide in an aqueous solution at room temperature. The product forms as a crystalline precipitate [6, 129, 132].

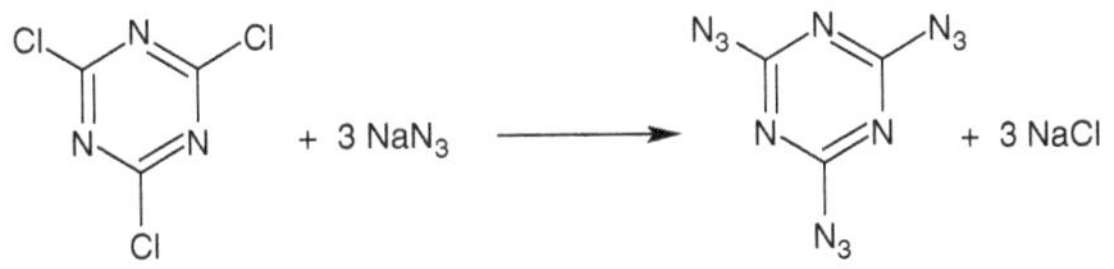

The substance is highly dangerous. Explosion during the manufacturing process, more specifically during drying, has been encountered [30]. The suitable solvent for recrystallization of cyanuric triazide is ethanol [6, 132]. Recrystallization of cyanuric triazide is a dangerous operation due to potential formation of large, extremely sensitive, crystals [45]. Long needles that form by evaporation of an organic solution can even explode spontaneously. Small crystals, formed by dilution of cyanuric triazide acetone solution, can explode when being dried at around 50 °C [136].

Cyanuric triazide also forms by reaction of 2,4,6-trihydrazino-1,3,5-triazine with sodium nitrite in an acidic environment [33, 137]. However, Bagal reported that this way of preparation is not suitable for making pure cyanuric triazide [46].

4.8.4 Uses

Cyanuric triazide was patented as an initiating explosive by Erwin Ott in 1919 [133] but it has not been used practically due to high sensitivity, volatility, and hygroscopicity [5, 8, 40, 45]. This substance has recently again been considered as an LA/LS replacement in initiator mixtures, an alternative to NOL-130, and an LA replacement in stab detonators [140]. Other authors again excluded it for practical use due to its low thermal stability [145, 146]. Some investigations of possible ways of preparation of carbon nanotubes via explosive decomposition of cyanuric triazide in confining cavity have been investigated [142, 143].

4.9 4,4′,6,6′-Tetra(azido)hydrazo-1,3,5-triazine and 4,4′,6,6′-Tetra(azido)azo-1,3,5-triazine

These two new high-nitrogen compounds were probably first reported by Huynh et al. in 2004 [147].

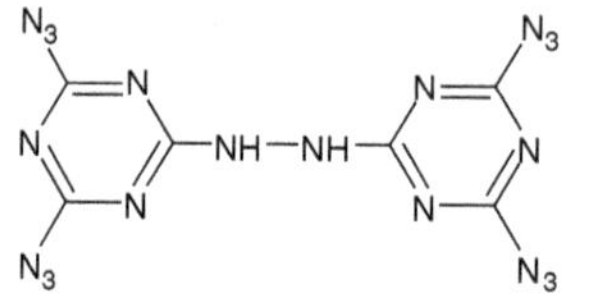

4,4′,6,6′-Tetra(azido)hydrazo-1,3,5-triazine
TAHT

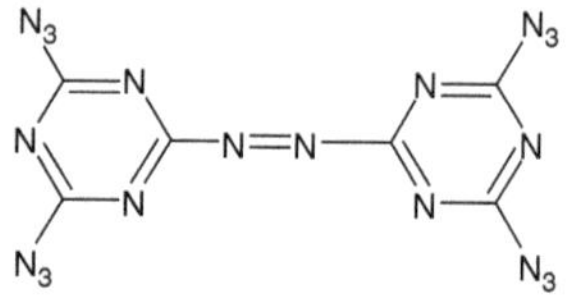

4,4′,6,6′-Tetra(azido)azo-1,3,5-triazine
TAAT

4.9.1 Physical and Chemical Properties

4,4′,6,6′-Tetra(azido)hydrazo-1,3,5-triazine (TAHT) has only one polymorph in which two 1,3,5-triazine rings are not co-planar but have a torsion angle of 105°. The density of this polymorph is 1.649 g cm^{-3}. On the other hand, 4,4′,6,6′-tetra(azido)azo-1,3,5-triazine (TAAT) crystallizes as an α or β polymorph with density 1.724 g cm^{-3} and 1.674 g cm^{-3}, respectively. The β polymorph has two conformers whose azido groups are oriented in different directions. The presence of hydrazo and azo linkages significantly decreases the volatility and dramatically increases the melting point relative to cyanuric triazide (m.p. is not observable up to decomposition at 200 and 202 °C, respectively). Both molecules have very high positive heat of formation 1,753 kJ mol^{-1} for TAHT and 2,171 kJ mol^{-1} for TAAT (it is believed that the ΔH_f for TAAT is the highest ever experimentally measured for an energetic high-nitrogen compound) [147, 148].

Pyrolysis of TAAT yields carbon nitrides such as C_2N_3 and C_3N_5 when decomposed under mild conditions (e.g., low temperature and without applied pressure) [149, 150].

4.9.2 Explosive Properties

It was to be expected that the hydrazo and azo linkages in TAHT and TAAT molecules would desensitize these compounds relative to cyanuric triazide. Reported values are summarized in Table 4.19 [147].

The impact sensitivity of TAHT is inferior to cyanuric triazide and even PETN, and sensitivity to friction is between the values for cyanuric triazide and PETN. TAAT is generally more sensitive than its hydrazo analogue and is comparable to cyanuric triazide itself (impact sensitivity is about the same, friction sensitivity is significantly lower) [147].

4.9 4,4′,6,6′-Tetra(azido)hydrazo-1,3,5-triazine and 4,4′,6,...

Table 4.19 Sensitivity of TAHT and TAAT to mechanical stimuli and their thermal stability compared with PETN and cyanuric triazide [147, 148]

Compound	Impact sensitivity (h_{50}, type 12) (cm)	Friction sensitivity (BAM) (kg)	Spark sensitivity (J)	DSC fast decomp. (°C)
Cyanuric triazide	6.2	<0.5	<0.36	187
Hydrazo comp.	18.3	2.9	<0.36	202
Azo comp.	6.2	2.4	<0.36	200
PETN	14.5	5.4	>0.36	178

4.9.3 Preparation

4,4′,6,6′-Tetra(chloro)hydrazo-1,3,5-triazine (prepared from cyanuric triazide [151]) is the starting material for both compounds. Its reaction with excess of hydrazine hydrate in acetonitrile forms 4,4′,6,6′-tetra(hydrazino)hydrazo-1,3,5-triazine which undergoes diazotization to TAHT. TAAT forms by oxidation of TAHT by chlorine in a water/chloroform suspension [147, 148, 150]. Simplified method of preparation of the key intermediate TAHT was published by Li et al. [152]. Instead of hydrazinolysis and diazotation they used nucleophilic substitution of 4,4′,6,6′-tetra(chloro)hydrazo-1,3,5-triazine with sodium azide for preparation of TAHT which they in the last step oxidized with N-bromosuccinimide.

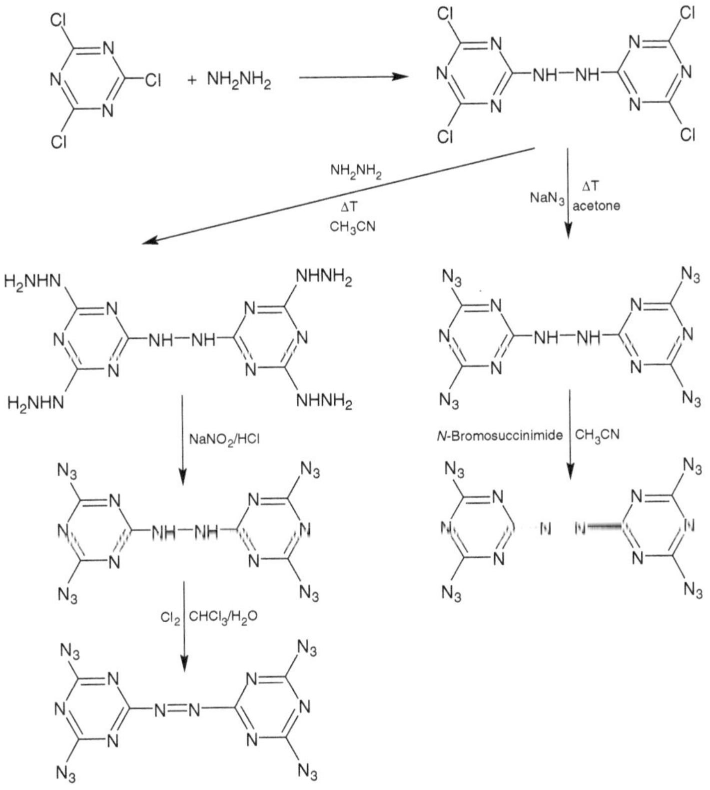

4.9.4 Uses

Huynh and Hiskey patented the use of TAAT for preparation of carbon nitrides, more specifically C_2N_3 and C_3N_5 [150].

4.10 1,3,5-Triazido-2,4,6-trinitrobenzene

1,3,5-Triazido-2,4,6-trinitrobenzene (TATNB, TNTAB) was first prepared by Oldřich Turek in 1924 by the action of sodium azide on 1,3,5-trichloro-2,4,6-trinitrobenzene [153, 154].

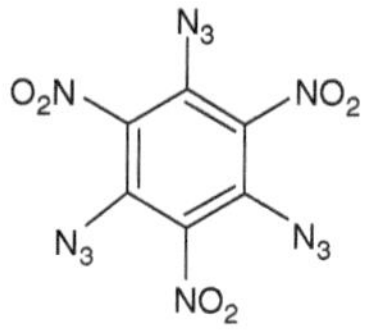

4.10.1 Physical and Chemical Properties

1,3,5-Triazido-2,4,6-trinitrobenzene is a substance which crystallizes in the form of bright yellow crystals. The melting point of TATNB is 131 °C and density 1.805 g cm^{-3} (pycnometric) [153, 154] or 1.84 g cm^{-3} (X-ray) [155]. The heat of formation is 765.8 kJ mol^{-1} or 2.28 MJ kg^{-1} (a value significantly higher than for TNT or HMX), TATNB is strongly endothermic compound. The experimentally determined heat of combustion is 3,200 kJ mol^{-1} [155]. TATNB is soluble in acetone and benzene, slightly soluble in alcohols, and insoluble in water. It decomposes in pyridine or monoethanolamine [153, 154, 156]. TATNB is stable in water (no change within 3.5 years of storage under water) [153, 154]. Turek reported the nonhygroscopic nature of TATNB [153, 154]; Ficheroulle and Kovache reported slight hygroscopicity (it absorbs 1.35 wt.% of water within 40 days of exposure to moist air) [156]. On exposure to light, the yellow color deepens and by very long exposure to sunlight, surface layers of TATNB crystals decompose to benzotrifuroxane [153, 154]. TATNB is compatible with most common metals and reactions do not take place in a moist atmosphere [153, 154]. Reaction was not observed for iron, steel, copper, and brass even when wet [33, 45]. Ficheroulle and Kovache, however, reported reaction with steel and lead on one hand and good resistance of aluminum, brass, and copper on the other, both in moist atmospheres [153, 154, 156]. TATNB can be chemically decomposed by action of aqueous sodium hydroxide when trinitrofluoroglucinol forms [156].

TATNB has low thermal stability. It decomposes by heating to benzotrifuroxane and gaseous nitrogen. This decomposition is quantitative and occurs even at low temperatures [153, 154].

4.10 1,3,5-Triazido-2,4,6-trinitrobenzene

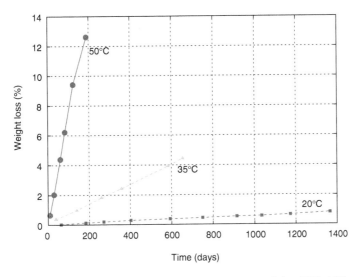

Fig. 4.17 Dependence of weight loss of TATNB on temperature and time [153, 154]

The dependence of the rate of decomposition (in weight loss) on temperature has been determined by Turek and is presented in Fig. 4.17. TATNB rapidly decomposes at 100 °C; full transformation is achieved within 14 h. The decomposition is not self-catalyzed [153, 154].

4.10.2 Explosive Properties

Sensitivity to impact is mostly mentioned significantly lower than for MF [5, 30, 40, 153, 154] (expect [33, 157]). The published values of sensitivity to impact are summarized below.

- 50 % probability of ignition at 2.5 J (1/2 kg hammer from 50 cm) [156].
- 0 % probability of ignition at 2 J and 100 % probability of ignition at 3.9 J vs. 0 % probability of ignition at 3.9 J and 50 % probability of ignition at 5.9 J for LA (2 kg hammer from relevant height) [33].
- Ignition at 5 J vs. 1–2 J for MF [40].
- Higher than 1.5 J (0.25 kg hammer from 60 cm) [155].
- Significantly lower than MF [3, 153, 154].

Table 4.20 Detonation velocity of TATNB

Density (g cm^{-3})	Detonation velocity (m s^{-1})	References
1.50	7,200	[33]
1.54	7,500	[30]
1.70	8,100	[33]

Sensitivity to friction is reported generally relatively low, but depends on crystal size [5]. Under water, TATNB is not sensitive to friction [153, 154].

TATNB is sensitive to flame and when lit, loose TATNB burns with a blinding blue flame without fumes [153, 154]. The burning behavior of the material pressed at 98, 196, and 294 MPa has been studied by Ficheroulle [156]. A thin layer of TATNB (0.5 mm thick) burns slowly with rate about 0.03 mm s^{-1} and a single crystal with burning speed about 0.05 m s^{-1}. These rates are slower than those of most other primary explosives (SA, LS, MF, or cyanuric triazide) [85]. However, if confined, for example pressed in a blasting cap, TATNB explodes violently [153, 154]. TATNB explodes when heated to 150 °C within 10 s [5].

TATNB is a powerful primary explosive. The dependence of detonation velocity on density is given in Table 4.20.

Brisance of TATNB is high; it is superior to that of TNT and tetryl in plate dent test. Power of TATNB determined by a Trauzl lead block test is 150 % TNT (vs. 27 % for LA and 38 % for MF) [153, 154] or 179 % TNT [5]. Ignition efficiency of TATNB is superior to that of both MF and LA:

- 0.02 g TATNB for TNT and 0.01 g for tetryl (TNT and tetryl were pressed at 49 MPa, TATNB at 29 MPa) [153, 154].
- 0.08 g TATNB for TNT and 0.05 g for tetryl (with reinforcing cap, TNT and tetryl was pressed at 49 MPa, TATNB at 29 MPa) vs. 0.12 g for LA and 0.25–0.30 g for MF (both for TNT) [33].

TATNB can be dead-pressed. It can be loaded into blasting caps by pressure up to 29 MPa; at higher pressures its brisance rapidly decreases. In combination with LA it can be pressed at significantly higher pressures (29–196 MPa) [153, 154, 158]. Fedoroff, Shefield any Kaye's encyclopedia mentions that pure TATNB becomes dead-pressed at pressures of about 290 MPa [5].

4.10.3 Preparation

1,3,5-Triazido-2,4,6-trinitrobenzene can be easy synthesized by the action of alkaline azide on 1,3,5-trichloro-2,4,6-trinitrobenzene. The product is slightly soluble in the reaction mixture from which it precipitates. The sodium azide can be used in the form of an aqueous ethanol solution while 1,3,5-trichloro-2,4,6-trinitrobenzene in the form of an acetone solution or alone [153, 154, 159, 160]. Purification of TATNB may be done by crystallization from solvents such as chloroform [153, 154, 159, 160], acetic acid, or mixture of acetic acid and acetone 1/1 (the last mixture is recommended for preparation of single crystals for structural studies) [155].

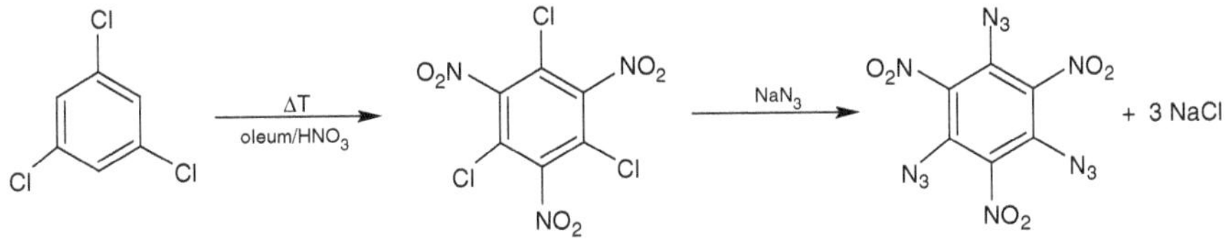

4.11 2,3,5,6-Tetraazido-1,4-benzoquinone

[Reaction scheme: 1,3,5-trichlorobenzene → (ΔT, oleum/HNO₃) → 1,3,5-trichloro-2,4,6-trinitrobenzene → (NaN₃) → 1,3,5-triazido-2,4,6-trinitrobenzene + 3 NaCl]

The starting substance, 1,3,5-trichloro-2,4,6-trinitrobenzene, can be prepared from 1,3,5-trichlorobenzene by nitration in a strong nitration mixture such as oleum/nitric acid at high temperatures [153, 154, 159, 160]. TATNB can also be prepared by nitration of 1,3,5-triazido-2,4-dinitrobenzene [155].

4.10.4 Uses

Turek suggested 1,3,5-triazido-2,4,6-trinitrobenzene as a replacement of mercury fulminate for detonators and blasting cartridges in 1927–1930 [158, 161, 162] and several years later again for filling in percussion caps [157, 163]. TATNB has high initiation efficiency; ideal sensitivity to mechanical stimuli; good chemical stability and it does not contain heavy metals. It would be an ideal candidate for a "green" primary explosive if it was not for its low thermal stability which disqualifies it from practical use. Some large-scale experiments have been carried out in the past to examine the possibilities of a practical application for TATNB, but without success being reported [30].

4.11 2,3,5,6-Tetraazido-1,4-benzoquinone

2,3,5,6-Tetraazido-1,4-benzoquinone (TeAzQ) was first prepared by Fries and Ochwat in 1923 by the action of sodium azide on 2,5-diazido-3,6-dichloro-1,4-benzoquinone [164].

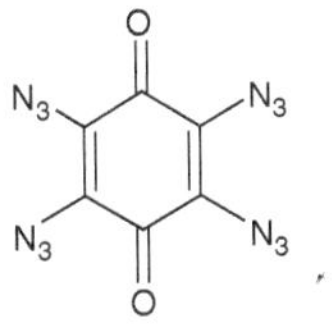

[Structure: 2,3,5,6-tetraazido-1,4-benzoquinone]

4.11.1 Physical Properties

2,3,5,6-Tetraazido-1,4-benzoquinone forms dark blue crystals with metallic luster [165] or nice brownish yellow prismatic crystals with a blue-black gloss [164]. It is soluble in acetone, slightly soluble in ethanol, and insoluble in water. The main

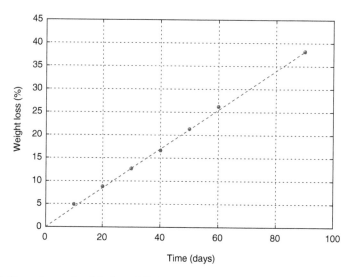

Fig. 4.18 Dependence of weight loss of TeAzQ on storage time [165]

drawback of this substance is its low thermal stability. It rapidly decomposes at laboratory temperature generating nitrogen. Nice blue crystals lose their gloss, break apart, and the substance quickly loses its initiating ability. The dependence of weight loss of TeAzQ on time (at room temperature) is shown in Fig. 4.18. To exclude the possibility that the substance is just volatile and that the weight loss is due to evaporation, Šorm encapsulated a sample in a closed vessel connected to a manometer and, from the pressure rise, calculated the amount of evolved nitrogen. He found that the substance is not volatile and that the weight loss is due to decomposition [165].

TeAzQ yields white 2,3,5,6-tetraazido-1,4-hydroquinone on reduction, but this is unstable and reoxidizes back to blue quinone while standing in air [165]. It decomposes under the action of sulfuric acid or sodium hydroxide liberating nitrogen [164].

4.11.2 Explosive Properties

The impact sensitivity of 2,3,5,6-tetraazido-1,4-benzoquinone is high; significantly higher than for LA. Values published in scientific papers are summarized in Table 4.21.

Sensitivity to friction is high as well; it explodes during gentle adjustment on sandpaper [166]. Ignition temperature is low, and it explodes at 91 °C (heating rate 5 °C min^{-1}). By action of flame it violently explodes, whereas 2,3,5,6-tetraazido-1,4-hydroquinone only ignites. TeAzQ has high ignition efficiency and excellent resistance to dead-pressing. Using pressures of 49 MPa or 98 MPa resulted in the same minimal amount of 0.02 g needed for initiating TNT [165].

Table 4.21 Impact sensitivity of 2,3,5,6-tetraazido-1,4-benzoquinone

	Converted to energy (J)	As reported in literature	Value reported	Reference
TeAzQ	~1	20 cm/0.5 kg	Not specified	[165]
	<0.4	<1.5 cm/2.5 kg	h_{50}	[166]
LA	1.13	4.6 cm/2.5 kg	h_{50}	[166]
LS	2.16	8.8 cm/2.5 kg	h_{50}	[166]

4.11.3 Preparation

2,3,5,6-Tetraazido-1,4-benzoquinone can be easily prepared by the action over several hours of sodium azide on 2,3,5,6-tetrachloro-1,4-benzoquinone (*p*-chloranil) in an ethanol suspension [165] or by the action of aqueous sodium azide on a methylene chloride solution of *p*-chloranil [166].

4.11.4 Uses

2,3,5,6-Tetraazido-1,4-benzoquinone has never been used due to its low thermal stability.

References

1. Curtius, T.: Ueber Stickstoffwasserstoffäure. Berichte der deutschen chemischen Gesellschaft **23**, 3023–3033 (1890)
2. Curtius, T., Radenhauser, R.: Zur Kenntniss der Stickstoffwasserstoffsäure. Journal für praktische Chemie **43**, 207–208 (1891)
3. Danilov, J.N., Ilyusin, M.A., Tselinskii I.V.: Promyshlennye vzryvchatye veshchestva; chast I. Iniciiruyushchie vzryvshchatye veshchestva. Sankt-Peterburgskii gosudarstvennyi tekhnologicheskii institut, Sankt-Peterburg (2001)
4. Bubnov, P.F.: Initsiruyushchie vzryvchatye veshchestva i sredstva initsirovaniya. Gosudarstvennoe izdatelstvo oboronnoi promyshlennosti, Moskva (1940)
5. Fedoroff, B.T., Sheffield, O.E., Kaye, S.M.: Encyclopedia of Explosives and Related Items. Picatinny Arsenal, New Jersey (1960–1983)
6. Krauz, C.: Technologie výbušin. Vědecko-technické nakladatelství, Praha (1950)
7. Evans, B.L., Yoffe, A.D., Gray, P.: Physics and chemistry of the inorganic azides. Chem. Rev. **59**, 515–568 (1959)
8. Urbański, T.: Chemistry and Technology of Explosives. Pergamon, Oxford (1984)
9. Hyronimus, F.: Improvements in and relating to the charge of ammunition primers. GB Patent 1,819, 1908

10. Rintoul, W.: Explosives. Rep. Prog. Appl. Chem. **5**, 523–565 (1920)
11. Audrieth, L.F.: Hydrazoic acid and its inorganic derivates. Chem. Rev. **15**, 169–224 (1934)
12. Miles, F.D.: The formation and characteristic of crystals of lead azide and of some other initiating explosives. J. Chem. Soc. 2532–2542 (1931)
13. Hattori, K., McCrone, W.: Lead azide, $Pb(N_3)_2$. Anal. Chem. **28**, 1791–1792 (1956)
14. Stettbacher, A.: Die Ermittlung des absoluten spezifischen Gewichts von Zündund Sprengstoffen. Chemisches Zentralblatt **113**, 366–367 (1942)
15. Fair, H.D., Walker, R.F.E.: Energetic Materials. Physics and Chemistry of Inorganic Azides. Plenum, New York (1977)
16. Miles, F.D.: The formation and constitution of crystals of lead salts containing water-soluble colloid. Philos. Trans. R. Soc. Lond. A Math. Phys. Sci. **235**, 125–164 (1935)
17. Davis, T.L.: The Chemistry of Powder and Explosives. Wiley, New York (1943)
18. Taylor, A.C., Thomas, A.T.: Spontaneous explosions during crystal growth of lead azide. J. Cryst. Growth **3**, 391–394 (1968)
19. Taylor, G.W.C.: The preparation of gamma lead azide. In: Proceedings of Symposium on Lead and Copper Azide, pp. A-3, 20-23, Waltham Abbey, 1966
20. Lamnevik, S., Soderquist, R.: On lead azides. 1 Refinement of the unit cell dimensions of alpha- and beta- lead azide. On a transformation of beta- to alpha- lead azide in the solid state. Report A 1105-F 110. FOA (1963)
21. Špičák, S., Šimeček, J.: Chemie a technologie třaskavin. Vojenská technická akademie Antonína Zápotockého, Brno (1957)
22. Jung, P.C.: Initiation and Detonation in Lead Azide and Silver Azide at Sub-millimeter Geometrics. Texas Technical University, Lubbock (2006)
23. Roth, J.: Initiation of lead azide by high-intensity light. J. Chem. Phys. **41**, 1929–1936 (1964)
24. Wöhler, L., Krupko, W.: Über die Lichtempfindlichkeit der Azide des Silbers, Quecksilberoxyduls, Bleis und Kupferoxyduls, sowie über basisches Blei- und Cupriazid. Berichte der deutschen chemischen Gesellschaft **46**, 2045–2057 (1913)
25. Todd, G., Eather, R., Heron, T.: The decomposition of lead azide under storage conditions. In: Proceedings of Symposium on Lead and Copper Azide, pp. B-2, 34–44, Waltham Abbey, 1966
26. McLaren, A.C.: The influence of preheating on the detonation velocity of lead azide. Research **10**, 409–410 (1957)
27. Gray, P., Waddington, T.C.: Thermochemistry and reactivity of azides. I. Thermochemistry of the inorganic azides. Proc. R. Soc. Lond. A Math. Phys. Sci. **A235**, 106–119 (1956)
28. Yoffe, A.D.: Thermal decomposition and explosion of azides. Proc. R. Soc. Lond. A Math. Phys. Sci. **A208**, 188–199 (1951)
29. Graybush, R.J., May, F.G., Forsyth, A.C.: Differential thermal analysis of primary explosives. Thermochim. Acta **2**, 153–162 (1971)
30. Urbański, T.: Chemistry and Technology of Explosives. PWN—Polish Scientific Publisher, Warszawa (1967)
31. Lamnevik, S.: Lead azide, the ideal detonant? In: Jenkins, J.M., White, J.R. (eds.) Proceedings of the International Conference on Research in Primary Explosives, vol. 2, pp. 9/1–9/17, Waltham Abbey (1975)
32. Brown, M.E., Swallowe, G.M.: The thermal decomposition of the silver(I) and mercury(II) salts of 5-nitrotetrazole and of mercury(II) fulminate. Thermochim. Acta **49**, 333–349 (1981)
33. Khmelnitskii, L.I.: Spravochnik po vzryvchatym veshchestvam. Voennaya ordena Lenina i ordena Suvorova Artilleriiskaya inzhenernaya akademiya imeni F. E. Dzerzhinskogo, Moskva (1962)
34. Urbański, T.: Chemie a technologie výbušin. SNTL, Praha (1959)
35. Taylor, G.W.C.: Technical requirements and prospects for new primary explosives. In: Jenkins, J.M., White, J.R. (eds.) Proceedings of the International Conference on Research in Primary Explosives, vol. 3, pp. 18/1–18/21, Waltham Abbey (1975)
36. Blay, N.J., Rapley, R.J.: In: Jenkins, J.M., White, J.R. (eds.) Proceedings of the International Conference on Research in Primary Explosives, vol. 3, pp. 20/1–20/19, Waltham Abbey (1975)

37. Wyatt, R.H.: Copper azide corrosion. In: Proceedings of Symposium on Lead and Copper Azide, pp. A-2, 15–19, Waltham Abbey, 1966
38. Wythes, P.E.: Accidents in manufacture of lead azide. In: Proceedings of Symposium on Lead and Copper Azide, pp. D-2, 99–102, Waltham Abbey, 1966
39. Military Explosives, Report TM-9-1300-214, Headquarters, Department of the Army (1984)
40. Meyer, R., Köhler, J., Homburg, A.: Explosives. Wiley-VCH, Weinheim (2002)
41. Šelešovský, J., Matyáš, R., Musil, T.: In: Using probit analysis for sensitivity tests - sensitivity curve and reliability. 14th Seminar on New Trends in Research of Energetic Materials, pp. 964–968, Pardubice, 2011
42. Matyáš, R., Šelešovský, J., Musil, T.: Sensitivity to friction for primary explosives. J. Hazard. Mater. **213–214**, 236–41 (2012)
43. Clark, A.K., Davies, N., Hubbard, P.J., Lee, P.R.: Cross sensitisation to impact between lead azide and tetryl. In: Jenkins, J.M., White, J.R. (eds.) Proceedings of International Conference on Research in Primary Explosives, vol. 3, pp. 22/1-22/15, Waltham Abbey (1975)
44. Medlock, L.E., Leslie, J.P.: Some aspects of the preparation and characteristics of lead azide precipitated in the presence of gelatin. In: Jenkins, J.M., White, J.R. (eds.) Proceedings of the International Conference on Research in Primary Explosives, vol. 2, pp. 10/1–10/16, Waltham Abbey (1975)
45. Tomlinson, W.R., Sheffield, O.E.: Engineering Design Handbook, Explosive Series of Properties Explosives of Military Interest. Report AMCP 706-177 (1971)
46. Bagal, L.I.: Khimiya i tekhnologiya iniciiruyushchikh vzryvchatykh veshchestv. Masinostroenie, Moskva (1975)
47. Kast, H., Haid, A.: Über die sprengtechnischen Eigenschaften der wichtigsten Initialsprengstoffe. Zeitschrift für das angewandte Chemie **38**, 43–52 (1925)
48. Wallbaum, R.: Sprengtechnische Eigenschaften und Lagerbeständigke der wichtigsten Initialsprengstoffe. Zeitschrift für das gesamte Schiess- und Sprengstoffwesen **34**, 197–201 (1939)
49. Kabik, I., Urman, S.: Hazards of copper azide fuzes. In: Proceedings of Minutes of the 14th Explosive Safety Seminar, pp. 533–552 (1972)
50. Blechta, F.: Dnešní stav otázky náhražek třaskavé rtuti. Chemický obzor **3**, 330–336 (1928)
51. Curtius, T.: Neues vom Stickstoffwasserstoff. Berichte der deutschen chemischen Gesellschaft **24**, 3341–3349 (1891)
52. Hanus, M.: Méně známé třaskaviny. Synthesia a.s., VÚPCH, Pardubice (1996)
53. Holloway, K.J., Taylor, G.W.C., Thomas, A.T.: Manufacture of dextrinated lead azide. US Patent 3,173,818, 1965
54. Thomas, A.T.: Spontaneous explosion during crystal growth of lead azide. In: Proceedings of Symposium on Lead and Copper Azide, pp. D-3, 103–111, Waltham Abbey, 1966
55. Garrett, W.L., Downs, D.S., Gora, T., Fair, H.D., Wiegand, D.A.: Preparation, characterization and electric field initiation of lead azide single crystal. In: Jenkins, J.M., White, J.R. (eds.) Proceedings of the International Conference on Research in Primary Explosives, vol. 1, pp. 4/1–4/9, Waltham Abbey (1975)
56. Garrett, W.L.: The growth of large lead azide crystals. Mater. Res. Bull. **7**, 949–954 (1972)
57. Herz, E.V.: A process for manufacture of detonating compositions for detonators and primers. GB Patent 187,012, 1922
58. Herz, E.V.: A process for manufacture of detonating compositions for detonators or primers. US Patent 1,498,001, 1924
59. Spear, R.J., Elischer, P.P.: Studies of stab initiation. Sensitization of lead azide by energetic sensitizers. Aust. J. Chem. **35**, 1–13 (1982)
60. Friederich, W.: Manufacture of primig compositions. GB Patent 180,605, 1922
61. Todd, G., Tasker, M.P.: The identity of the gamma modification of basic lead azide type I. Helv. Chim. Acta **54**(7), 2210–2212 (1971)

62. Sinha, S.K.: Study of basic azides of lead by thermometric titration. In: Hansson, J. (ed.) Proceedings of 3rd Symposium on Chemical Problems Connected with the Stability of Explosives, pp. 16–32. Sektionen for detonik och Forbranning, Ystad (1973)
63. Sinha, S.K., Srivastava, R.C., Surve, R.N.: Basic azides of lead as safer primary explosives. In: Jenkins, J.M., White, J.R. (eds.) Proceedings of the International Conference on Research in Primary Explosives, vol. 2, pp. 11/1–11/18, Waltham Abbey (1975)
64. Agrawal, J.P.: High Energy Materials—Propellants, Explosives, Pyrotechnics. Wiley-VCH, Weinheim (2010)
65. Taylor, C.A., Rinkenbach, W.H.: Silber azide: An initiator of detonation. Army Ordnance **5**, 824–825 (1925)
66. Piechowicz, T.: Solubilité de l'azoture d'argent dans l'ammoniaque et dans la pyridine. Bulletin de la Societe Chimique de France **5**, 1566–1567 (1971)
67. Taylor, A.C., Nims, L.F.: The standard potential of the silver-silver azide electrode. J. Am. Chem. Soc. **60**, 262–264 (1938)
68. Dennis, L.M., Isham, H.: Hydronitric acid V. J. Am. Chem. Soc. **29**, 18–31 (1907)
69. Merwe, L.: The preparation and chemical and physical characterization of silver azide. In: Proceedings of 12th Symposium on Explosives and Pyrotechnics, vol. 12, 1984
70. Lamnevik, S.: Prevention of copper azide formation in ammunition. In: Proceedings of Symposium on Lead and Copper Azide, pp. C-6, 92–96, Waltham Abbey, 1966
71. Taylor, G.W.C.: The Manufacture of Silver Azide RD 1336. Report 2/R/50 (Accession No. ADA 474242), Explosives Research and Development Establishment, Waltham Abbey (1950)
72. Field, J.E.: The mechanisms of initiation and propagation in primary explosives: a review. In: Jenkins, J.M., White, J.R. (eds.) Proceedings of the International Conference on Research in Primary Explosives, vol. 1, pp. 1/1–1/24, Waltham Abbey, Essex (1975)
73. Taylor, G.W.C., Jenkins, J.M.: Progress toward primary explosives on improved stability. In: Proceedings of 3rd Symposium on Chemical Problems Connected with the Stability of Explosives, pp. 43–46, Ystad, Sweden (1973)
74. Bates, L.R., Jenkins, J.M.: Search for new detonator. In: Jenkins, J.M., White, J.R. (eds.) Proceedings of the International Conference on Research in Primary Explosives, vol. 2, Waltham Abbey (1975)
75. Bekk, J.: The photographic behaviour of silver azoimide. J. Chem. Soc. **108**, II200–II201 (1915)
76. Evans, B.L., Yoffe, A.D.: Structure and stability of inorganic azides. II. Some physical and optical properties, and the fast decomposition of solid monovalent inorganic azides. Proc. R. Soc. Lond. A Math. Phys. Sci. **A250**, 346–366 (1959)
77. Courtney-Pratt, J.S., Rogers, G.T.: Initiation of explosion by light and by flying fragments. Nature **175**, 632–633 (1955)
78. Wöhler, L., Krupko, W.: Action of light on silver, mercurous, lead and cuprous azoimides; basic azoimides of lead and copper. J. Chem. Soc. **104II**, 703 (1913)
79. Gray, P.: Chemistry of the inorganic azides. Quart. Rev. **17**, 441–473 (1963)
80. Hitch, A.R.: Thermal decomposition of certain inorganic trinitrides. J. Am. Chem. Soc. **40**, 1195–1204 (1918)
81. Millar, R.W.: Lead free initiator materials for small electro explosive devices for medium caliber munitions. Final Report 04 June 2003. Report QinetiQ/FST/CR032702/1.1, QuinetiQ, Farnborough (2003)
82. Taylor, A.C., Rinkenbach, W.H.: Sensitivities of detonating compounds to frictional impact, impact, and heat. J. Franklin Inst. **204**, 369–376 (1927)
83. Wöhler, L., Martin, F.: Azides; Sensitiveness of. J. Soc. Chem. Ind. **36**, 570–571 (1917)
84. Roux, J.J.P.A.: The dependence of friction sensitivity of primary explosives upon rubbing surface roughness. Propellants Explosives Pyrotechnics **15**, 243–247 (1990)
85. Evans, B.L., Yoffe, A.D.: The burning and explosion of single crystals. Proc. R. Soc. Lond. **A238**, 325–333 (1956)
86. Bowden, F.P., Williams, R.J.E.: Initiation and propagation of explosion in azides and fulminates. Proc. R. Soc. Lond. A Math. Phys. Sci. **1951**, A176–A188 (1951)

References

87. Yoffe, A.D., Evans, B.L., Deb, S.K.: Foreign cations in silver azide. Nature **180**, 294–295 (1957)
88. Wöhler, L., Martin, F.: Die Initialwirkung von Aziden und Fulminaten. Zeitschrift für das gesamte Schiess- und Sprengstoffwesen, (1917); **12**, 18–21
89. Darier, G.E., Goudet, C.: Preventing explosions in handling azides and other explosives. US Patent 1,349,411, 1920
90. Gray, P., Waddington, T.C.: Detonation and decomposition of silver azide sensitized by the cyanamide ion. Chem. Ind. 1255–1257 (1955)
91. Blechta, F.: Verfahren zur Herstellung von Initialzündsätzen, welche kolloidale Silber- und Quesksilberazide in lockerer, nicht zusammenbackender Form enthalten. AT Patent 126,150, 1932
92. Blechta, F.: Une nouvelle amorce a l'azothydrure. Chimie et industrie 921–925 (1933)
93. Taylor, G.W.C.: Improvements in the manufacture of silver azide. GB Patent 781,440, 1957
94. Williams, E., Peyton, S.V., Harris, R.C.: Improvements in or relating to the manufacture of silver azide. GB Patent 887,141, 1962
95. McGuchan, R.: Improvements in primary explosive compositions and their manufacture. In: Proceedings of 10th Symposium on Explosives and Pyrotechnics, San Francisco, 1979
96. Costain, T.S.: Process for producing silver azide. US Patent 3,943,235, 1976
97. Hirlinger, J.M., Bichay, M.: New Primary Explosives Development for Medium Caliber Stab Detonators. Report SERDP PP-1364, US Army ARDEC, Washington DC (2004)
98. Hirlnger, J., Fronabarger, J., Williams, M., Armstrong, K., Cramer, R.J.: Lead azide replacement program. In: Proceedings of NDIA, Fuze Conference, 2005
99. Hodgkinson, W.R.: Improvements in and relating to the production of azides. GB Patent 128,014, 1919
100. Turrentine, J.W.: Contributions of electrochemistry of hydronitric acid and its salts. I. The corrosion of some metals in sodium trinitride solution. J. Am. Chem. Soc. **33**, 803–828 (1911)
101. Cirulis, A.: Die explosiven Eigenschaften des Kupferazids $Cu(N_3)_2$, Zeitschrift fur das gesamte Schiess- und Sprengstoffwesen (1943); **38**, 42–45
102. Straumanis, M., Cirulis, A.: Das Kupfer(II)-azid. Darstellungsmethoden, Bildung und Eigenschaften. Zeitschrift für anorganische und allgemeine Chemie **251**, 315–331 (1943)
103. Colton, R.J., Rabalais, J.W.: Electronic structure of some inorganic azides from X-ray electron spectroscopy. J. Chem. Phys. **64**, 3481–3486 (1976)
104. Wilsdorf, H.: Die Kristallstruktur des einwertigen Kupferazids, CuN_3. Acta Crystallogr. **1**, 115–118 (1948)
105. Duke, J.R.C.: The crystallography of copper azides. In: Proceedings of the Symposium on Lead and Copper Azides, pp. C-5, 87–91, Waltham Abbey, 1966
106. Harris, R.C.: The formation and detection of copper azides corrosions. In: Proceedings of Symposium on Lead and Copper Azide, pp. C-1, 70–71, Waltham Abbey, 1966
107. Lamnevik, S.: Copper azide corrosion. In: Proceedings of Symposium on Lead and Copper Azide, pp. C-2, 72–77, Waltham Abbey, 1966
108. Holloway, K.J.: The preparation, identification and sensitiveness of copper azide. In: Proceedings of Symposium on Lead and Copper Azide, pp. C-3, 78–83, Waltham Abbey, 1966
109. Turrentine, J.W., Moore, R.L.: The action of hydronitric acid on cuprous chloride and metallic copper. J. Am. Chem. Soc. **34**, 382–384 (1912)
110. Singh, K.: Sensitivity of cuprous azide towards heat and impact. Trans. Faraday Soc. **55**, 124–129 (1959)
111. Curtius, T., Rissom, J.: Azoimide. J. Chem. Soc. **76**, B92 (1899)
112. Senise, P., Neves, E.F.A.: Solubility study of copper(II) azide in aqueous sodium azide solutions of low ionic strength. J. Inorg. Nucl. Chem. **33**, 351–358 (1971)
113. Medlock, L.E.: Corrosion of copper detonator tubes in the presence of lead azide. In: Proceedings of Symposium on Lead and Copper Azide, pp. C-4, 84–86, Waltham Abbey, 1966

114. Okubo, S., Shindo, K., Oinuma, S.: Copper azide detonators. I. Preparation of copper azide and its impact sensitivity test. CAN **52**, 8559 (1958)
115. Wöhler, L., Martin, F.: Über neue Fulminate und Azide. Berichte der deutschen chemischen Gesellschaft **50**, 586–596 (1917)
116. Nockemann, P., Cremer, U., Ruschewitz, U., Meyer, G.: Mercurous azide, $Hg_2(N_3)_2$. Zeitschrift für anorganische und allgemeine Chemie **629**, 2079–2082 (2003)
117. Belomestnykh, V.N.: Uprugie cvoistva neorganicheskikh azidov pri ctandartnykh usloviyakh. Neorganicheskie materialy **29**, 210–215 (1993)
118. Birckenbach, L.: Cadmium azide (at the same time a warning). Zeitschrift für anorganische und allgemeine Chemie **214**, 94–96 (1933)
119. Wöhler, L., Martin, F.: Sensitiveness of azides. Angew. Chem. **30**(I), 33–39 (1917)
120. Rathsburg, H.: Über die Bestimmung der Reibungsempfindlichkeit von Zünstoffen. Zeitschrift für das angewandte Chemie **41**, 1284–1286 (1928)
121. Curtius, T., Rissom, J.: Neue Untersuchungen über den Stickstoffwasserstoff N_3H. J. für praktische Chemie **58**, 261–309 (1898)
122. Klapötke, T.M., Geissler, P.: Preparation and characterization of the first binary arsenic azide species: $As(N_3)_3$ and $[As(N_3)_4][AsF_6]$. J. Chem. Soc. Dalton Trans. 3365–3366 (1995)
123. Klapötke, T.M., Schulz, A., McNamara, J.: Preparation, characterization and *ab initio* computation of the first binary antimony azide, $Sb(N_3)_3$. J. Chem. Soc. Dalton Trans. 2985–2987 (1996)
124. Haiges, R., Vij, A., Boatz, J.A., Schneider, S., Schroer, T., Gerken, M., Christe, K.O.: First structural characterization of binary As^{III} and Sb^{III} azides. Chemistry **10**, 508–517 (2004)
125. Villinger, A., Schulz, A.: Binary bismuth(III) azides: $Bi(N_3)_3$, $[Bi(N_3)_4]^-$, and $[Bi(N_3)_6]^{3-}$. Angew. Chem. Int. Ed. **49**, 8017–8020 (2010)
126. Haiges, R., Boatz, J.A., Vij, A., Vij, V., Gerken, M., Schneider, S., Schroer, T., Yousufuddin, M., Christe, K.O.: Polyazide chemistry: Preparation and characterization of $As(N_3)_5$, $Sb(N_3)_5$, and $[P(C_6H_5)_4][Sb(N_3)_6]$. Angew. Chem. **116**, 6844–6848 (2004)
127. Boyer, J.H., Canter, F.C.: Alkyl and aryl azides. Chem. Rev. **54**, 1–57 (1954)
128. Finger, H.: Über Abkömmlinge des Cyanurs. J. für praktische Chemie **75**, 103–104 (1907)
129. Taylor, C.A., Rinkenbach, W.H.: Preparation and detonating properties of cyanuric triazide. J. Franklin Inst. **196**, 551 (1923)
130. Hart, C.V.: Carbonic acid azides. J. Am. Chem. Soc. **50**, 1922–1930 (1928)
131. Ott, E., Ohse, E.: Zur Kenntnis einfacher Cyan- und Cyanurverbindungen. II. Über das Cyanurtriazid, (C_3N_{12}). Berichte der deutschen chemischen Gesellschaft **54**, 179–186 (1921)
132. Ott, E.: Explosive and process of making same. US Patent 1,390,378, 1921
133. Ott, E.: Verfahren zur Herstellung von Initialzündmitteln und von Treib- und Sprengmitteln. DE Patent 350,564, 1922
134. Hughes, E.W.: The crystal structure of cyanuric triazide. J. Chem. Phys. **3**, 1–5 (1935)
135. Sutton, T.C.: Structure of cyanuric triazide $(C_3N_3)(N_3)_3$. Philos. Mag. **15**, 1001–1018 (1933)
136. Ficheroulle, H., Kovache, A.: Contribution à l'étude des explosifs d'amorçage. Memorial des poudres **41**, 1–22 (1959)
137. Imray, O.: Manufacture of a new explosive. GB Patent 170,359, 1921
138. Gillan, E.G.: Synthesis of nitrogen-rich carbon nitride networks from an energetic molecular azide precursor. Chem. Mater. **12**, 3906–3912 (2000)
139. Muraour, H.: Sur la théorie des réactions explosives. Cas particulier des explosifs d'amorçage. Memories présentés a la Société chimique **51**, 1152–1166 (1932)
140. Mehta, N., Cheng, G., Cordaro, E., Naik, N., Lateer, B., Hu, C., Stec, D., Yang, K.: Performance testing of lead-free stab detonator. In: Proceedings of NDIA Fuze Conference, 2006
141. Pepekin, V.I.: Limiting detonation velocities and limiting propelling powers of organic explosives. Dokl. Phys. Chem. **414**(2), 159–161 (2007)
142. Kroke, E., Schwarz, M., Buschmann, V., Miehe, G., Fuess, H., Riedel, R.: Nanotubes formed by detonation of C/N precursors. Adv. Mater. **11**, 158–161 (1999)

143. Utschig, T., Schwarz, M., Miehe, G., Kroke, E.: Synthesis of carbon nanotubes by detonation of 2,4,6-triazido-1,3,5-triazine in the presence of transition metals. Carbon **42**, 823–828 (2004)
144. Fogelzang, A.E., Egorshev, V.Y., Pimenov, A.Y., Sinditskii, V.P., Saklantii, A.R., Svetlov, B.S.: Issledovanie statsionarnogo goreniya initsiiruyushchikh vzryvchatykh veshchestv pri vysokykh davleniyakh. Dokl. Akad. Nauk SSSR **282**, 1449–1452 (1985)
145. Huynh, M.V., Coburn, M.D., Meyer, T.J., Wetzer, M.: Green primaries: Environmentally friendly energetic complexes. Proc. Natl. Acad. Sci. **103**, 5409–5412 (2006)
146. Huynh, M.V., Coburn, M.D., Meyer, T.J., Wetzer, M.: Green primary explosives: 5-nitrotetrazolato-N^2-ferrate hierarchies. Proc. Natl. Acad. Sci. **103**, 10322–10327 (2006)
147. Huynh, M.H.V., Hiskey, M.A., Hartline, E.L., Montoya, D.P., Gilardi, R.: Polyazido high-nitrogen compounds: hydrazo- and azo-1,3,5-triazine. Angew. Chem. Int. Ed. **43**, 4924–4928 (2004)
148. Huynh, M.H.V., Hiskey, M.A., Pollarod, C.J., Montoya, D.P., Hartline, E.L., Gilardi, R.: 4,4′,6,6′-Tetrasubstituted hydrazo- and azo-1,3,5-triazines. J. Energetic Mater. **22**, 217–229 (2004)
149. Huynh, M.H.V., Hiskey, M.A., Archuleta, J.G., Roemer, E.L.: Preparation of nitrogen-rich nanolayered, nanoclustered, and nanodendritic carbon nitrides. Angew. Chem. Int. Ed. **44**, 737–739 (2005)
150. Huynh, M.H.V., Hiskey, M.A.: Preparation of high nitrogen compound and materials therefrom. US Patent 2006/0211565, 2006
151. Loew, P., Weis, C.D.: Azo-1,3,5-triazines. J. Heterocyc. Chem. **13**, 829–833 (1976)
152. Li, X.T., Li, S.H., Pang, S.P., Yu, Y.Z., Luo, Y.J.: A new efficient route for the synthesis of 4,4′,6,6′-tetra(azido)azo-1,3,5-triazine. Chin. Chem. Lett. **18**, 1037–1039 (2007)
153. Turek, O.: Le 2,4,6-trinitro-1,3,5-triazido-benzene, nouvel explosif d'amorçage. Chimie et industrie **26**, 781–794 (1931)
154. Turek, O.: 1,3,5-Triazido-2,4,6-trinitrobenzen, nová iniciálná výbušina. Chemický obzor **7**, 76–79; 97–104 (1932)
155. Adam, D., Karaghiosoff, K., Klapotke, T.M., Hioll, G., Kaiser, M.: Triazidotrinitro benzene: 1,3,5-$(N_3)_3$-2,4,6-$(NO_2)_3C_6$. Propellants Explosives Pyrotechnics **27**, 7–11 (2002)
156. Ficheroulle, H., Kovache, A.: Contribution à l'étude des explosifs d'amorçage. Mémorial des poudres **31**, 1–27 (1949)
157. Zielinski, B.: Ignition mixture for percussion caps of all kind, small munitions, and primers. US Patent 2,111,719, 1938
158. Turek, O.: Blasting cartridge, percussion cap, detonator, detonating fuse, and the like. US Patent 1,743,739, 1930
159. Turek, O.: A method of producing 2,4,6-trinitro-1,3,5-triazidobenzene. GB Patent 298,981, 1928
160. Turek, O.: Verfahren zur Herstellung von 1,3,5-Trinitro-2,4,6-triazidobenzol. DE Patent 498,050, 1930
161. Turek, O.: Improvements in and connected with explosive charges for detonators, percussion caps, boosters, detonating fuses, projectiles and the like. GB Patent 298,629, 1927
162. Turek, O.: Verfahren zur Herstellung von Sprengladungen für Sprengkapseln, Zündkapseln, Detonationszündschnüre u. dgl. DE Patent 494,289, 1928
163. Improvements in ignition mixtures for percussion caps of all kind, small munitions and primers. GB Patent 465,768, 1936
164. Fries, K., Ochwat, P.: Neues über Dichlor-2.3-naphthochinon-1.4. Berichte der deutschen chemischen Gesellschaft **56**, 1291–1304 (1923)
165. Šorm, F.: O tetrazidobenzchinonu (1,4). Chemický obzor **14**, 37–39 (1939)
166. Gilligan, W.H., Kamlet, M.J.: On the explosive properties of tetraazido-p-benzoquinones. Tetrahedron Lett. **19**, 1675–1676 (1978)

Chapter 5
Salts of Polynitrophenols

Many heavy metal salts of polynitrophenols (e.g., phenol, resorcinol, or phloroglucinol) have the character of primary explosives while alkaline salts are secondary explosives. Primarily lead and barium salts are used or recommended for practical applications. In general, lead and barium salts of nitrophenols have high sensitivity to flame but low initiation efficiency. These properties make them unsuitable in pure form and they are therefore mainly used in mixtures with other substances. The increasing number of nitrogroups in the molecule increases the overall energetic content, which is reflected in its higher sensitivity. The number of nitro groups may be varied depending on the requirements for particular applications. For example, lead mononitroresorcinol is used in fuse heads while lead trinitroresorcinol is used in mixtures with other primary explosives, such as LA, in detonators. Salts of polynitrophenols containing an azido group in the molecule (primarily lead salts of mono and dinitroazidophenol) have also been proposed as primary explosives but have not yet found a practical application [1].

5.1 Salts of Picric Acid

Picric acid forms normal and basic metallic salts. Out of all the salts of picric acid so far investigated, only the normal lead salt is worth mentioning as it was for some time used as a primary explosive in various initiating compositions. Investigations and some limited use were also reported for the potassium salt; however, its low sensitivity (F of I only 64–67) prevented it from fulfilling the role of a full-fledged primary explosive and served rather as a energetic component of priming compositions, igniferous compositions, or as a component of mixtures for fuseheads [2, 3].

5.1.1 Normal Lead Picrate

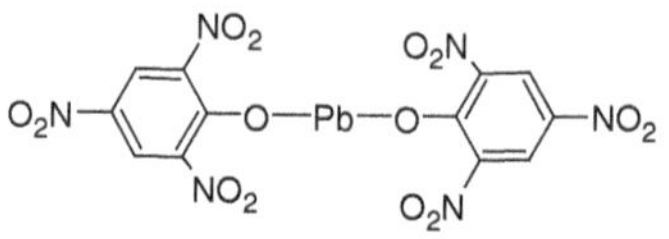

$$\text{O}_2\text{N}-\underset{\underset{\text{NO}_2}{|}}{\overset{\overset{\text{NO}_2}{|}}{\text{C}_6\text{H}_2}}-\text{O}-\text{Pb}-\text{O}-\underset{\underset{\text{O}_2\text{N}}{|}}{\overset{\overset{\text{O}_2\text{N}}{|}}{\text{C}_6\text{H}_2}}-\text{NO}_2$$

Normal lead picrate forms several hydrates containing from 1 to 5 molecules of crystal water. The tetrahydrate of lead picrate is obtained when an aqueous solution of picric acid is treated with lead carbonate. The monohydrate results when the tetrahydrate is retained at 80 °C and also when the anhydride is exposed to air. The anhydrous salt is obtained when the tetrahydrate or monohydrate is heated to 150 °C [4].

The tetrahydrate forms silky yellow needles while the anhydride forms a yellow powder [4]. Lead picrate is insoluble in water, ether, chloroform, benzene, and toluene; sparingly soluble in acetone and alcohol [5].

Lead picrate is considered highly sensitive to mechanical impact and thermal stimuli [6]. The anhydride is more sensitive to mechanical stimuli than the hydrates. Impact sensitivity of anhydride is significantly higher than the sensitivity of mercury fulminate (4 cm/0.5 kg vs. 24 cm for MF) [7, 8]. Handling of lead picrate anhydride represents the same level of risk as handling of lead styphnate. The ignition temperature is 281 °C (explosion takes place instantaneously or within 1 s) [7]. The formation of lead picrate by reaction of tetryl (which decomposes to picric acid) with lead azide is reported as a possible reason for the higher sensitivity of this mixture compared to pure LA [6].

5.1.1.1 Preparation

Neutral lead picrate is prepared by reaction of aqueous solutions of picric acid [5, 9, 10] or mixture of picric acid and its sodium salt [5] with lead nitrate or acetate. Lead picrate is isolated as a yellow precipitate.

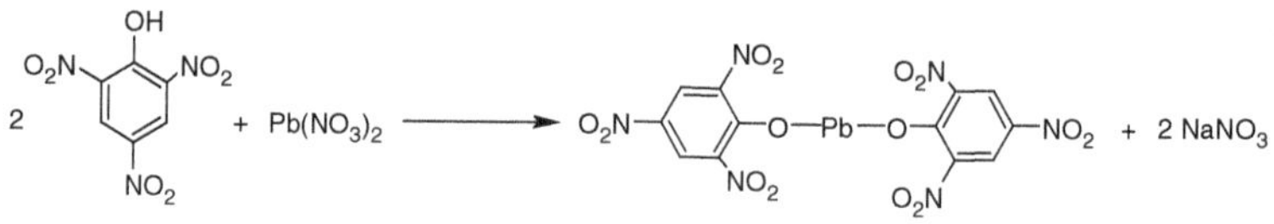

$$2\ \text{C}_6\text{H}_2(\text{OH})(\text{NO}_2)_3 + \text{Pb}(\text{NO}_3)_2 \longrightarrow (\text{O}_2\text{N})_3\text{C}_6\text{H}_2\text{-O-Pb-O-C}_6\text{H}_2(\text{NO}_2)_3 + 2\ \text{NaNO}_3$$

5.1.1.2 Use

Lead picrate was used in applications similar to those for lead styphnate, for example, as part of ignition mixtures in fuses during World War II [8, 9]. Lead picrate can be used as an active component in initiating mixtures, e.g., for electrical squibs in bridge-wire detonators. It has been also used in mixtures for electric fuseheads [11].

5.1.2 Basic Lead Picrate

Lead picrate forms several basic salts with varying lead content. They form citric yellow crystals practically insoluble in water or alcohol. Its explosive properties are similar to those of lead styphnate but its flammability is lower [10]. Sensitivity to electrostatic discharge is high (at the level of LS) and is the reason for the many accidents which occurred during its manufacture [12]. Sensitivity to impact is 2.5 J (it explodes by 5 kg hammer from 5 cm) [13].

Basic lead picrate is prepared by reaction of picric acid with yellow lead oxide in water. The mixture is stirred and boiled during the reaction. Product forms as a precipitate [11]. Basic lead picrate has similar explosive properties to those of LS and was therefore used in similar applications (e.g., as a component of percussion mixtures) [10–12]. Due to its lower flammability, it was, however, replaced in practical use by LS [12].

5.2 Salts of Dinitroresorcinol

Dinitroresorcine (DNR) forms two isomers: 2,4 and 4,6. Unlike in the nitration of some other aromatic molecules (toluene, phenol), it is possible to prepare practically pure dinitro isomers. The position of nitro groups in the ring depends on the reaction conditions. The 2,4-isomer of DNR can be easily prepared by dinitrosation of resorcinol followed by alkaline oxidation of 2,4-dinitrosoresorcinol [8, 14]. 2,4-DNR cannot be prepared by sulfonation of resorcinol followed by reaction with nitric acid (method used for phenol) because this method yields the trinitro compound. The 4,6-DNR isomer can be prepared in two ways (a) by nitration of 4,6-diacetylresorcinol and (b) directly by nitration of resorcinol using 98 % nitric acid at low temperatures (between -20 and $-15\ °C$) [8].

Among the various possible lead salts of DNR only two have found industrial application as primary explosives: normal lead salt of 2,4-dinitroresorcine (2,4-LDNR) and basic lead salt of 4,6-dinitroresorcine (basic 4,6-LDNR).

5.2.1 Lead salts of 2,4-Dinitroresorcinol

2,4-Dinitroresorcinol (2,4-DNR) forms a variety of lead salts (2,4-LDNRs) including acid, normal, and a large number of basic ones. The most commonly used and best characterized is the normal 2,4-LDNR.

Table 5.1 Comparison of impact sensitivity and brisance for LS and LDNR isomers [8]

	LS	Normal 2,4-LDNR	Basic 4,6-LDNR
N content (%)	9.0	6.9	4.3
Pb content (%)	44.2	51.1	64.1
Impact sensitivity (cm for 1 kg hammer)	17	30	60
Brisance (sand test in % TNT)	50	41.7	31.3

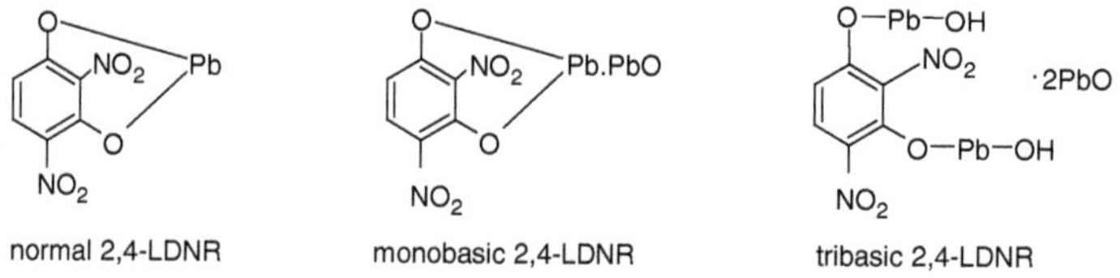

normal 2,4-LDNR   monobasic 2,4-LDNR   tribasic 2,4-LDNR

Normal 2,4-LDNR exists in two forms: yellow and orange described in one source [8] or orange and red described in another source [15]. The crystal density is 3.2 g cm^{-3} [8, 15]. These forms have differing physical but similar explosive properties. Lead 2,4-dinitroresorcinol is insoluble in water, acetone, benzene, and other common organic solvents [8].

5.2.1.1 Explosive Properties

Normal 2,4-LDNR is a less powerful explosive with even lower initiation efficiency than lead styphnate. According to Payne, 0.4 g is not sufficient for initiation of tetryl [8, 14]. 2,4-LDNR is less sensitive to impact and friction than lead styphnate. Sensitivity and brisance of practically useful lead salts of polynitroresorcines generally decrease both with increasing amount of lead and decreasing amount of nitrogen (Table 5.1) [8]. The main reason for the practical application of normal 2,4-LDNR, despite its poorer explosive properties compared to LS, is increased handling safety (particularly less susceptible to static electricity) [8, 9, 16].

5.2.1.2 Preparation

2,4-LDNR is generally prepared from 2,4-dinitroresorcinol and soluble lead salt (nitrate or acetate) via the sodium salt (introduced as sodium carbonate). The reaction is carried out in hot or boiling water [8–10, 14, 15].

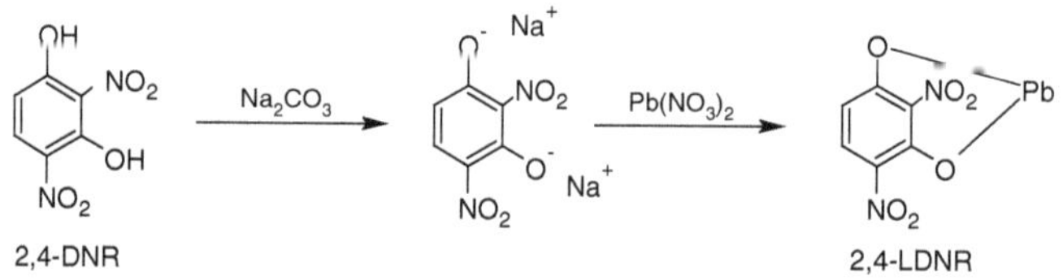

2,4-DNR   2,4-LDNR

The sodium salt can be replaced by other salts (e.g., magnesium) as well [17, 18]. The reaction conditions (pH, temperature regime, rate, and order of addition of

5.2 Salts of Dinitroresorcinol

reactants) influence not only the type of the prepared salt but also the shape of the crystals and free-flowing properties of the final product [14, 17].

Basic 2,4-LDNR (mono-, tri-, and tetra-) are prepared from the corresponding equimolar amounts of the sodium salt of DNR and lead nitrate in a basic environment [14]:

$$C_6H_2(NO_2)_2(ONa)_2 + 4\ Pb(NO_3)_2 + 6\ NaOH \longrightarrow C_6H_2(NO_2)_2(OPbOH)_2 \cdot 2PbO + 8\ NaNO_3 + 2\ H_2O$$
Na salt of 2,4-DNR → tribasic 2,4-LDNR

$$C_6H_2(NO_2)_2(ONa)_2 + 5\ Pb(NO_3)_2 + 8\ NaOH \longrightarrow C_6H_2(NO_2)_2(OPbOH)_2 \cdot 3PbO + 10\ NaNO_3 + 3\ H_2O$$
tetrabasic 2,4-LDNR

5.2.1.3 Use

Normal 2,4-LDNR was discovered at the end of nineteenth century and has been employed since 1940 [19]. Its application is similar to that of LS due to their similar explosive properties, particularly as an ingredient of priming mixtures. It has also been used in compositions for electric detonators [8, 9, 16, 19].

5.2.2 Lead Salts of 4,6-Dinitroresorcinol

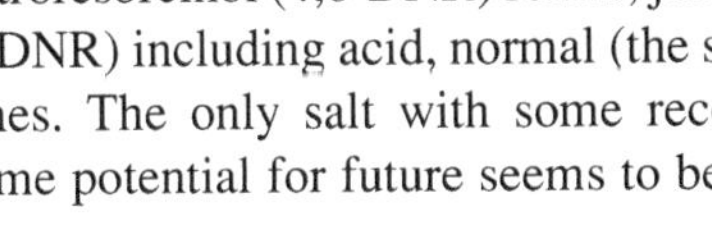

acid 4,6-LDNR 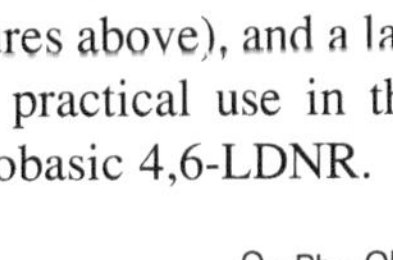   normal 4,6-LDNR

4,6-Dinitroresorcinol (4,6-DNR) forms, just like the 2,4-isomer, a variety of lead salts (4,6-LDNR) including acid, normal (the structures above), and a large number of basic ones. The only salt with some recorded practical use in the past and possibly some potential for future seems to be monobasic 4,6-LDNR.

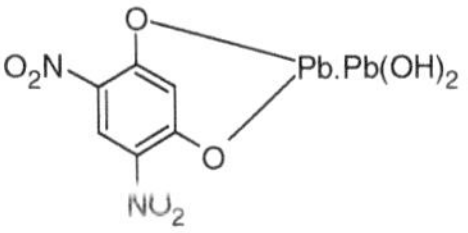

monobasic 4,6-LDNR formula I   monobasic 4,6-LDNR formula II   monobasic 4,6-LDNR formula III

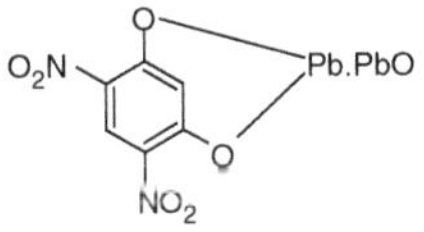

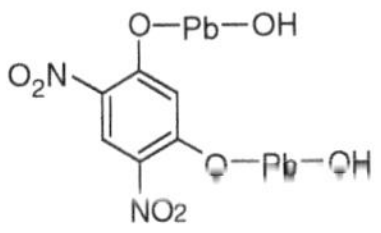

Three structures of monobasic 4,6-LDNR are reported in the literature (a) formula I—is used by [20, 21], (b) formula II—is used by [22], and (c) formula

Table 5.2 Lead salts of 4,6-DNR

Composition	Pb/DNR	Comments	References
Acid salt	1/2	Reported	[24, 25]
Normal salt	1/1	Not yet isolated	[24]
2/3 Basic salt	5/3	Reported	[21, 24]
Monobasic salt (referred as basic salt)	2/1	Not yet isolated in pure form	[24]
		Reported	[8, 14, 20, 23]
Dibasic salt	3/1	Reported	[14, 20, 24]
Tribasic salt	4/1	Doubtful existence reported	[14, 20, 24]

Table 5.3 Lead salts of 4,6-DNR

Composition	Color	Comments	References
2/3 Basic salt	–	Density: crystal 3.65 g cm^{-3}, bulk 1.3–1.4 g cm^{-3}	[21, 24]
Monobasic salt	Yellow	Less dense "micro-crystalline"	[8, 15]
	Brick-red	Denser red form	[8, 14, 15]
Dibasic salt	Dark red	–	[14]
Tribasic salt	Deep orange	–	[14]

III—is used by [8, 14, 15, 23]. For di- and tri-basic 4,6-LDNR, Payne proposed the following structures [14]:

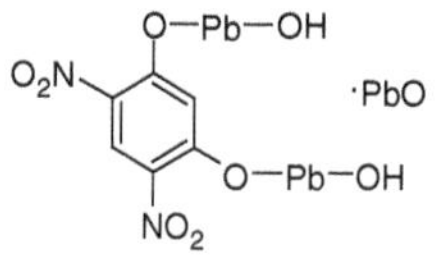

dibasic 4,6-LDNR 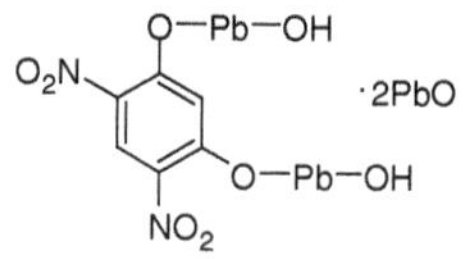   tribasic 4,6-LDNR

The problems related to the correctness of each of the structures are similar to those encountered with basic lead styphnate and are therefore addressed there.

5.2.2.1 Physical and Chemical Properties

The overview of reported 4,6-LDNR salts is outlined in Table 5.2 and their crystal form and color in Table 5.3 [20, 22, 24].

The form prepared by Taylor et al. [21] and mentioned in [24] is made of basic salt containing 3 molecules of lead 4,6-dinitroresorcinol and 2 molecules of lead hydroxide – 3[PbDNR]·2Pb(OH)$_2$ further reported as 2/3 salt. The ratio of lead to DNR in this substance is 5:3. The 2/3 basic lead salt has a density 3.65 g cm^{-3} and it is practically insoluble in water (0.01 g/100 ml). The salt is compatible with common metals and oxidizers and stable under water for a long time (no change was observed over a period of 1 year) [24].

5.2 Salts of Dinitroresorcinol

The most frequently described, and probably the only lead salt that has been practically used, is monobasic lead 4,6-dinitroresorcinol, mostly called just basic lead 4,6-dinitroresorcinol. It can be prepared in two forms—yellow and red (see Table 5.3) both having the same chemical and explosive properties [15].

5.2.2.2 Explosive Properties

All of the lead salts of DNR have worse explosive properties than LS. The initiation efficiency of monobasic 4,6-LDNR is poor (0.4 g does not initiate tetryl pressed to 20.7 MPa) [8]. The comparison of 4,6-LDNR with LS and normal 2,4-LDNR was mentioned previously (see Table 5.1).

The sensitivity data of 2/3 basic lead salt of 4,6-DNR were published by Jenkins. The figure of insensitiveness (F of I) is 20 (compared to RDX = 80) and ignition temperature is 248 °C at heating rate 5 °C min^{-1} [24].

5.2.2.3 Preparation

Preparation of various basic salts is based on the reaction of either (a) lead nitrate in a basic environment (NaOH) or (b) freshly prepared lead hydroxide (precipitated from aqueous solutions of lead acetate and sodium hydroxide) with 4,6-dinitroresorcinol or its salts.

$C_6H_2(NO_2)_2(ONa)_2$ + 2 $Pb(OH)_2$ $\longrightarrow$ $C_6H_2(NO_2)_2(OPbOH)_2$ + 2 H_2O
Na salt of 4,6-DNR   monobasic 4,6-LDNR

$C_6H_2(NO_2)_2(ONa)_2$ + 3 $Pb(NO_3)_2$ + 4 NaOH $\longrightarrow$ $C_6H_2(NO_2)_2(OPbOH)_2 \cdot PbO$ + 6 $NaNO_3$ + H_2O
                                                  dibasic 4,6-LDNR

$C_6H_2(NO_2)_2(ONa)_2$ + 4 $Pb(NO_3)_2$ + 6 NaOH $\longrightarrow$ $C_6H_2(NO_2)_2(OPbOH)_2 \cdot 2PbO$ + 8 $NaNO_3$ + 2 H_2O
                                                  tribasic 4,6-LDNR

The form of the product (Table 5.4) depends on the reaction conditions (amount of lead compound, temperature, reaction period, pH, rate and order of addition of reactants) [24]. The following forms are reported based on the temperature of the reaction mixture: acid 50–60 °C preferably 55 °C, normal 40–50 °C preferably 45 °C, and 2/3 basic at 18–28 °C preferably 23 °C [25]. The particular basic salts are formed from exact molar proportions of lead salts in a basic environment (NaOH) [14, 15, 22]. The manufacturing process of 2/3 basic salt (RD 1353) has been described by Jenkins [24]. The normal salt is prepared from 4,6-dinitroresorcinol in an aqueous suspension of freshly prepared lead hydroxide [25]. Crystal modifiers (e.g., carboxymethyl cellulose) are recommended for preparation of the product in a more suitable form for processing [26, 27].

Table 5.4 Content of Pb in various salts of 4,6-DNR

Composition	Pb content (%)	References
Acid salt	34.1	[25]
Normal salt	51.1	[25]
2/3 Basic salt	61.0	[21, 24, 25]
Monobasic salt	64.1	[8, 14, 15, 22]
Dibasic salt	71.5	[14, 22]
Tribasic salt	75.9	[14, 22]

5.2.2.4 Uses

Large-scale manufacturing of basic lead 4,6-dinitroresorcinol was developed in Great Britain during World War II [8, 23].

5.3 Salts of Trinitroresorcine

5.3.1 Lead Styphnate

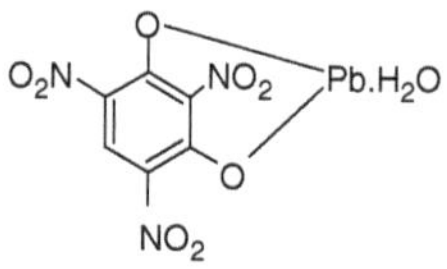

Lead styphnate (lead salt of 2,4,6-trinitrobenzene-1,3-diol, LS, LTNR, tricinate) was, according to the well-established reference [8], first prepared by Edmund von Herz in 1914 by reaction of magnesium styphnate with lead acetate in presence of nitric acid. However, in reality, preparation of LS was published more than 40 years earlier by Stenhouse [28]. This discrepancy in historical data is probably caused by the fact that Herz patented the substance as a component of a detonator primary charge and was therefore the first one who found some real application [29, 30]. The preparation route used by Stenhouse is based on the reaction of lead acetate with an aqueous solution of styphnic acid [28].

5.3.1.1 Physical and Chemical Properties

Normal lead styphnate forms a monohydrate with crystal density from 3.06 to 3.1 g cm^{-3} [31, 32]; heat of formation of LS is -835 kJ mol^{-1} [1]. LS exists as two polymorphs, α and β [24]. Many authors, however, do not distinguish between these two forms [5, 8, 9, 12].

LS crystallizes as monohydrate in the form of gold, orange, or reddish-brown monoclinic crystals (Fig. 5.1). The reason for the various colors of LS crystals has

5.3 Salts of Trinitroresorcine

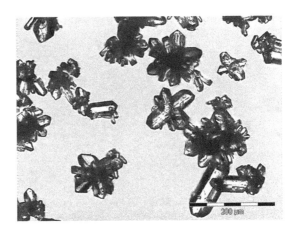

Fig. 5.1 Crystals of lead styphnate

not been successfully explained [33]. Crystals of LS turn red on heating with concurrent fracture of the crystals into irregular particles of greatly decreased dimensions. The loss of water leads to formation of a sensitive anhydride with density 2.9 g cm^{-3}. Upon rehydration, the red color of the anhydrous salt reverts back to the characteristic orange. The crystals containing crystal water are slightly hydroscopic. Hydroscopicity is mostly reported significantly higher than MF and LA [34, 35] even though some authors (e.g., Kast and Haid) reported lower hydroscopicity than LA and MF [36]. LS does not lose crystal water easily at ambient temperatures and its loss is not significant even up to about 100 °C [10, 37]. According to Wallbaum, LS does not change during 2 months of storage at 65–70 °C [33, 38]. Above 100 °C LS loses crystal water [1]; it takes 16 h at 115 °C, 7 h at 135 °C, and under 4 h at 145 °C as can be seen from Fig. 5.2 [39]. During this process, crystals of LS break apart releasing water from the crystals. Hailes [37], however, reports that crystal rupture occurs only at rapid heating above 200 °C which is in disagreement with our experience and today's technological processes. The phenomenon of reversible dehydratation/rehydratation coupled with crystal rupture is industrially used (at 140 °C) for modification of crystal size leading to elimination of graining, which is a risky operation. The industrial product before and after the heat treatment is shown at the same magnification in Fig. 5.3. Efforts aimed at removing the crystal water by storing LS in a desiccator above calcium chloride at normal temperature failed [33].

LS is slightly soluble in water (0.04 g per 100 g of water at 15 °C [10] or 0.09 g per 100 g of water at 17 °C [34]), methanol, pyridine, and amyl acetate; it is practically insoluble in ether, acetone, and common chlorinated solvents [34]. Lead styphnate is soluble in aqueous ammonium acetate, formamide [1, 20, 33], acetic acid [28], and ethanolamine (30 g per 100 g of ethanolamine at 17 °C.) It can be precipitated from acetic acid solution by addition of alcohol [28].

LS can be kept under water or a mixture of water and isopropyl alcohol until used, without decomposition [40]. LS, unlike LA, does not react with carbon dioxide. Dry LS does not react with common metals; wet LS reacts only with

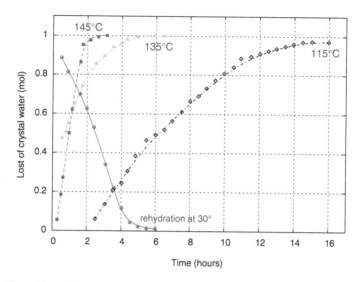

Fig. 5.2 Rate of loss of water from normal LS mono hydrate at various temperatures [39]

Fig. 5.3 Effect of heat treatment on the crystal structure of LS; *left*—as synthesized, *right*—the same product after heat treatment (by kind permission of Pavel Valenta, Austin Detonator, Vsetín, Czech Republic)

lead forming basic salt [34]. LS is decomposed by action of acids (e.g., sulfuric, nitric) giving styphnic acid and the relevant lead salt. With bases, LS also decomposes forming its basic salts and lead hydroxide [1, 10]. The laboratory destruction of LS may be done using bases. The procedure is based on dissolving LS in 40× its weight of an aqueous solution of 20 % sodium hydroxide and adding a solution of sodium dichromate in an amount equal to half the weight of LS [8]. Ficheroulle and Kovache [34] recommended nitric acid (10 %) as a decomposition agent in which case lead nitrate and styphnic acid are formed. LS is compatible with most common explosives when dry (e.g., RDX, tetryl, TNT) [20].

The influence of γ irradiation (cobalt-60, at 25 °C) upon the decomposition of LS has been published by Flanagan [41]. He observed that LS is many times more

5.3 Salts of Trinitroresorcine

resistant to ionizing radiation than LA. The rate of thermal decomposition of LS increases with the dose of previously applied irradiation. Irradiations in air and in vacuum gave the same results [41].

5.3.1.2 Explosive Properties

LS is extremely sensitive to flame and electric spark; moreover, various handling operations, or container characteristics, can easily create an electrostatic charge on the surface of the crystals. Its sensitivity is higher than MF, LA, DDNP and most other primary explosives. This extreme sensitivity presents a significant problem which resulted in the past in many accidents during LS production or handling. The critical operations include those where movement of the material can result in electrostatic charge formation, especially mechanical grain size and shape modifications and sieving [33]. Many accidents occur during production and handling. Significant safety precautions are therefore necessary to eliminate sources of static electric.

The extreme sensitivity of LS to electrostatic discharge was the reason behind the work on desensitized LS development and manufacture. One of the methods of desensitization is based on washing the final product with a solution of a desensitizing agent (e.g., benzene solution bee wax) [33]. The LS acquires the form of uniform granules, less sensitive to mechanical stimuli and electrostatic discharge, however, still keeping the same desired sensitivity to flame as the non-desensitized variety [10].

During World War II, Germany developed a method of coating LS with graphite producing so-called "black LS." The idea was to make the compound more conductive and hence less sensitive but the practical experience showed no improvement as to the number of explosions in production [12].

LS is often used in bridge-wire compositions where it is in direct contact with wire heated by an electric current. An interesting study of the effect of loading density on the hot wire initiation of lead and barium styphnates was carried out by Naval Ordnance Laboratory [42]. It was found that the energy required for the initiation remained constant and the time to ignition decreased as the loading density increased.

The sensitivity to mechanical stimuli (impact and friction) is not that critical, as can be seen from Figs. 2.15 and 2.19. The sensitivity of LS to impact is mostly reported lower than LA [15, 36, 43–46]; however, some authors report sensitivity being higher [47, 48]. The same discrepancy in results is published for friction sensitivity where lower values are reported by [5, 7, 38], while higher sensitivity is reported by Danilov [1]. Our own results of friction sensitivity testing at room temperature indicate that LS is less sensitive than LA (see Fig. 2.19). Sensitivity of LS to mechanical stimuli significantly increases with increasing temperatures (even at 75 °C) [38], while it decreases at low temperatures [20]. Increasing of impact sensitivity of LS during storage at higher temperatures can be demonstrated by trials conducted over 60 days at 75 °C when it increased from 23 cm to 5 cm using 1 kg hammer [10].

Table 5.5 Detonation velocity of lead styphnate

Density (g cm^{-3})	Detonation velocity (m s^{-1})	References
0.93	2,100	[9]
2.6	4,900	[36]
2.9	5,200	[36]
3.1	5,600	[50]

Table 5.6 The power of LS compared with LA and MF measured in lead block

	Trauzl block for 10 g (cm^3) [1]	Trauzl block for 10 g (cm^3) [38]	Lead block for 2 g (cm^3) [36]	Trauzl block for 10 g (%TNT) [15]
LS	130	122	29.0	40
MF	110	133	33.0	51
LA	110	113	25.6	39

According to Danilov et al. [1], the heat of explosion of LS is 725 kJ mol^{-1}. The course of explosive decomposition of LS is difficult to characterize precisely, but the decomposition is best described by the following equation [33, 36, 49]:

$$80 \text{ [LS]} \rightarrow 80 \text{ Pb}_{(g)} + 172 \text{ CO}_2 + 294 \text{ CO} + 14 \text{ HCN} + 31 \text{ H}_2 + 81 \text{ H}_2\text{O}_{(g)} + 113 \text{ N}_2$$

The volume of gaseous product is 270 dm^3 kg^{-1} [33]. The detonation velocity of LS at various densities and conditions is summarized in Table 5.5.

The initiating efficiency of LS is very poor and in its pure form it is not able to initiate any other common high explosives except uncompressed PETN [38]. Bubnov, for example, reported that even 2 g LS are not able to initiate tetryl [51]. This is the reason why it is considered as a poor primary explosive or even not considered at all as a typical primary explosive [12]; this type of primary explosive is sometimes described as a pseudo initiating substance [33].

Power of LS is relatively low. The exact values determined by lead block are listed in Table 5.6 together with MF and LA for comparison [15, 38].

Very interesting results of TNT equivalency for LS were published by Swatosh et al. [52] from blast wave measurements particularly from max. peak pressure and positive impulse. Depending on the form of the test and scaled distance, TNT equivalence was in the range of 30–65 %.

The ignition temperature of LS is 255 °C at heating rate 5 °C min^{-1} [3] and 275–277 °C at heating rate 20 °C min^{-1} [36]. The temperature at which explosion occurred within 5 s was 280 °C according to Danilov et al. [1] and instantaneous explosion occurred at 270 °C when measured by Kast and Haid [36]. The dependence of time to detonation of LS on temperature has been published by Hailes [37]. The results are in the original paper presented for varying sample amounts (from 0.5 to 6.7 mg), which makes the data very difficult to compare as the size influences the

5.3 Salts of Trinitroresorcine

Fig. 5.4 Time to explosion of LS as a function of temperature [37]

time to explosion. We have therefore summarized only the results from 0.5 to 2.0 mg and the results are presented in Fig. 5.4.

5.3.1.3 Preparation

The earliest published preparation method found by the current authors is based on addition of a solution of lead acetate to an aqueous solution of styphnic acid [28]. The reaction yielded a yellow gelatinous form of LS when simply mixing the reactants and deep yellow needles if boiled with slight excess of lead carbonate instead of lead acetate.

The usual way of LS preparation is based on adding a solution of a soluble styphnate salt (Na, K, Mg, NH_4^+), most commonly magnesium styphnate, to a well-stirred lead nitrate or lead acetate solution. The lead styphnate precipitates from the reaction mixture.

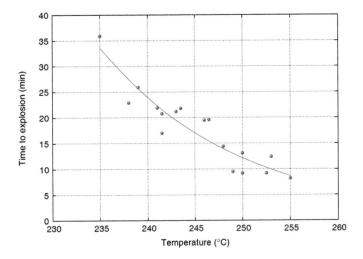

The reaction conditions (especially pH and temperature) are the key factors influencing the type of LS formed. Control of pH is essential to avoid basic styphnate formation (addition of weak acetic acid).

Magnesium styphnate is prepared by addition of magnesium oxide to trinitroresorcine water suspension [8, 10, 12, 53–55]. Precipitation of pure LS requires a higher temperature (mostly 65–80 °C) than that required to produce material for wax coating (around 45–50 °C) [10]. Some more complicated temperature profiles employed in production of LS are summarized in [55]. If the temperature drops below 40 °C during preparation, an amorphous form unsuitable for further technological processing is formed. The technological details of LS production are detailed in many books [1, 10, 12, 33, 56]. The addition of crystal modifiers (e.g., carboxymethyl cellulose) is recommended for preparation of LS in a more suitable form for processing [26, 27]; addition of PVA to the reaction mixture very strongly suppresses formation of β-LS [57]. The purity of styphnic acid has an impact on the properties of LS as well. If it is prepared from very pure styphnic acid, it has a markedly increased tendency to hydrolyze in water [3].

The influence of reaction conditions on properties, yield, and purity of LS was studied in detail by Stettbacher (LS was prepared by reaction of sodium styphnate and lead nitrate). His observations were summarized by Bagal as follows [33]:

- The yield of LS increases with purity of styphnic acid.
- The order of batching does not have significant influence on quality of LS. It does not matter which reactant's pH is adjusted, the quality and quantity of LS stay the same.
- LS prepared at high temperature forms a precipitate; the gelatinous form of LS forms when reaction takes place at room temperature. The gelatinous form of LS only deflagrates by action of flame rather than detonation.
- The amount of acetic acid that is added for modification of pH has to be low, the yield of LS decreasing with increasing amount of acetic acid.
- The yield of LS is higher with more concentrated solutions.

Free styphnic acid is not a suitable starting material due to its low solubility in water and formation of nitric acid as a by-product of the reaction (nitric acid decomposes LS). When the sodium or magnesium salt of styphnic acid is used, innocuous sodium or magnesium nitrate forms as a by-product [33]. The reagents for LS production are used in equimolar quantity. However, when excess lead acetate is used as a starting substance, the double salt of LS and lead acetate is formed. But the double salt of LS and an initial lead salt only forms when the lead salt of a weak organic acid is used (e.g., lead acetate) [58].

An unusual dihydrate of LS can be prepared via magnesium styphnate in a similar way to that for normal LS. However, the solutions used for preparation of the dihydrate must be diluted, reaction temperature must be low (about 9 °C), and the reaction mixture must not be stirred. The dihydrate forms within several hours (about 18 h) as red needles [33].

5.3 Salts of Trinitroresorcine   145

Lead styphnate can also be prepared as a jelly [58, 59]. The jelly product forms by mixing a concentrated solution of lead nitrate or acetate with styphnic acid at low pH (2.8–5.5) at room temperature. The sequence of addition and the concentration of reagents are critical. At pH >5.5 LS forms in the crystalline form [59].

5.3.1.4 Use

The application of lead styphnate is dictated by its high sensitivity to flame and its low brisance. Initiating efficiency of lead styphnate is poor but it easily initiates less flame-sensitive primary explosives. The combination of the high flammability of lead styphnate and the high initiating efficiency of lead azide has been the reason for using a mixture of these two substances in blasting caps since 1920 [60, 61]. Mixtures of LS and GNGT were used in percussion and stab-initiated applications [10]. Another function of lead styphnate in this mixture is coating lead azide and hence protecting from mechanical stimuli and the chemical influence of moisture and carbon dioxide from the environment [1, 33, 51]. A drawback, not frequently considered, is the acid reaction of LS in presence of moisture that accelerates hydrolysis of LA [3]. The phenomenon of compatibility of LA/LS has already been discussed in Chap. 4.

LS is also the main component of many pyrotechnic mixtures for primers often in combination with GNGT (e.g., SINOXIDE composition [62]). It is moreover used as a component of fusehead composition (Valenta Private Communication). Despite its excellent properties for primer applications and plenty of experience, the presence of lead is causing its replacement by non-toxic alternatives.

5.3.2 Basic Lead Styphnate

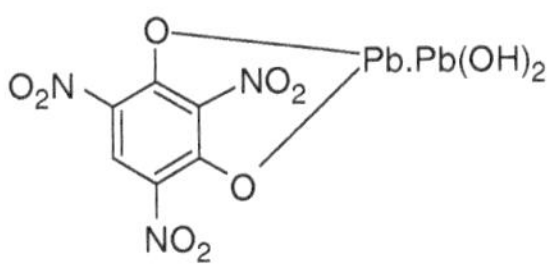

monobasic LTNR
formula I

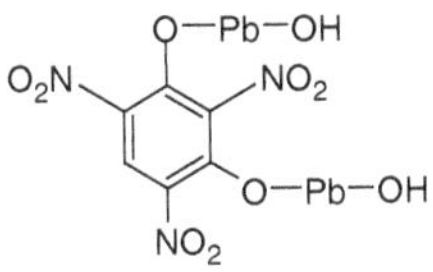

monobasic LTNR
formula II

Lead styphnate forms a variety of basic salts of which only the monobasic salt (simply called basic) is of practical significance. Monobasic LS was, according to Fedoroff and Shefield [8], first prepared by Griess in 1874 [63]; however, reference to preparation of this substance was not found by the current authors in the original article.

Table 5.7 Differing forms of mono basic lead styphnate

Form	Polymorph	ρ (g cm^{-3})	References
Yellow orthorhombic	α	4.05	[67]
Red diamond	β	4.12	[65, 67]
		4.06	[24]
Yellow amorphous	γ	4.13	[67]

Two structures of basic lead styphnate are commonly used. The first structure represents a double salt of lead styphnate and lead hydroxide (formula I) [20, 33, 39, 64]. In the second structure each lead atom is bonded with one oxygen atom of the trinitroresorcinol molecule (formula II) [8, 65, 66].

5.3.2.1 Physical and Chemical Properties

Basic lead styphnate exists in three polymorphic forms listed in Table 5.7 [20, 64, 67].

Some literature resources list only two forms, red prisms (4.06 g cm^{-3}) and yellow needles (3.88 g cm^{-3}) [8, 33, 40, 65]. The bulk density of the red form is 0.6–0.65 g cm^{-3} while for the yellow version it is only 0.25 g cm^{-3} [33].

Basic lead styphnate is practically insoluble in water (red form slightly more than practically insoluble yellow form—0.0008 g/100 ml at 25 °C) [24, 33]. Both of them are soluble in aqueous ammonium acetate and practically insoluble in organic solvents that do not dissolve lead styphnate [40]. The hygroscopicity of both forms is not high; however, when comparing the two forms, the red is significantly less hydroscopic than the yellow [33]. In presence of moisture it is almost neutral unlike the acidic normal salt. Basic LS is thermally stable in moist air (90 %) at higher temperatures (74 °C) for a considerable time [20]. The basic salt has superior thermal stability compared to the normal salt [64]. It is compatible with aluminum and copper metals and also with oxidizers [24]. It is stable in water (no changes under water for 2 months) [24] and can be chemically decomposed in similar way to that for normal lead styphnate [8]. Normal LS is formed on treating styphnic acid with basic LS in 50 % ethanol at pH = 4 [59].

5.3.2.2 Explosive Properties

Basic lead styphnate is quite sensitive to electrostatic discharge which makes its handling and loading particularly hazardous. The sensitivity of each form is different and varying degrees have been found by various researchers. The sensitivities to impact are reported according to the following sequence:

Yellow amorphous ≪ yellow orthorhombic < red prism diamond [67]
Red hexagonal < yellow needles [68]

The sensitivity to impact of β form (red) is reported to be the same as or higher than that for LS (F of I is 10 while 20 for LS); sensitivity to friction is similar to LS and temperature of ignition is 248–250 °C at heating rate 5 °C min^{-1} [24]. The temperature of ignition of basic LS is 240 °C (heating rate 5 °C min^{-1}) [69]. Bagal reported 280 °C for the yellow form and 287 °C for the red form (explosion within 5 s) [33].

The sensitivity to electrostatic discharge is reported in the following order [24]:

Yellow amorphous (γ) < yellow (α) < red (β)

The sensitivity of basic LS to flame is about the same as for LS [33]. The brisance of the orthorhombic yellow form is higher than that of other forms based on the results of a sand test [67]. Sensitivity of basic LS to flame rapidly decreases in presence of water. It cannot be ignited by flame if the water content is higher than 10 % [20].

5.3.2.3 Preparation

The basic lead styphnate is prepared by reaction of the sodium or magnesium salt of 2,4,6-trinitroresorcine in an alkaline reaction medium with lead nitrate [8].

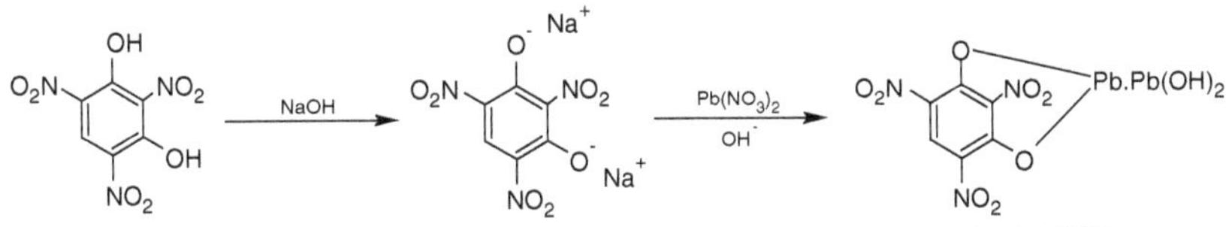

An alternative preparation method involves the ammonia salt of 2,4,6-trinitroresorcine [24, 64, 66]. The formation of the desired form and crystal shape depends on reaction conditions. The addition of 2-mononitroresorcinol in an amount between 0.1 and 5 % is sometimes employed in the preparation of the red form to improve its bulk density and flowability [24, 64]. The presence of foreign substances in the starting 2,4,6-trinitroresorcine (added in the sulfonation operation in the preparation of styphnic acid) has also a significant influence on the shape of basic LS crystals [70]. Various procedures are patented for preparation of red [24, 64, 65, 71] and yellow [39, 64–67] basic lead styphnate. Just like LS, the basic LS can also be prepared as a jelly (see lead styphnate).

5.3.2.4 Use

Basic lead styphnate has been used as a component of priming compositions [24] for percussion primers and stab detonators (e.g., component of NOL 130 primer mixture that contains 40 % basic LS, 20 % LA, 2 % GNGT, 15 % Sb$_2$S$_3$, and 20 % BaNO$_3$) [8]. It is also used as a component of fusehead compositions [24].

5.3.3 Double Salts of Lead Styphnate

Lead styphnate also forms double salts. Mainly salts with tetrazole compounds were examined as brisant primary explosives. The main attention was aimed at double salts of styphnic acid with 1,3-di(5-tetrazoyl)triazene (LDDS) and styphnic acid with 5-nitraminotetrazole (PbNATNR). The study of these salts was done in the 1960s and 1970s.

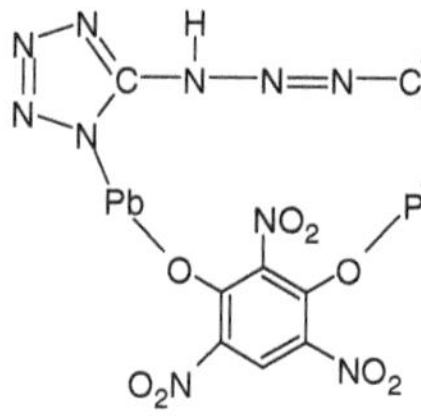

5.3.3.1 Physical and Chemical Properties

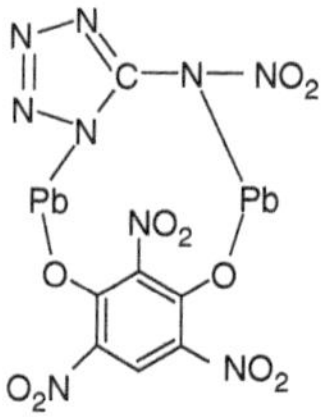

The double salt of styphnic acid and 5-nitraminotetrazole (PbNATNR) forms lemon yellow crystals with crystal density 3.6 g cm^{-3} (density of LS is 3.03 g cm^{-3}). It is practically insoluble in cold water, slightly soluble in hot water (0.032 g/100 ml at 25 °C and 0.175 g/100 ml at 70 °C). PbNATNR is stable under hot water for a long time, no decomposition being observed after several months stored at 65 °C. Thermal stability is superior to that of LS, and weight loss after 24 h at 210 °C is 0.5 % (compared with 24.3 % for LS) [72].

According to Sinha, LDDS shows higher thermal stability than GNGT, LS and even higher than LA. The thermal stability is better even at high humidity. The compound does not lose weight up to 280 °C. It is interesting that the double salt LDDS has higher thermal stability than either of the two components, lead 1,3-di (5-tetrazoyl)triazene and lead styphnate [72, 73].

5.3.3.2 Explosive Properties

The impact sensitivity of LDDS is higher than LS, sensitivity to friction is reported to be of the same order as that of service LA or higher (see Table 5.8) [74].

The impact sensitivity of the double salt of styphnic acid and 5-nitramino-tetrazole (PbNATNR) is higher than LS, and it explodes with far greater force and vigor. Brisance of PbNATNR is high, it is superior to that of LS, LA, and MF. The ignition temperature is 340 °C (compared with 320 °C for LS) [72].

5.3 Salts of Trinitroresorcine

Table 5.8 Explosive properties of LDDS and PbNATNR

	LDDS	PbNATNR	LS	LA	MF	References
Impact sensitivity (cm)[a]	–	14	16	–	–	[72]
Impact sensitivity (kg m cm^{-2})	0.107	–	0.210	0.202	0.112	[73, 74]
Explosion temperature (heating rate 5 °C min^{-1}, TGA)	290, 312	–	255, 270	–	–	[73]
Sand test (g)	–	24.0	16.1	14.2	18.9	[72]

[a]100 g weight.

5.3.3.3 Preparation

Double salts are prepared by action of aqueous soluble lead salt on an equimolar mixture of styphnic acid (in the form of the magnesium salt) and soluble salt of the tetrazole compound [72, 73].

5.3.3.4 Use

Double salts were suggested as a lead azide replacement. Both salts are reported as powerful as lead azide and show neither low sensitivity to flame nor a tendency to hydrolyze [72, 73]. Information about their practical application is not available.

5.3.4 Barium Styphnate

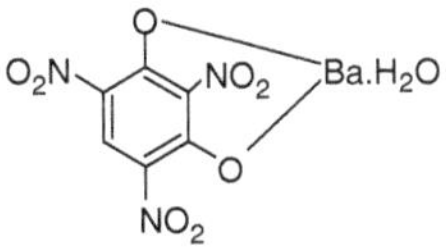

The scientific literature is a bit puzzling regarding the history of this substance. Griess is sometimes mentioned as the first chemist who prepared barium styphnate [63] in 1874. It is, however, not true as the article clearly references work of Stenhouse [28] who prepared the barium, lead, and silver salts of styphnic acid for comparison with orcin (5-methylbenzene-1,3-diol) salts of the same metals four years earlier. In addition, the article of Griess is surprisingly often erroneously referred to as the first work publishing preparation of lead styphnate.

Table 5.9 Differing forms of barium styphnate monohydrate [33]

Aspect	Form	ρ (g cm^{-3})
Yellow labile	Enol form of TNR	–
Red labile	Acid form of TNR	2.581
Yellow stable	Diketo form of TNR	2.625

Table 5.10 Temperature of ignition (explosion within 5 s) and sensitivity to flame [33]

	Temperature of ignition (°C)	Sensitivity to flame (cm)[a]
Monohydrate—yellow	356	1
Monohydrate—red	361	1.5
Anhydride	346	5

[a]Height at which 100 % initiation occurs (without pressing) measured on pendulum for investigation of sensitivity to flame.

5.3.4.1 Physical and Chemical Properties

Barium styphnate (barium salt of 2,4,6-trinitrobenzene-1,3-diol; BaS) forms an anhydride and two hydrates, monohydrate and trihydrate; only the anhydride and the monohydrate are used in practical applications [20, 75, 76]. According to Bagal the monohydrate of barium styphnate exists in three polymorphic forms listed in Table 5.9 [33].

The red form is not stable and easily changes to the yellow labile form in (a) acidic environments, (b) large amount of residues of solvent or inorganic compounds from precipitation in crystals, or (c) in presence of DNR in TNR. In solution, the red form also changes into the yellow labile form. The red form is slightly soluble in cold water, better in boiling water, and insoluble in ethanol, ether, and other organic solvents. The yellow stable form is less soluble in water than red BaS (about 2.5 times less) [33].

Barium styphnate trihydrate forms yellow-orange needles. It is more soluble in water than the monohydrate. Trihydrate loses two crystal waters at 105 °C and red needles of monohydrate are formed [33]. The original work of Griess reports the trihydrate formed as rhombohedric prisms with low solubility in water [63].

The monohydrate is more difficult to dehydrate than LS and it only loses its molecule of water after heating to 170 °C [2] or 160 °C at pressure 91 kPa [33].

The anhydride is, however, not stable in air and quickly absorbs moisture reverting back to its monohydrate form. The monohydrate is slightly hydroscopic, less than LA [33], and it is thermally stable as it loses only 0.1 % wt. within 2 h at 160 °C [2].

5.3.4.2 Explosive Properties

Barium styphnate is not very sensitive to impact; less than LA and LS. The F of I of this compound is only 40 (while 20 for LS and 15 for LA) [2]. Thermal sensitivity and sensitivity to flame depends on the form. The values for the anhydride and the two monohydrate forms are summarized in Table 5.10. Brisance of BaS monohydrate is lower than for LS [33]. It is reported to lose water of crystallization accompanied by a change of color to orange-red followed by explosion "with extreme violence" when gently heated on platinum foil [28].

5.3.4.3 Preparation

The earliest method for BaS is described by Stenhouse who dissolved trinitroresorcine in 100 parts of boiling water and added excess of barium carbonate [28]. On cooling he obtained pale yellow rhomboidal plates. Analysis showed presence of three waters of crystallization.

Direct precipitation of BaS anhydride is not possible as one of three polymorphic forms of the monohydrate is produced depending on reaction conditions. The red form is created when freshly prepared styphnic acid is used for precipitation of barium styphnate. It is formed by direct reaction of equivalent amounts of TNR and barium carbonate. The yellow stable salt is formed when "old TNR" is used for BaS preparation, in an acidic environment or in excess precipitating agent. It is commonly prepared via the sodium salt of styphnic acid. The trihydrate of BaS forms in reaction mixture with low concentration of reagents and without stirring. The trihydrate also sometimes contaminates the monohydrate [33].

Preparation of barium styphnate monohydrate for practical applications is similar to preparation of LS. Magnesium styphnate is used as a starting material which reacts with barium nitrate, chloride, or acetate resulting in precipitation of barium styphnate. A reaction temperature of 60–85 °C is recommended. Product precipitates in form of the trihydrate, which dehydrates to the monohydrate by the slow addition of sufficiently dilute nitric acid [75]. It is difficult to re-crystallize in a suitable form and it is therefore recommended to use a crystal-modifying agent (e.g., carboxymethyl cellulose) [2, 26, 27].

Preparation of BaS with physical properties suitable for volumetric loading using crystal growth modifiers has been described by Orbovic and Codoceo [76]. From the published results it seems that the best product is obtained when using a mixture of water-soluble cationic copolymer Pel (polyectrolyte AS-3605) and CMC (carboxymethyl cellulose).

5.3.4.4 Use

Barium styphnate has been used for firearms in Britain since World War II [2]. The type of production process and shape of prepared crystals determines the application of this primary explosive. It is used in delay composition, as an ingredient of primary compositions, as a component of compositions for electric primers or igniters, delay compositions, and so on [2, 3, 75, 76].

5.3.5 Other Salts of Styphnic Acid

The current authors have not been able to find if styphnate salts other than lead, basic lead, and barium have ever been practically used. Information about silver and cupric salts is sparse.

The silver salt of styphnic acid has been proposed as a primary explosive in several patents [77, 78]. Silver styphnate forms an anhydrous salt or a monohydrate. The monohydrate loses water of crystallization at 65 °C [33]. It is sensitive to flame; sensitivity to impact is about the same as for LS [33] or higher, approaching MF [7]. The temperature of ignition is 286 °C (heating rate 5 °C min^{-1}) [33]. Silver styphnate can be prepared in a similar way to that used for LS by the action of the soluble silver salt on the soluble styphnate (e.g., magnesium, sodium) or by the action of silver carbonate on styphnic acid [33, 77, 78]. Probably the earliest recorded method of preparation is based on short time boiling of an aqueous solution of pure styphnic acid with slight excess of silver oxide followed by filtration of the precipitate deposited on cooling [28]. The silver salt is precipitated in the form of long yellowish-brown needles easily re-crystallizable from water.

The cupric salt of styphnic acid forms a tetrahydrate that changes to the monohydrate by heating at 100 °C. According to Bagal, the cuprous salt of styphnic acid has never been prepared [33]. Styphnic acid also forms many double salts [33].

References

1. Danilov, J.N., Ilyusin, M.A., Tselinskii, I.V.: Promyshlennye vzryvchatye veshchestva; chast I. Iniciiruyushchie vzryvshchatye veshchestva. Sankt-Peterburgskii gosudarstvennyi tekhnologicheskii institut, Sankt-Peterburg (2001)
2. Collins, P.H., Holloway, K.J., Williams, R.J.: The development of less-sensitive primary explosives. In: Jenkins, J.M., White, J.R. (eds.) Proceedings of International Conference on Research in Primary Explosives, vol. 2, pp. 12/1–12/20. Waltham Abbey, England (1975)
3. Taylor, G.W.C.: Technical requirements and prospects for new primary explosives. In: Jenkins, J.M., White, J.R. (eds.) Proceedings of the International Conference on Research in Primary Explosives, vol. 3, pp. 18/1–18/21. Waltham Abbey, England (1975)
4. Sillberrad, O., Phillips, H.A.: The metallic picrates. J. Chem. Soc. **93**, 485 (1908)
5. Meyer, R., Köhler, J., Homburg, A.: Explosives. Wiley-VCH Verlag GmbH, Weinheim (2002)
6. Clark, A.K., Davies, N., Hubbard, P.J., Lee, P.R.: Cross sensitization to impact between lead azide and tetryl. In: Jenkins, J.M., White, J.R. (eds.) Proceedings of the International Conference on Research in Primary Explosives, vol. 3, pp. 22/1–22/15. Waltham Abbey, England (1975)
7. Taylor, A.C., Rinkenbach, W.H.: Sensitivities of detonating compounds to frictional impact, impact, and heat. J. Franklin Inst. **204**, 369–376 (1927)
8. Fedoroff, B.T., Sheffield, O.E., Kaye, S.M.: Encyclopedia of Explosives and Related Items. Picatinny Arsenal, New Jersey (1960–1983)
9. Urbański, T.: Chemie a technologie výbušin. SNTL, Praha (1959)
10. Špičák, S., Šimeček, J.: Chemie a technologie třaskavin. Vojenská technická akademie Antonína Zápotockého, Brno, Czechoslovakia (1957)
11. Straka, A.: Speciální technika. FMVS Praha a GRt ZVS Brno, Praha (1976)

References

12. Krauz, C.: Technologie výbušin. Vědecko-technické nakladatelství, Praha (1950)
13. Urban, L., Šenkýř, V. (eds.): Speciální technika, Federální ministerstvo všeobecného strojírenství Praha, závody všeobecného strojírenství, generální ředitelství Brno, Praha, p. 536 (1976)
14. Payne, J.R.: Thermochemistry of lead 2,4- and 4,6-dinitroresorcinol. Thermochim. Acta **265**, 73–87 (1995)
15. Tomlinson, W.R., Sheffield, O.E.: Engineering Design Handbook, Explosive Series of Properties Explosives of Military Interest. Report AMCP 706-177 (1971)
16. Peyton, S.V., Williams, E.: Improvements in or relating to ignitory compositions. GB Patent 892,741, 1962
17. Rubenstein, L.: Improvements in or relating to the production of normal lead dinitroresorcinate. US Patent 616,456, 1949
18. Rubenstein, L.: Production of normal lead dinitroresorcinate. US Patent 2,493,551, 1950
19. Tang, T.B.: Decomposition of lead(II) 2,4-dinitroresorcinate. Thermogravimetry, calorimetry, microscopy and time-resolved mass spectroscopy. Thermochim. Acta **61**, 341–356 (1983)
20. Khmelnitskii, L.I.: Spravochnik po vzryvchatym veshchestvam. Voennaya ordena Lenina i ordena Suvorova Artilleriiskaya inzhenernaya akademiya imeni F. E. Dzerzhinskogo, Moskva (1962)
21. Taylor, G.W.C., Thomas, A.T.T., Williams, R.J.E.: Lead compounds of 4:6-dinitroresorcinol. GB Patent 1,374,235, 1974
22. Jones, D.T., Rubenstein, L.: Improvements in or relating to explosive salts and applications thereof. GB Patent 582,976, 1946
23. Urbański, T.: Chemistry and Technology of Explosives. Pergamon Press, Oxford (1984)
24. Jenkins, J.M.: New processes for manufacture of lead salts of nitrophenols. In: Jenkins, J.M., White, J.R. (eds.) Proceedings of the International Conference on Research in Primary Explosives, vol. 3, pp. 21/1–21/25. Waltham Abbey, England (1975)
25. Taylor, G.W.C., Napier, S.E.: Improvements in or relating to the manufacture of lead 4:6-dinitroresorcinate. GB Patent 1,094,921, 1967
26. Taylor, G.W.C., Napier, S.E.: Improvements in and relating to the preparation of explosive substances. GB Patent 849,101, 1960
27. Taylor, G.W.C., Napier, S.E.: Preparation of explosive substances containing carboxymethyl cellulose. US Patent 3,291,664, 1966
28. Stenhouse, J.: Contributions to the history of Orcin. No. I. Nitro-substitution compounds of the Orcins. Proc. R. Soc. Lond. **19**, 410–417 (1870–1871)
29. Herz, E.R.: Verfahren zur Herstellung von Zündsätzen. DE Patent 285,902, 1914
30. Herz, E.R.: Method of manufacturing detonators, detonating caps and the like. GB Patent 142,823, 1921
31. McCrone, W.C., Adams, O.W.: Crystallographic data. Lead styphnate (normal). Anal. Chem. **27**, 2014–2015 (1955)
32. Miles, F.D.: The formation and characteristic of crystals of lead azide and of some other initiating explosives. J. Chem. Soc. 2532–2542 (1931)
33. Bagal, L.I.: Khimiya i tekhnologiya iniciiruyushchikh vzryvchatykh veshchestv. Mashinostroenie, Moskva (1975)
34. Ficheroulle, H., Kovache, A.: Contribution à l'étude des explosifs d'amorçage. Mémorial des poudres **31**, 7–27 (1949)
35. Stettbacher, A.: Bleitrinitroresorcinat. Nitrocellulose **8**, 21–27 (1937)
36. Kast, H., Haid, A.: Über die sprengtechnischen Eigenschaften der wichtigsten Initialsprengstoffe. Zeitschrift für das angewandte Chemie **38**, 43–52 (1925)
37. Hailes, H.R.: The thermal decomposition of lead styphnate. Trans. Faraday Soc. **29**, 544–549 (1933)
38. Wallbaum, R.: Sprengtechnische Eigenschaften und Lagerbeständigke der wichtigsten Initialsprengstoffe. Zeitschrift für das gesamte Schiess- und Sprengstoffwesen **34**, 197–201 (1939)

39. Zingaro, R.A.: Lead salt of 2,4,6-trinitroresorcinol. J. Am. Chem. Soc. **76**, 816–819 (1954)
40. Military Explosives. Report TM 9-1300-214, Washington, DC (1984)
41. Flanagan, T.B.: Effect of nuclear irradiation upon the decomposition of lead styphnate monohydrate. Nature **181**, 42–43 (1958)
42. Leopold, H.S.: Effect of loading density on the hot wire initiation of normal lead styphnate and barium styphnate. Report NOLTR-70-96, AD716024, U.S. Naval Ordnance Laboratory, White Oak, Silver Spring (1970)
43. Fronabarger, J., Sanborn, W.B., Bichay, M.: An investigation of some alternatives to lead based primary explosive. In: Proceedings of 37th AIAA 2001–3633 Joint Propulsion Conference and Exhibit, pp. 1–9. Salt Lake City, Utah (2001)
44. Hiskey, M.A., Huynh, M.V.: Primary explosives. US Patent 2006/0030715A1, 2006
45. Huynh, M.H.V., Coburn, M.D., Meyer, T.J., Wetzer, M.: Green primaries: Environmentally friendly energetic complexes. Proc. Natl. Acad. Sci. **103**, 5409–5412 (2006)
46. Clark, L.V.: Diazodinitrophenol, a detonating explosive. J. Ind. Eng. Chem. **25**, 663–669 (1933)
47. Millar, R.W.: Lead-free initiator materials for small electro-explosive devices for medium caliber munitions. Report QinetiQ/FST/CR032702/1.1, QINETIQ Ltd, Farnborough (2003)
48. Wallbaum, R.: Sprengtechnische Eigenschaften und Lagerbeständigke der wichtigsten Initialsprengstoffee. Zeitschrift für das gesamte Schiess- und Sprengstoffwesen **34**, 161–163 (1939)
49. Kast, H.: Weitere Untersuchungen über die Eigenschaften der Initialsprengstoffe. Zeitschrift für das Gesamte Schiess- und Sprengstoffwesen **21**, 188–192 (1926)
50. Novotný, M., Mečíř, R., Sedláček, S., Tamchyna, V.: Teorie výbušin. VŠCHT v Pardubicich, Pardubice (1981)
51. Bubnov, P.F.: Initsiruyushchie vzryvchatye veshchestva i sredstva initsirovaniya. Gosudarstvennoe izdatelstvo oboronnoi promyshlennosti, Moskva (1940)
52. Swatosh, Jr. J.J., Napadensky, H.S., Cook, J.R., Levmore, S.: Blast parameters of lead styphnate, lead azide, and tetracene. Report AD-A021410, Picatiny Arsenal, dist. NTIS (1975)
53. Davis, T.L.: The Chemistry of Powder and Explosives. Wiley, New York (1943)
54. Hill, S.G.: Comercial detonators; Manufacture of lead styphnate and lead azide. Can. Chem. Process Ind. **26**(18), 43 (1942)
55. Holloway, K.J., Jenkins, J.M., Taylor, G.W.C.: Improvements in or relating to the manufacture of lead styphnate. GB Patent 1,380,437 (1975)
56. Urbański, T.: Chemistry and Technology of Explosives. PWN—Polish Scientific Publisher, Warszawa (1967)
57. Taylor, G.W.C.: The preparation of gamma lead azide. In: Symposium on Lead and Copper Azide, pp. A–3, 20–23, Waltham Abbey (1966)
58. Garfield, F.M.: Manufacture of normal lead trinitroresorcinate and double salts thereof. US Patent 2,295,104, 1942
59. Rosen, B.: The lead styphnate jelly. J. Am. Chem. Soc. **77**, 6517–6518 (1955)
60. Friederich, W.: Process for the manufacture of detonating caps for mining and military purposes. GB Patent 138,083, 1920
61. Grotta, B.: Development and application of initiating explosives. Ind. Eng. Chem. **17**, 134–138 (1925)
62. Hagel, R., Redecker, K.: Sintox – a new, non-toxic primer composition by Dynamit Nobel AG. Propellants Explosives Pyrotechnics **11**, 184–187 (1986)
63. Griess, P.: Ueber Einwirkung von Salpeter-Schwafelsäure auf Orthonitrobenzoësäure. Berichte der deutschen chemischen Gesellschaft **7**, 1223–1228 (1874)
64. Taylor, G.W.C., Thomas, A.T.: Basic lead styphnate. GB Patent 1,381,257, 1975
65. Brün, W.: Priming mixture. US Patent 1,942,274, 1934
66. Tauson, H.: Process of making basic lead trinitroresorcinol. US Patent 2,020,665, 1935
67. Hitchens, A.L.J., Garfield, F.M.: Basic lead styphnate and process of making it. US Patent 2,265,230, 1941

References

68. Brün, W.: Manufacture of styphnic acid. US Patent 2,275,169, 1942
69. Maksacheff, M., Whelan, D.J.: Thermochemistry of normal and basic lead styphnates using differential scanning calorimetry. Report MRL-R-1000, Materials Research Laboratories, Melbourne (1986)
70. Brün, W.: Manufacture of styphnic acid salts. US Patent 2,275,170–2,275,173 (1942)
71. Brün, W.: Basic lead styphnate and process of making it. US Patent 2,137,234 (1938)
72. Staba, E.A.: Crystalline double salt of lead nitroaminotetrazole and lead styphnate. US Patent 3,310,569 (1967)
73. Bahadur, K.: Study of some double salts of lead useful as initiatory explosives by thermo gravimetry and thermometric titration. In: Hansson, J. (ed.) Proceedings of Symposium on Chemical Problems Connected with the Stability of Explosives, pp. 1–15. Ystad, Sweden (1973)
74. Sinha, S.K., Srivastava, R.C., Surve, R.N.: Metallic double salts—a new concept of primary explosive. In: Jenkins, J.M., White, J.R. (eds.) Proceedings of the International Conference on Research in Primary Explosives, vol. 2, pp. 13/1–13/10. Waltham Abbey, England (1975)
75. Taylor, G.W.C., Napier, S.E., Stansfield, F.: Improvements in and relating to the production of barium styphnate monohydrate. GB Patent 1,094,981, 1967
76. Orbovic, N., Codoceo, C.L.: Production of exploding materials for detonators: Control of crystal growth of lead and barium 2,4,6-trinitroresorcinate. Propellants Explosives Pyrotechnics **33**, 459–466 (2008)
77. Brün, W.: Priming mixture. US Patent 2,097,510, 1937
78. Brün, W.: Silver salt of styphnic acid. CA Patent 373,990, 1938

Chapter 6
Diazodinitrophenol

6.1 Introduction

The 2-diazo-4,6-dinitrophenol (dinol, diazol, DDNP, DDNPh, or DADNPh) is the first-ever synthesized diazonium compound. Its synthesis is attributed to Griess [1] who prepared it by introducing nitric oxides into an alcoholic solution of 2-amino-4,6-dinitrophenol (picramic acid) [1, 2]. Its explosive character was, however, first reported more than 30 years later in 1892 by Lenze [3].

6.2 Structure

Although DDNP has been known for almost 150 years, its structure is still debated and general consensus does not exist. The most obsolete suggested structures are not mentioned here but may be found in [4].

The properties (chemical, physical, and spectral) of general ortho (structure I, benzo [d][1,2,3]oxadiazole) and para (structure II, 2-oxa-3,4-diazabicyclo[3.2.2] nona-1(7),3,5,8-tetraene) isomers of diazophenols are very similar indicating that the structures should also be similar. Even though the cyclization and formation of two rings does not seem to be unreasonable for the ortho-isomer, it definitely does not look probable for the para-isomer. It is therefore assumed that the structures I and II are rather improbable for diazophenols [5]. This assumption was confirmed when structure I was prepared and shown to be unstable even in a solid argon matrix. The meta-(3-diazo) isomer has not been reported [5, 6].

R. Matyáš and J. Pachman, *Primary Explosives*,
DOI 10.1007/978-3-642-28436-6_6, © Springer-Verlag Berlin Heidelberg 2013

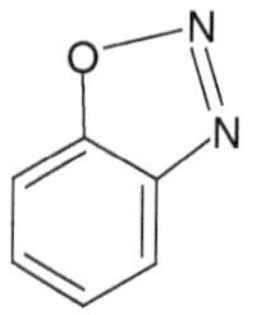

structure I 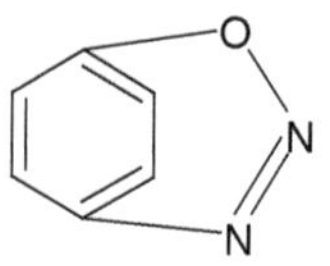   structure II

The possible cyclization of the diazo structure III of DDNP and formation of cyclic isomer IV is improbable based on the property-related assumption mentioned above. The theoretical studies employing the AM1 level of theory further support this conclusion [6].

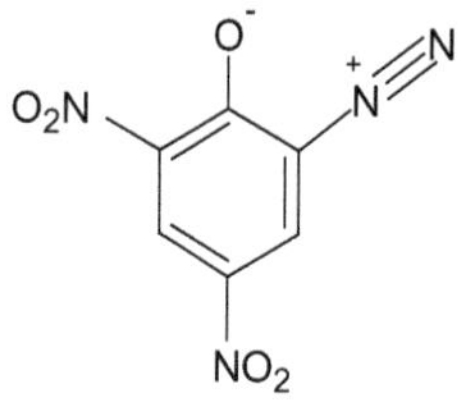

structure III 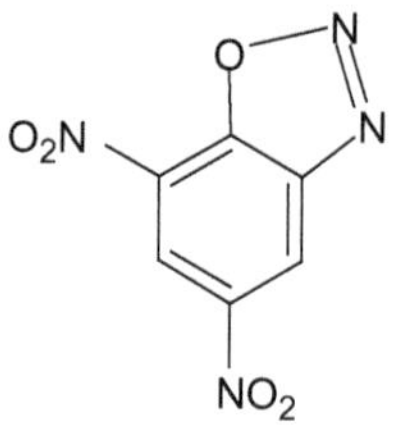   structure IV 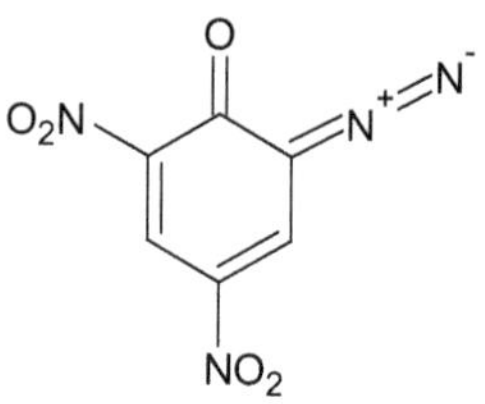   structure V

X-ray diffraction of DDNP was first carried out by Lowe-Ma et al. [5] and later confirmed by Holl et al. [6]. The bonding of DDNP has been discussed on the basis of theoretical calculations and compared to the results of X-ray diffraction, IR, and NMR [5, 6].

According to Lowe-Ma et al. [5], neither the zwitterionic (structure III) nor the quinonoid (structure V) structure provides a satisfactory model for the ortho-diazophenols including DDNP. Some possible tautomeric structure intermediate between these two also does not seem to be probable since DDNP exhibits discrete features of each of them [5]. In later work, these authors incline to structure VI based on the comparison of X-ray crystallography, NMR, and MO calculations. This is closely in agreement with the more recent findings of Holl et al. [6] who proposed diazo structure VII, which is suggested as the best match with single crystal X-ray diffraction and computational studies.

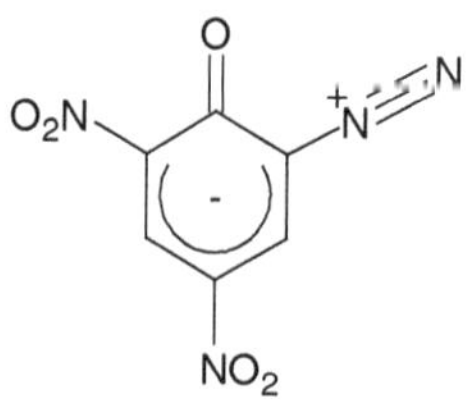

structure VI 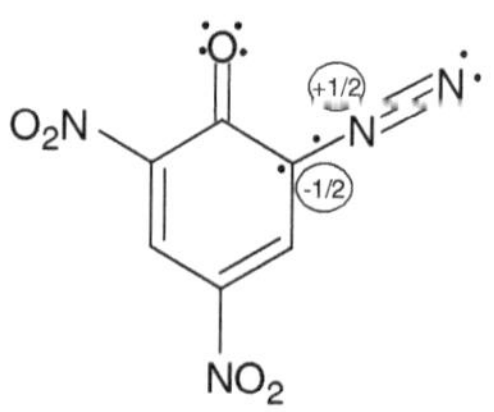   structure VII

Fig. 6.1 Crystals of DDNP prepared by re-crystallization from acetone

6.3 Physical and Chemical Properties

DDNP crystals are yellow (Fig. 6.1). However, shades of the technical product can range from dark yellow or green to dark brown [4]. The density of DDNP is most often reported as 1.63 g cm^{-3} [7–11]; however, according to Lowe-Ma et al. the density is 1.719 g cm^{-3} (X-ray) [5, 12]. Bagal reports the density of DDNP to be 1.71 g cm^{-3} and states that the value 1.63 g cm^{-3} is for the technical product. The bulk density of DDNP is 0.5–0.9 g cm^{-3} and only 0.27 g cm^{-3} for DDNP in fine powder form [4]. Heat of formation of DDNP is 321 kJ mol^{-1} [13]. It melts at 157–158°C with decomposition; explodes violently at higher temperatures [9, 14, 15].

DDNP is mostly reported as a very slightly hygroscopic compound [11, 16]. Špičák and Šimeček [10] reported 1 % weight gain by storing in 70 % relative humidity (RH) for 40 days; Ficheroulle and Kovache [14] a weight gain of 2.25 % in 100 % RH compared to 1.22 % for LS; TM9-1300-214 reports 0.04 % in 90 % RH at 30 °C [15].

DDNP is slightly soluble in water (0.08 g in 100 ml at 25 °C) [15] but the explosive properties are not affected as the material does not show any signs of reaction at ordinary temperatures. The storage under water does not affect the color or sensitivity of later isolated DDNP [16] and it shows "unimpaired brisance for 24 months at ordinary temperatures or 12 months at 50 °C" when stored under water [15].

The solubility of DDNP in various solvents is shown in Table 6.1. DDNP is readily soluble in acetic acid, hot acetone, and nitrobenzene, less soluble in methanol, ethanol, and ethyl acetate, poorly in other usual organic solvents [4, 7, 16].

It is photosensitive and turns dark when irradiated by light. The samples exposed to sunlight show signs of decomposition after only 40 h of irradiation [13]. However, the photolysed crystal surface becomes a protective layer against further action of the light on the inner portion of the crystal [17]. DDNP is stable in cold mineral acids; it is decomposed by hot concentrated sulfuric acid and by cold diluted hydroxide solution, liberating nitrogen [13, 16]. Reaction with alkalis is

Table 6.1 Solubility of DDNP

	Solubility (g per 100 g of solvent)	
	25°C [15]	50°C [7]
Water	0.08	–
Ethyl acetate	–	2.45
Methanol	0.57	1.25
Ethanol	0.84	2.43
Trichloromethane	–	0.11
Tetrachloromethane	–	Trace
Benzene	0.09	0.23
Toluene	–	0.15
Petrol ether	–	Insol. (at 20°C)
Ethyl ether	0.04	0.08 (at 30°C)
Carbon disulfide	–	Trace (at 30°C)
Acetic acid	1.40	–
Acetone	6.0	–

used for non-explosive disposal of diazidodinitrophenol (0.5 % aqueous solution of sodium hydroxide) [13]. It also reacts with sodium azide forming the sodium salt of 2-azido-4,6-dinitrophenol and liberating nitrogen. This reaction may be used for analysis. The quantity of DDNP in the sample is determined from the volume of nitrogen produced [4].

DDNP is compatible with most usual explosives but is incompatible with LA [8, 14]. Compatibility of wet DDNP with common metals was examined by Ficheroulle and Kovache [14]. They observed that DDNP in a moist atmosphere does not corrode metals, and only attacks them very slightly. The specifications for military type of DDNP are summarized in MIL-D-82885.

Thermal stability of DDNP is higher than stability of MF but significantly lower than that of LA. This is a property which may be limiting its use for some applications [11]. DDNP does not change when stored at 65 °C and it loses 1.25 % of its weight over 96 h at 100 °C [4]. Extensive study of thermal stability of DDNP and kinetics of its decomposition was carried out by Kaiser and Ticmanis. DDNP is stable at moderate temperatures (~60 °C) for a long period; however, it quickly decomposes at temperatures of more than 100 °C [18].

6.4 Explosive Properties

Sensitivity of DDNP to impact is about the same as that for mercury fulminate [4] or slightly less [7, 19] (see Fig. 2.15). Its sensitivity to impact significantly depends on the size of DDNP particles—fine crystals of DDNP are twice as sensitive as the coarse product [3].

Its sensitivity of DDNP to friction is reported lower than that of MF and about the same as that of LA [20]; however, according to our own experiments its

6.4 Explosive Properties

Table 6.2 Detonation velocity of DDNP

Density (g cm^{-3})	Detonation velocity (m s^{-1})	References
0.9	4,400	[8]
1.5	6,600	[9]
1.6	6,900	[8]

sensitivity relatively low – it is between MF and PETN (see Fig. 2.19) [21]. Sensitivity to electric discharge and its comparison with other primary explosives is illustrated in Fig. 2.21. Upon ignition, unconfined and unpressed DDNP burns like nitrocellulose even in quantities of several grams. It does not ignite or detonate under water even when initiated by a blasting cap [4, 16, 22].

Smoleński and Pluciński reported that explosive decomposition of DDNP results mainly in the formation of carbon monoxide, carbon, hydrogen cyanide, and nitrogen. The reaction proceeds according to the following equation [13]:

$$10\ C_6H_2N_4O_5 \longrightarrow 42\ CO + 2.52\ CO_2 + 2.94\ H_2O + 3.15\ H_2 + 7.67\ C + 7.82\ HCN + 16.1\ N_2$$

The dependency of detonation velocity of DDNP on density is summarized in Table 6.2. Its initiating efficiency based on tetryl acceptor charge is better that of MF and LA [12]. The values in TM9-1300-214 [15] confirm superiority of DDNP to MF but the efficiency with respect to LA differs based on the type of acceptor as shown in Table 6.3. It indicates that DDNP is a better initiator for less-sensitive secondary explosives.

Initiating efficiency of DDNP was recently measured by Vala and Valenta. According to their results, it is about the same as of LA when measured using PETN acceptor charge when strong confinement is used (Vala and Valenta private communication).

DDNP belongs to the group of highly brisant explosives being superior to mercury fulminate and also, in some areas, to lead azide. The results of a sand test are summarized in Table 6.3 [7]. Sand-crushing strength of DDNP is not affected by storing under water for 50 days. The power of DDNP measured by a small Trauzl block is significantly superior to LA and MF (Table 6.3).

The ignition temperature is 185 °C [7, 13] or 177 °C [23] (within 5 s); 200 °C for explosion within 1 s [7]. According to our experiments, ignition temperatures of DDNP prepared by Garfield's method [24] are in the range 161–163 °C (heating rate 5 °C min^{-1}, static air atmosphere, sample weight 5 mg). As mentioned previously, sunlight affects the quality of DDNP. The ignition temperature of irradiated DDNP is somewhat lower than for a non-irradiated sample [13].

There has been some disagreement about dead-pressing of DDNP. In older literature, DDNP is reported as a primary explosive, which cannot be dead-pressed

Table 6.3 Initiating efficiency and brisance of LA, MF, and DDNP

	Initiating efficiency [15] (gram of primary explosive)			Brisance[a] [7] (gram of sand)	Brisance[b] [7] (gram of sand)	Power by small [7] Trauzl block (cm^3)
	Tetryl	TNT	Ammonium picrate			
DDNP	0.12	0.15	0.28	19.3	90.6	25.0
MF	0.19	0.24	No detonation	6.5	48.4	8.1
LA	0.10	0.26	No detonation	7.2	36.0	7.2

[a]Sand test values are for 0.2 g charge measured in detonator shell and pressed under a reinforcing capsule at 23.4 MPa
[b]Sand test values are for 1 g charge measured in detonator shell and pressed under a reinforcing capsule at 23.4 MPa

even at high pressures. The attempts to dead-press DDNP reported by Clark indicate that he could not dead-press it when using pressures up to 690 MPa. Two attempts using greater pressures (800 and 900 MPa) both ended by explosion [3, 8, 9, 11]. However, neither the densities obtained at 690 MPa nor other details have been reported. It is therefore difficult to explain why the explosions occurred.

Later experimental works clearly show that it is easily possible to dead-press DDNP. The maximum initiation efficiency is reached at densities from 1.2 to 1.3 g cm^{-3}. At density 1.4 g cm^{-3} it is completely dead-pressed. The density at which DDNP becomes dead-pressed does not depend on the specific surface of the material to be pressed as much as it does in the case of MF [25, 26]. Bagal reported that DDNP becomes dead-pressed by pressure above 19.6 MPa [4].

6.5 Preparation

DDNP is prepared by diazotation of an aqueous solution of 2-amino-4,6-dinitrophenol (picramic acid) or its sodium salt with sodium nitrite and hydrochloric acid [11].

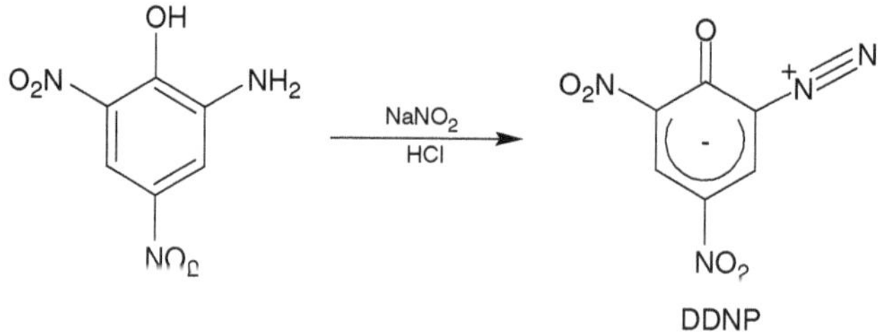

Needle- or leaf-shaped crystalline DDNP is obtained when pure reactants are used at common reaction conditions (Fig. 6.2). The addition of acid to the reaction mixture is the more common way of DDNP preparation [13, 27–29] although addition of nitrite into acidic solution of picraminate acid is also published [7, 24].

6.5 Preparation

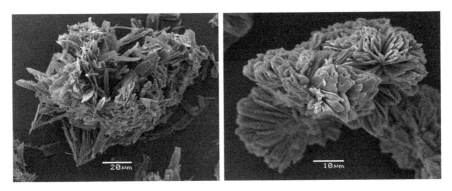

Fig. 6.2 Crystals of DDNP prepared by authors following Alexander's method [27] (*left*) and Garfield and Dreher's method [24] (*right*)—pure reactants were used at both methods; dyes and phenols were not used

The free-flowing product can be obtained by both optimization of the reaction conditions (rate of addition of hydrochloric acid to the mixture) [28] or by use of some crystal shape modifying substances during reaction. Rounded grains of DDNP are reported when pyrogallol, hydroquinone, dinitroresorcinol, and some other substances are added [27] and tabular crystals when adding triphenylmethane dyes to the reaction mixture [24]. Slow mixing of reagents is one of the most important factors for producing free-flowing DDNP according to [28, 29]. Babcock and Kenvil proposed use of an injector or atomizer for acid batching with addition of the acid under the surface of the diazoting mixture. This way acid is added in the form of fine droplets. The process eliminates formation of local high acidity regions and hence prevents the formation of very fine particles of DDNP [29].

Our own experience makes us a bit uncertain about the above-mentioned methodologies for obtaining technologically acceptable product. After numerous unsuccessful trials, we have deviated from the published routes and developed our own which enables us to produce large-size round particles (Fig. 6.3, left). Similar spherical DDNP with a little bit different structure was obtained by Nesveda (Fig. 6.3, right).

Another way of DDNP preparation is oxidation of picramic acid with chromic acid [30]. It is believed by authors that a part of picramic acid is oxidized to complete decomposition forming nitrogen dioxide and subsequently nitrous acid which then diazotizes picramic acid. Strong foaming due to a gas evolution (CO, CO_2, and NO_2) and low yield (32.5 % by weight) support proposed mechanism including partial decomposition of picramic acid.

Re-crystallization may be done from nitrobenzene or from acetone solution by adding the solution to cold ether. DDNP may be purified by dissolving in hot acetone and precipitation by addition of iced water. The purification turns the

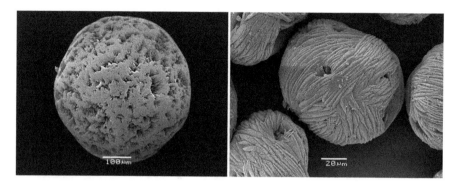

Fig. 6.3 SEM photographs of spherical DDNP prepared by authors (*left*) and the same prepared by Jiří Nesveda, Sellier and Bellot, Vlašim (*right*)

dark brown raw product a brilliant yellow [7, 19]. Several alternative ways of DDNP preparation are summarized in the literature [7].

6.6 Use

Although first prepared in 1858, DDNP was first proposed for use as an explosive in 1892 when Lenze highlighted its valuable explosive properties [13]. Probably the first patent treating DDNP as an explosive was by Dehn in 1922 [16]. DDNP found its application as an initiating explosive in both military and commercial detonators, particularly in USA and Japan [8, 26, 31, 32]. It is also used in stab and percussion primer mixtures and in these applications it has been the subject of many patents including, for example, [31–33]. Due to the absence of heavy metals in the molecule it is widely used in non-toxic types of primer in which it serves as the energizing component (e.g. SINTOX [34]). It is practically the only easily available replacement of LS in today's percussion priming mixtures [35]. A mixture containing DDNP and tetrazene has been proposed as an explosive filler for explosive riveting [36].

Despite its reported use, DDNP exhibits several disadvantages which make it not an ideal primary explosive:

- It causes allergic reactions in some workers
- It can be dead-pressed
- It has low density and therefore it requires a relatively large volume
- It is easily electrified resulting in difficulties during handling and filling [4]
- Problems of waste water treatment
- Relatively low thermal stability for some applications (Vala and Valenta private communication)

The use of DDNP in military applications was further questioned by Fronabarger and Williams pointing out its inability to fulfill shelf-life and reliability requirements [37]. On the other hand, the American technical manual TM 9-1300-214 [15] considers DDNP satisfactory for both commercial and military use.

References

1. Griess, P.: Vorläufige Notiz über die Einwirkung von salpetriger Säure auf Aminitrophenylsäure. Annalen der Chemie und Pharmacie **106**, 123–125 (1858)
2. Griess, P.: Ueber eine neue Klasse organischer Verbindungen, welche Wasserstoff durch Stickstoff vertreten enthalten. Annalen der Chemie und Pharmacie **113**, 200–217 (1860)
3. Urbański, T.: Chemie a technologie výbušin. SNTL, Praha (1959)
4. Bagal, L.I.: Khimiya i tekhnologiya iniciiruyushchikh vzryvchatykh veshchestv. Mashinostroenie, Moskva (1975)
5. Lowe-Ma, C.K., Nissar, R.A., Wilson, W.S.: Diazophenols – their structure and explosive properties, report AD-A197439. Naval Weapons Centrum, China Lake, USA (1987)
6. Holl, G., Klapötke, T.M., Polborn, K., Rienäcker, C.: Structure and bonding in 2-dizo-4,6-dinitrophenol (DDNP). Propellants, Explosives, Pyrotechnics **28**, 153–156 (2003)
7. Clark, L.V.: Diazodinitrophenol, a detonating explosive. J. Ind. Eng. Chem. **25**, 663–669 (1933)
8. Fedoroff, B.T., Sheffield O.E., Kaye S.M.: Encyclopedia of Explosives and Related Items. Picatinny Arsenal, New Jersey (1960–1983)
9. Meyer, R., Köhler, J., Homburg, A.: Explosives. Wiley-VCH Verlag GmbH, Weinheim (2002)
10. Špičák, S., Šimeček, J.: Chemie a technologie třaskavin. Vojenská technická akademie Antonína Zápotockého, Brno, Czechoslovakia (1957)
11. Danilov, J.N., Ilyusin, M.A., Tselinsky, I.V.: Promyshlennye vzryvchatye veshchestva; chast I. Iniciiruyushchie vzryvshchatye veshchestva. Sankt-Peterburgskii gosudarstvennyi tekhnologicheskii institut, Sankt-Peterburg (2001)
12. Lowe-Ma, C.K., Nissan, R.A., Wilson, W.S., Houk, K.N., Wang, X.: Structure of diazophenols: ^{13}C NMR spectroscopy, and molecular orbital studies. J. Chem. Res. (S), 214–215 (1988)
13. Urbański, T.: Chemistry and Technology of Explosives. PWN – Polish Scientific Publisher, Warszawa (1967)
14. Ficheroulle, H., Kovache, A.: Contribution à l'étude des explosifs d'amorçage. Mémorial des pouders **31**, 7–27 (1949)
15. Military explosives. Report TM 9-1300-214, Washington DC (1984)
16. Dehn, W.M.: Process of increasing the sensitiveness and power of explosive compositions and product thereof. US Patent 1,428,011, 1922
17. Yamamoto, K.: Primary explosives. VIII. Photolysis of diazodinitrophenol. Chem. Abstr. **64**, 3273d; CAN 3264:26627 (1966)
18. Kaiser, M., Ticmanis, U.: Thermal stability of diazodinitrophenol. Thermochim. Acta **250**, 137–149 (1995)
19. Tomlinson, W.R., Sheffield, O.E.: Engineering design handbook, explosive series of properties explosives of military interest. Report AMCP 706-177 (1971)
20. John, H.J., Yeager, C., Pile, D., Webb, T.: Non-toxic primer mix. US Patent 6,478,903 B1, 2002
21. Matyáš, R., Šelešovský, J., Musil, T.: Sensitivity to friction for primary explosives. J. Hazard. Mater. 213–214, 236–241 (2012)
22. Davis, T.L.: The Chemistry of Powder and Explosives. Wiley, New York (1943)
23. Henkin, H., McGill, R.: Rates of explosive decomposition of explosives. Ind. Eng. Chem. **44**, 1391–1395 (1952)
24. Garfield, F.M., Dreher, H.W.: Manufacture of diazodinitrophenol. US Patent 2,408,059, 1946

25. Strnad, J.: Primary explosives and pyrotechnics – lecture notes. Institute of Energetic Materials, University of Pardubice, Pardubice (1999)
26. Strnad, J.: Iniciační vlastnosti nejpoužívanějších třaskavin a vývoj nových metodik jejich měření. Dissertation thesis, Institute of Chemical Technology, Pardubice (1972)
27. Alexander, H.B.: Method of producing diazodinitrophanol and product thereof. US Patent 2,103,926, 1937
28. Hancock, R.S., Pritchett, L.C.: Method for producing diazodinitrophenol. US Patent 1,952,591, 1934
29. Babcock, L.W., Kenvil, N.J.: Diazodinitrophenol and the method of the preparation of the same. US Patent 2,155,579, 1939
30. Urbański, T., Szyc-Lewańska, K., Bednarczyk, M., Ejsmund, J.: On formation of 2,4-dinitro-6-diazoxide by oxidation of picramic acid. Bulletin de l'Académie polonaise des sciences Séries des sciences chimiques **8**, 587–590 (1960)
31. Kaiser, H.E.: Percussion cap. US Patent 1,852,054, 1932
32. Lopata, F.G., Mei, G.C.: Non-toxic, non-corrosive rimfire cartridge. US Patent 4,674,409, 1987
33. Burns, J.E.: Ammunition. US Patent 1,862,295, 1932
34. Hagel, R., Redecker, K.: Sintox – a new, non-toxic primer composition by Dynamit Nobel AG. Propellants, Explosives, Pyrotechnics **11**, 184–187 (1986)
35. Jung, S.M., Lee, H.S., Son, J.W., Pak, C.H.: Non-toxic primer powder compositions for small-caliber ammunition. US Patent S 2005 224147, 2005
36. Tsukii, T., Kikuchi, S.: Explosive for riveting. Chem Abst **49**, 10628 (1955)
37. Fronabarger, J.W., Williams, M.D.: Lead-free primers. US Patent 2009/0223401, 2009

Chapter 7
Salts of Benzofuroxan

7.1 Introduction

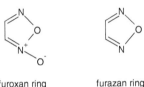

furoxan ring   furazan ring

The furoxan ring (furazan oxide; 1,2,5-oxadiazole 2-oxide) in a molecule is a sensitizing structure which produces a similar sensitizing effect as, for example, an azido group. The presence of the furoxan ring itself, however, is not sufficient to sensitize any compound to a point where it would be useful as a primary explosive. Even the sensitivity of benzotrifuroxan (BTF), representing a molecule with the highest possible number of furoxan groups present on an aromatic ring, is insufficient and places the substance among secondary explosives—between PETN and RDX to be exact [1]. Metallic salts are much more sensitive and possess the necessary properties to be considered as primary explosives. The sensitivity of furoxan salts can further be increased by incorporation of explosophores (energetic groups like nitro or azido) onto an aromatic ring. Careful variation of the number of the furoxan and energizing groups gives the organic chemist the possibility to design molecules with just the correct sensitivity necessary for a primary explosive [2].

benzotrifuroxane (BTF)   5,7-diamino-4,6-dinitrobenzofuroxan (CL-14)

On the other hand, the presence of an amino group on the aromatic ring decreases the sensitivity of the molecule. This can be illustrated by the example of 4,6-dinitrobenzofuroxan (4,6-DNBF). 7-Amino-4,6-dinitrobenzofuroxan with one amino group is significantly less sensitive than 4,6-DNBF itself and the sensitivity is further dramatically reduced by introduction of another amino group. 5,7-Diamino-4,6-dinitrobenzofuroxan (CL-14) is half as sensitive to impact as TNT [3, 4].

The same trend in sensitivity applies also to furoxan salts. The alkaline salts of CL-14 (sodium, potassium, rubidium, and cesium) are less sensitive to impact and friction than relevant salts of 4,6-DNBF due to the presence of amino groups in the aromatic ring [5]. The incorporation of amino groups also significantly increases the thermal stability of these salts (ignition in the range 240–280 °C) compared with its 4,6-DNBF analogs (ignition in the range 160–220 °C) [5]. In practical terms, the most promising for future applications are potassium salts (lithium and sodium salts are hygroscopic; rubidium and cesium are expensive).

An alternative to furoxans are furazans which have a similar molecular structure and also form metallic salts. Far less attention, however, has been paid to this class of substance. Nitrobenzofurazanes are among the rare cases where at least some information has been published. Like the nitrobenzofuroxans, they are not sufficiently sensitive, but they do form salts, which in some cases may fulfill the criteria for use in some specific applications. The most promising substances appear to be alkaline salts particularly the potassium salt of 4,6-dinitrobenzofurazan. It is, however, too insensitive for priming applications [6]. Other nitrobenzofurazanes and their salts were originally investigated as possible LA replacements, but they failed to provide sufficient performance and were therefore reconsidered as potential LS replacements. Based on the information available in open literature today, it is not possible to predict their usefulness in future applications.

7.2 Salts of 4,6-Dinitrobenzofuroxan

4,6-Dinitrobenzofuroxan (4,6-DNBF) was first prepared in 1892. The structure of the furoxan ring remained unclear for a long time. The furoxans were until the 1960s assumed to be *o*-dinitroso compounds. Only later was benzofuroxan's structure firmly established and confirmed by X-ray crystallography [7, 8].

4,6-DNBF is acidic and easily forms stable salts [9]. Based on the type of metal, 4,6-DNBF forms σ Meisenheimer complex salts [10, 11] with resonance structure corresponding to either formula I or II as shown below [12].

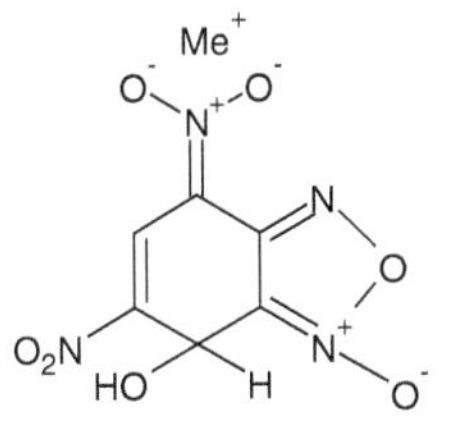

7.2 Salts of 4,6-Dinitrobenzofuroxan

structure I

Me = Li$^+$, Na$^+$, K$^+$, Mg^{2+}, Ca^{2+}, Sr^{2+}, Mn^{2+}, Co^{2+}, Ni^{2+}, Zn^{2+}, Cd^{2+}

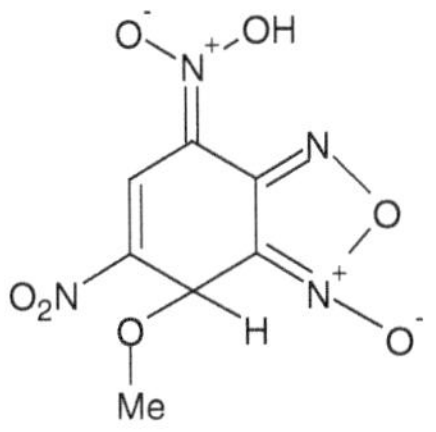

structure II

Me = Al^{3+}, Pb^{2+}, Fe^{3+}, Cr^{3+}, Cu^{2+}

Sinditskii et al. [12] performed IR analysis of several of these metallic salts and based on the results proposed two types of Meisenheimer complex structure depending on the nature of the metal cation. Both types have nitro groups in the *aci*-form, however the cations of the first group (structure I) have the metal atom bonded to the oxygen atoms of the nitro-group in the *aci*-form by an ionic type of bond. The second group comprises adducts in which the metal atom is covalently bonded to the oxygen atom of the hydroxyl group (structure II) [12]. The latter type of bond was confirmed by IR and NMR analysis for the Cr^{3+}, Fe^{3+}, and Cu^{2+} salts [13]. The IR spectrum of the barium salt does not enable ascribing it unambiguously to either one of the groups [12].

7.2.1 Physical and Chemical Properties

4,6-DNBF itself melts at 174 °C and decomposes at 273 °C. The physical properties of several metallic salts of 4,6-DNBF are summarized in Table 7.1. The potassium salt of 4,6-DNBF (KDNBF) forms small golden to orange platelets with density which have poor pouring and mixing properties. Crystal density is 2.21 g cm^{-3} [14]. KDNBF is slightly soluble in water (0.245 g/100 ml at 30 °C) [17] and the solubility increases with temperature allowing it to be re-crystallized at 70 °C [18]. Sodium salt (NaDNBF) is soluble in water [18] and its solution is used for preparation of other salts. Exact details of its solubility are however not reported.

7.2.2 Explosive Properties

4,6-Dinitrobenzofuroxan itself has the characteristics of a secondary explosive. Its metallic salts, some of which are summarized in Table 7.1, have on the other hand primary explosive characteristics.

Table 7.1 Properties of metallic salts of 4,6-dinitrobenzofuroxan

Physical properties	Potassium salt (KDNBF)	Sodium salt (NaDNBF)	Barium salt (BaDNBF)	Silver salt (AgDNBF)	Rubidium salt (RbDNBF)	Cesium salt (CsDNBF)	References
	Golden orange platelets	Red-brown crystals	Lustrous black-brown crystals	Red-brown crystals	Brick-red crystals	Buff crystals	[14] [15]
Ignition temperature (°C)	199–201	174–175	257–259	159–160	–	–	[14]
	220	160	–	–	188	166	[15]
	208–211	172–175	249–254	–	–	–	[12]
	207–210[a]						[16]
Impact sensitivity 2 kg[b] (cm)	35.0 (50 %) 29 (F of I)	33.5 (50 %) 28 (F of I)	–	–	35.0 (50 %) 29 (F of I)	30.0 (50 %) 24 (F of I)	[5, 15]
Friction sensitivity (kg)[c]	3.8	5.4	–	–	1.0	0.2	[5, 15]
Sensitivity to electric discharge (mJ)	26.5	24.0	–	–	28.9	24.5	[15]
	Fires at 4500, 450 and 45	Fires at 4500, 450 and 45	Fires at 4500 No fires at 450	Fires at 4500 No fires at 450	–	–	[14]

[a] With rapid decomposition from 190 °C
[b] Sensitivity to impact: RDX 35 cm; tetryl 85 cm; TNT 110 cm [5]
[c] Sensitivity to friction: RDX 12 kg; tetryl and TNT 32.4 kg [5]

7.2 Salts of 4,6-Dinitrobenzofuroxan

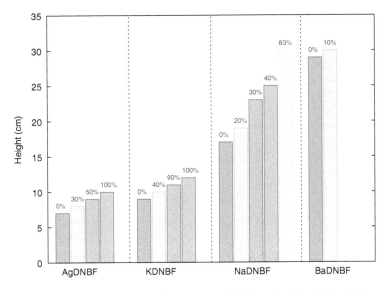

Fig. 7.1 Sensitivity of 4,6-DNBF salts to impact as a probability of ignition (ball and disc method, 6 or 10 trials at one height, maximum test height 30 cm) [14]

Table 7.2 Impact sensitivity of KDNBF

Sensitivity to impact				
Converted to energy (J)	As reported in literature	Value reported	Apparatus	Reference
0.67	15.2 cm/0.45 kg	h_{min}	Picatinny arsenal apparatus	[17]
2.17	6.1 cm/3.63 kg	h_{50}	Bureau of expl. machine	[19]
2.47	12.6 cm/2 kg	h_{50}	Drop hammer	[20]
6.9	35 cm/2 kg	h_{50}	Fall hammer	[15]
Between MF and LA		–		[16, 21, 22]

The impact sensitivity of the potassium salt is slightly lower than that of a silver salt but significantly higher than that of sodium and barium salts as demonstrated by Fig. 7.1. A discrepancy exists in the literature concerning the absolute values of impact sensitivity of KDNBF. Some of the values are summarized in Table 7.2. The sensitivity to friction is reported about the same as the sensitivity of LS [16]. According to our own results (see Fig. 2.19), KDNBF is less sensitive to friction than most other commonly used primary explosives (e.g., LA, LS) but more than DDNP [23]. Initiating efficiency is high and according to Zhang et al. it reaches 0.02 g for RDX (both pressed 52.37 MPa) [20]. Brisance of KDNBF by sand test is 92 % TNT [17].

The ignition temperature of KDNBF is 200 °C (heating rate 5 °C min^{-1}; DSC) according to Whelan et al. [24]. KDNBF does not melt and when heated it decomposes directly from the solid state. A detailed study of the thermal decomposition of KDNBF was carried out by Jones et al. [19, 25–27] and Li et al. [28].

The rubidium salt is slightly more sensitive to impact than LA (8.7 cm for 0.8 kg hammer compared to 10.3 cm for LA) and slightly more sensitive to impact than KDNBF. Sensitivity to flame is similar to KDNBF; sensitivity to electrostatic discharge is lower than LA and LS but higher than KDNBF. The exothermic decomposition starts at 189 °C (heating rate 10 °C min^{-1}) [29].

The sensitivity of the cesium salt to impact, friction, and electrostatic discharge is reported higher than for KDNBF [30]. As in the case of KDNBF, discrepancies exist in the literature about the absolute values of the impact sensitivity of the cesium salt. Mehilal et al. [15] reported 35 cm for 2 kg hammer (h_{50}) whilst Zhang et al. [30] reported 8.9 cm/0.8 kg (h_{50}). The sensitivity to flame is the same as for KDNBF [30].

Sinditskii [12] studied the burning rate characteristics of metallic salts in the pressure range 0.1–30 MPa. Based on his observations, it is possible to divide salts of 4,6-DNBF by their burning behavior into two groups which correspond well with the two previously mentioned structural groups. The substances in the first group (identical to structural group I) showed fast burning rates regardless of the metal type. With these substances, the burning rate did not depend significantly on the surrounding pressure and all of the salts began to burn at atmospheric pressure. The thermally stable KDNBF has the highest burning rate of all of the salts in the first group.

The burning rate of the salts from the second group (structure II) significantly depended on the surrounding pressure and the behavior of these salts differed only slightly from 4,6-DNBF itself. Some of the substances in this group did not burn at atmospheric pressure. The barium salt, which could not be previously placed into either of the structural groups, falls into the second group according to its burning behavior [12].

7.2.3 Preparation of 4,6-Dinitrobenzofuroxan

The most common general procedure for preparation of benzofuroxans is based on thermolysis of the benzene compound containing adjacent nitro and azido groups in the ortho position [31, 32].

4,6-Dinitrobenzofuroxan can be prepared by thermal decomposition of 1-azido-2,4,6-trinitrobenzene (prepared from 1-chlor-2,4,6-trinitrobenzene) [13, 15] or more effectively, without isolation of 1-azido-2,4,6-trinitrobenzene, directly from 1-chlor-2,4,6-trinitrobenzene [31]. The other method of 4,6-dinitrobenzofuroxan synthesis is based on nitration of benzofuroxan (prepared from o-nitroaniline) [31–33].

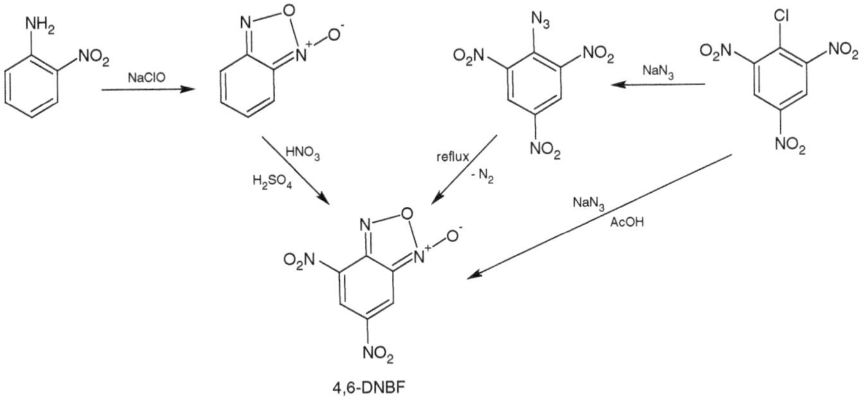

7.2 Salts of 4,6-Dinitrobenzofuroxan

7.2.4 Preparation of 4,6-Dinitrobenzofuroxan Salts

The first salts of 4,6-dinitrobenzofuroxan (potassium, sodium, ammonium, and silver) were prepared by Drost in 1899. The product of precipitation of 4,6-DNBF by potassium bicarbonate was however misidentified as the potassium salt of *m*-dinitro-*o*-dinitrosobenzene. The other salts were also thought to be salts of the same compound [18].

KDNBF is produced by neutralization of an aqueous solution of 4,6-dinitrobenzofuroxan or its sodium salt with a soluble potassium salt (bicarbonate, carbonate, nitrate, chloride) [14, 15, 17, 18, 34]. Various crystal modifiers may be used such as Tween 80 (polyoxyethylenesorbitan monooleate), PVA, dextrin, etc., for preparation of spherical KDNBF [22, 33, 35]. Temperature, stirring speed, dripping speed of reactants, and cooling time all have significant influence on crystal shape and size [22]. The product prepared at our department is shown in Fig. 7.2. It is further possible to prepare KDNBF from an acetone/water environment yielding golden platelets melting at 209 °C with explosion [36]. It is recommended to prepare KDNBF by precipitation from its aqueous solution [17, 37].

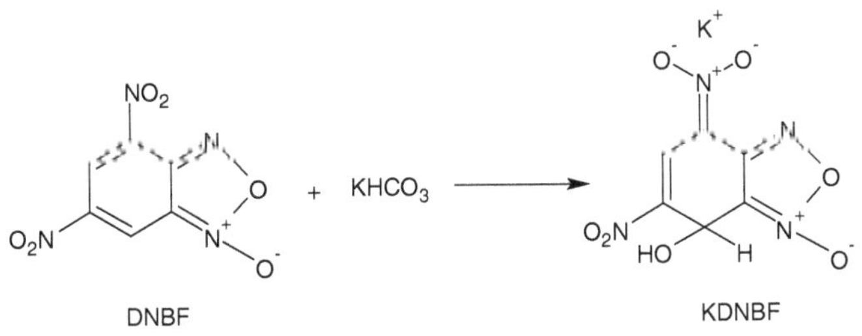

When methanol is used instead of an aqueous environment, potassium salts of methoxy derivates of dinitrobenzofuroxan or dinitrobenzofurazan (depending on reaction conditions) form instead of KDNBF [7].

Fig. 7.2 SEM photography of KDNBF

The sodium salt of 4,6-DNBF may be prepared in the same way as described above when a sodium salt (bicarbonate) is used for the reaction. Other salts of 4,6-DNBF may be prepared by precipitation of aqueous solutions of the sodium salt of 4,6-DNBF by relevant soluble metallic inorganic salts (e.g., nitrate, chloride) [13–15, 30].

One of the problems related to the preparation of 4,6-DNBF by nitration of benzofuroxan is its variable yield and the amount of side products. It was found by HPLC that up to 10 % of the isolated product may be 5,6-DNBF. This negatively influences the color and the shape of the resulting potassium salt [38]. Spear et al. however disagree with this and report that the presence of the 5,6-DNBF isomer in the nitrated product is below 1 % [31].

7.2.5 Uses

Although use of several metallic salts of 4,6-DNBF is protected by patent literature (e.g., barium salt [39]) only the potassium salt has been practically used in explosion initiating compositions. It has been reported in both military and commercial applications in the USA since the early 1950s [40]. The potassium salt has been proposed as a nontoxic replacement of lead styphnate for primers for hunting and sporting ammunition [16].

Patents usually refer to the application of KDNBF in primer composition for percussion caps. It is recommended in combination with tetrazene [41–45] or sometimes with DDNP. A mixture with the latter was reported to have good stability at wide temperature ranges and to withstand even very humid environments [46].

The use of KDNBF as the only primary explosive in priming mixtures is environmentally quite attractive due to the absence of any toxic heavy metal or DDNP. The absence of GNGT further makes the primer more stable at higher temperatures and humidity [47]. Pure KDNBF was also patented as a priming charge in detonators [37] and for use in gas generators for automobile safety devices (airbags, seat belt tensioners) [44].

7.3 Potassium Salt of 7-Hydroxylamino-4,6-dinitro-4,7-dihydrobenzofuroxan

Table 7.3 Stab sensitivity of potassium salt of 7-hydroxylamino-4,6-dinitro-4,7-dihydrobenzofuroxan/LA and GNGT/LA mixtures [48]

Composition	Ratio	Energy (50 % prob. of initiation, mJ)
Potassium 7-hydroxylamino-4,6-dinitro-4,7-dihydrobenzofuroxan/LA	1/10	6.5–7.6
	1/20	9.2
GNGT/LA	1/10	3.3
GNGT/LA	1/20	3.5

7.3 Potassium Salt of 7-Hydroxylamino-4,6-dinitro-4,7-dihydrobenzofuroxan

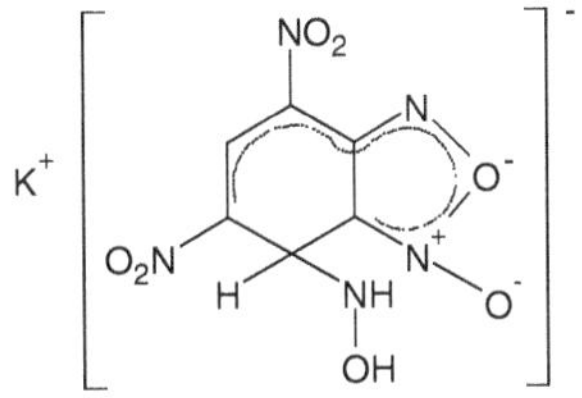

Another interesting substance containing a furoxan ring is the potassium salt of 7-hydroxylamino-4,6-dinitro-4,7-dihydrobenzofuroxan. It forms dark red crystals with density 1.92 g cm^{-3} and melting point 165 °C. It decomposes during melting with some gas being released. The ignition temperature is however reported in the same patent to be 152 °C (probably due to a different heating rate). The sensitivity to impact is 2.7 J (11 cm for 2.5 kg hammer; 50 % probability). This compound was examined as a potential sensitizer of LA in stab priming mixtures. The common mixtures of LA/GNGT were tested for comparison. The results are summarized in Table 7.3. The stab energies of mixtures of potassium salt of 7-hydroxylamino-4,6-dinitro-4,7-dihydrobenzofurxan and LA are twice of those corresponding 1/10 tetrazene compositions and three times of those of 1/20 [48].

Potassium salt of 7-hydroxylamino-4,6-dinitro-4,7-dihydrobenzofuroxan can be prepared in the same way as KDNBF. It is formed by the reaction of 4,6-dinitrobenzofuroxan with hydroxylamine in presence of a methanolic solution of potassium bicarbonate. The yield is nearly quantitative [48].

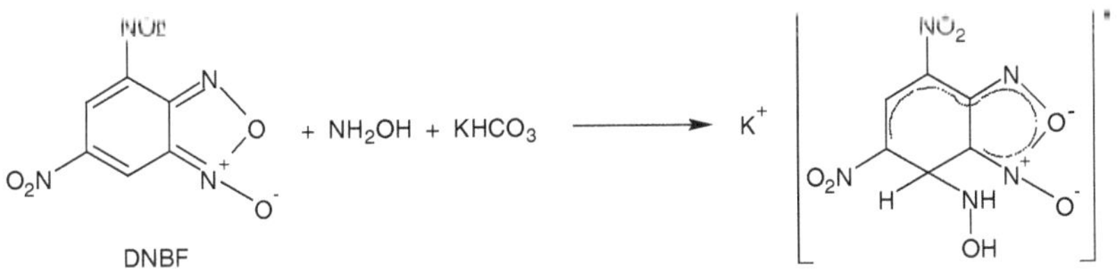

7.4 Potassium Salt of 7-Hydroxy-4,6-dinitrobenzofuroxan

$$K^+ \left[\begin{array}{c} \text{structure of 7-hydroxy-4,6-dinitrobenzofuroxan anion with } NO_2, O_2N, O^-, N^+, O, N, O^- \text{ groups} \end{array} \right]^-$$

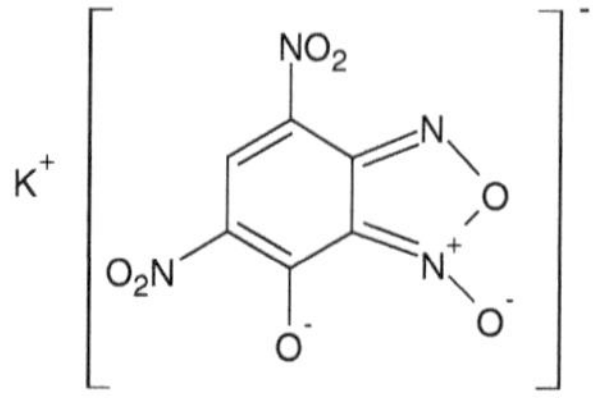

The potassium salt of 7-hydroxy-4,6-dinitrobenzofuroxan (KDNP) has a very similar chemical structure to KDNBF. However, as opposed to KDNBF which forms Meisenheimer adducts, this compound is aromatic. The aromaticity of KDNP has been confirmed by X-ray analysis [46].

7.4.1 Physical and Chemical Properties

The potassium salt (KDNP) exists in two forms. The first is a monohydrate which is prepared in aqueous solution. The second form is an anhydrous salt resulting from using a nonaqueous reaction medium [46].

The density of the anhydrous salt is 1.94–2.13 g cm^{-3}. The monohydrate forms brown crystals with density depending on crystal morphology. The potassium salt is only slightly hygroscopic compared with its more hygroscopic sodium analog [46]. KDNP is moderately soluble in water [49].

7.4.2 Explosive Properties

Sensitivity to impact and friction of the various forms of KDNP is summarized in Table 7.4. The sensitivity of crystalline KDNP to impact is equivalent to milled LS while the needles and amorphous compound are half as sensitive. The sensitivity of KDNP to friction is much lower than that of LS of similar crystal size; the needle crystals are significantly more sensitive than either the crystalline or amorphous form [46].

KDNP is more thermally stable than KDNBF; it remains stable at 120 °C for 90 days. The higher thermal stability compared with KDNBF is probably due to loss of aromaticity of the KDNBF molecule. The DSC data indicate that KDNP is roughly equivalent to LS in terms of onset and peak temperatures [46].

The comparison of KDNP's performance in terms of the ignition time was carried out in a steel closed bomb. The results were compared to results obtained for standard LS under the same conditions. Samples were loaded at 103 MPa and

Table 7.4 Sensitivity to impact, friction, and DSC of various crystal morphologies of KDNP [46]

Compound	Impact sensitivity Ball drop (mJ)	Friction sensitivity Small BAM (g)		DSC at 20 °C min^{-1} (°C)	
		No-fire level	Low fire level	Onset	Peak
KDNP—needles	51	175	200	278	283
KDNP—crystalline	25	1,300	1,400	280	284
KDNP—amorphous	51	1,400	1,500	266	275
Milled LS	25	40	50	280	305

fired in a closed bomb. The ignition time of KDNP is comparable to LS. The time from first indication of pressure increase to its maximum rise is faster for the needle and crystalline material. The amorphous form of KDNP performed worse than LS. The reason of this divergence is not known [46].

Some other results of KDNP investigations were summarized by Fronabarger et al. [49]. The report unfortunately does not specify the crystal form of the material. Nevertheless, the reported properties are as follows: impact sensitivity by ball drop method—51 ± 22 mJ, friction sensitivity by BAM 175 g (no-fire level) 200 g (low fire level), thermal properties by DSC at 20 °C min^{-1}—small endo at 145 °C, decomposition onset 278 °C, solubility in water—moderately soluble at normal temperature. Reactivity with aluminum, stainless steel, brass, and cadmium was not observed. Output measured as an impetus is reported better than LS in a closed bomb test.

7.4.3 Preparation

Two preparation procedures for KDNP are published: one by Norris et al. [50] and the second by Fronabarger et al. [46].

The starting material in the synthesis of KDNP according to Norris is commercially available 5-chlorobenzofuroxan. Nitration of this material gives 5-chloro-4,6-dinitrobenzofuroxan which is thermally unstable and spontaneously changes to 7-chloro-4,6-dinitrobenzofuroxan (undergoes a Boulton-Katritzky rearrangement [51]). The alkaline salt forms by reaction of this compound with the relevant carbonate (potassium, sodium, cesium [46]) in aqueous environment [50].

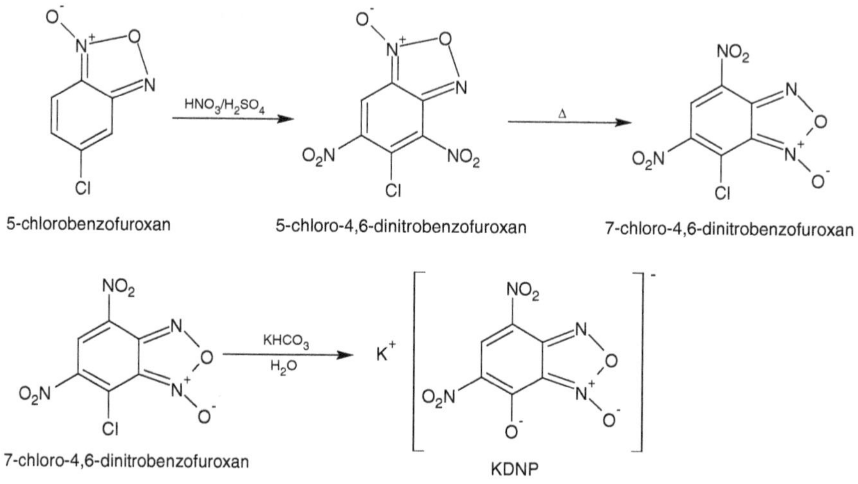

5-chlorobenzofuroxan →(HNO$_3$/H$_2$SO$_4$) 5-chloro-4,6-dinitrobenzofuroxan →(Δ) 7-chloro-4,6-dinitrobenzofuroxan

7-chloro-4,6-dinitrobenzofuroxan →(KHCO$_3$, H$_2$O) KDNP

The product of this reaction is monohydrate KDNP in the shape of brown needles. This form is not suitable for handling and loading in standard initiators due to particle morphology. The anhydride can be obtained by re-crystallization of monohydrate from 2-methoxyethanol by slow addition of isopropanol. Re-crystallized KDNP gives a fine, amorphous material more suitable for initiator loading [46].

The above reaction resembles previously mentioned preparation of Meisenheimer adducts by reaction of benzofuroxan with potassium carbonate in aqueous solution. It is interesting that using 7-chlorofuroxan instead of furoxan as a starting material, under otherwise similar conditions, leads to a nucleophilic aromatic substitution and formation of fully aromatic phenol salts. The same is true for furazans (Read, Personal Communication).

An alternative synthetic procedure for preparation of KDNP was recently proposed by Fronabarger et al. [46]. The starting compound is readily available 3-bromoanisole, which is nitrated with nitric acid/oleum in ethylene dichloride forming 3-bromo-2,4,6-trinitroanisole [52]. The bromine in the molecule is substituted for the azido group (reaction with sodium or preferably with potassium azide in methanol solution) giving furoxan ring by following reflux. Exchanging methanol for the higher boiling point diethyl carbonate allows reduction of the reaction time. This synthetic route provides KDNP in its anhydrous form [46].

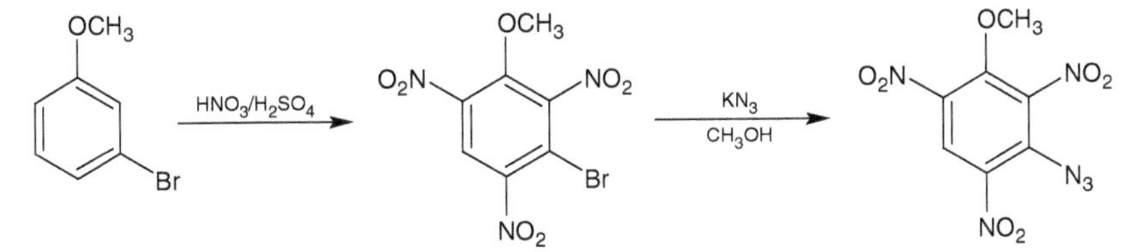

7.5 Salts of Bis(furoxano)-2-nitrophenol

[Reaction scheme: 3-bromoanisole → 3-bromo-2,4,6-trinitroanisole (HNO₃/H₂SO₄) → 3-azido-2,4,6-trinitroanisole (KN₃, CH₃OH) → KDNP (ΔT, CH₃OH or diethyl carbonate)]

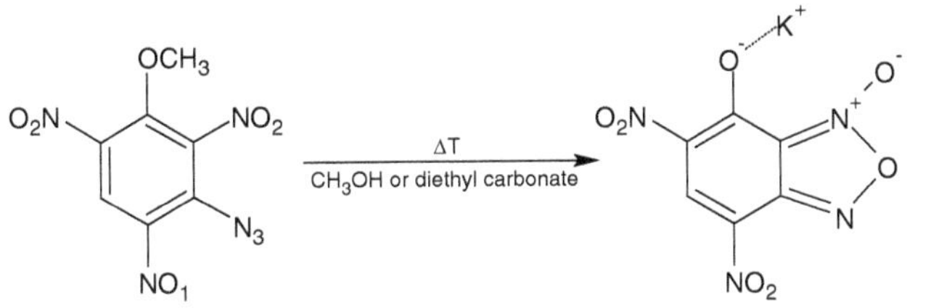

The sodium salt can be prepared in the same way as the potassium salt. It appears in an anhydrous form (as opposed to the Norris method [50]), which is, however, highly hygroscopic and rapidly absorbs water on standing (18 % water uptake on standing at 25 °C for 5 h in humid conditions 92 % RH). The potassium salt did not demonstrate a tendency to adsorb water even under these conditions [46].

7.4.4 Uses

The potassium salt of 7-hydroxy-4,6-dinitrobenzofuroxan is proposed as a low toxicity LS replacement with application in hot-wire igniter systems [51, 53].

7.5 Salts of Bis(furoxano)-2-nitrophenol

[Structures I and II of Me⁺ salts of bis(furoxano)-2-nitrophenol]

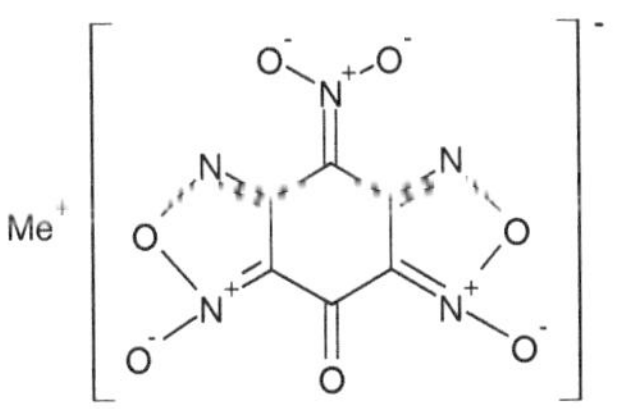

structure I 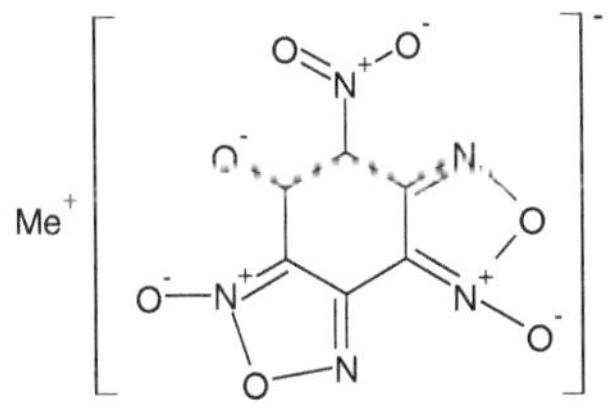   structure II

The salts of bis(furoxano)-2-nitrophenol (BFNP, hydroxynitrobenzodifuroxan) have been recently reported by Fronabarger et al. [54] and Sitzmann et al. [55]

emerging from a program focused on development of drop in replacements of LS and LA.

An interesting discovery was made in the molecular structure of the potassium salt of BFNP (KBFNP). The crystal structure analysis showed that two isomers co-crystallized together (structure I and II). Relative occupancy for structure I and structure II was determined to be 51:49. It is interesting that other salts (rubidium and guanidinium) were found to exist in discrete forms. To make it even more interesting, the rubidium salt exists in *"trans"* form (structure I, carbonyl and nitro group in para position) while the guanidinium salt exists in *"cis"* form (structure II, oxy and nitro groups in para to each other). NMR spectra indicate that all of the salts exist as mixtures of both isomers in solution [2, 54–56].

7.5.1 Physical and Chemical Properties

Alkaline salts of bis(furoxano)-2-nitrophenol are red-brown while the guanidinium salt is light yellow. Fronabarger et al. assume that the red-brown color is due to structure I. It is supported by the fact that when yellow crystals of the guanidinium salt (type II structure) are dissolved in dimethyl sulfoxide the color of the solution is red-brown, the same as for the other salts. The content of both forms of the guanidinium salt in solution was proved by NMR analysis. Removing solvent leads to a light yellow residue [2].

Density of the potassium salt is 2.106 g cm^{-3} [55] and it explodes at about 165 °C on heating. The sodium salt forms a monohydrate and is highly water soluble [2, 55]. The rubidium salt has properties similar to the potassium salt; it forms red-brown crystals with density 2.399 g cm^{-3} and ignition temperature 200 °C by DSC at 20 °C min^{-1} [2, 56].

7.5.2 Explosive Properties

The comparison of the explosive properties of KBFNP with LA and LS are summarized in Table 7.5. KBFNP is relatively highly impact and friction-sensitive, exceeding LS. The ignition time and time to peak pressure of KBFNP were examined in a closed bomb and compared with KDNBF. It was observed that KBFNP has a much shorter ignition time and a much shorter time to peak pressure and a substantially higher impetus relative to LS (ignited with ZPP—zirconium/potassium perchlorate mixture). In another study, ignition of ZPP by KBFNP and KDNBF was investigated. Slightly faster ignition and significantly faster pressure rise times to peak pressure were observed for KBFNP than for KDNBF (thin film bridges, capacitor discharge firing mode) [57, 58].

7.5 Salts of Bis(furoxano)-2-nitrophenol

Table 7.5 Sensitivity and DSC of KBFNP compared with LA and LS [53, 55]

Compound	Impact sensitivity Ball drop (mJ)	Friction sensitivity Small BAM (g)		DSC at 20 °C min^{-1} (°C)	
		No-fire level	Low fire level	Onset	Peak
KDNBF	9 ± 2	20	30	203	209
LA (RD1333)	–	–	10	332	341
LS	25 ± 1	40	50	290	305

KBFNP performance in a witness plate test is almost imperceptible in comparison with LA (0.8 mm dent in aluminum block vs. 37.3 mm for LA; both compounds pressed at 10 MPa) [53, 55].

KBFNP lacks good long-term thermal stability at higher temperatures (weight loss at 120 °C was about 34.7 % within 90 days), however it seems good at temperatures below 100 °C at which the weight loss was only 0.6 % after 150 h. The thermal stability of other alkaline, ammonium, and guanidinium salts is lower than that of the potassium salt (DSC onset was found to be 189 °C for Rb, 194 °C for Cs, 149 °C for Na salt) [2, 54–56].

7.5.3 Preparation

The preparation of bis(furoxano)-2-nitrophenol salts is a four-step procedure developed by Fronabarger and his collaborators. It is based on formation of a diazidotrinitrobenzene intermediate containing a hydrolysable group. This intermediate gives bis(furoxano)-2-nitrophenol by hydrolysis [54].

3,5-Dichloroanisole is suggested as a starting material thanks to its commercial availability and due to a presence of a readily hydrolyzed methoxy group. 3,5-Dichloroanisole is nitrated in the first step to 3,5-dichloro-2,4,6-trinitroanisole. The procedure was patented by Ott and Benziger and gives nearly a quantitative yield [57, 58]. 3,5-Dichloro-2,4,6-trinitroanisole is then converted to 3,5-diazido-2,4,6-trinitroanisole by reaction with sodium azide in water/dimethylcarbonate under phase transfer conditions. The intermediate 3,5-diazido-2,4,6-trinitroanisole is thermally labile and can be easily converted to bis(furoxano)nitroanisole by reflux in toluene. The methoxy group is readily hydrolyzed under mild conditions to form the desired bis(furoxano)nitrophenol [2, 54–56].

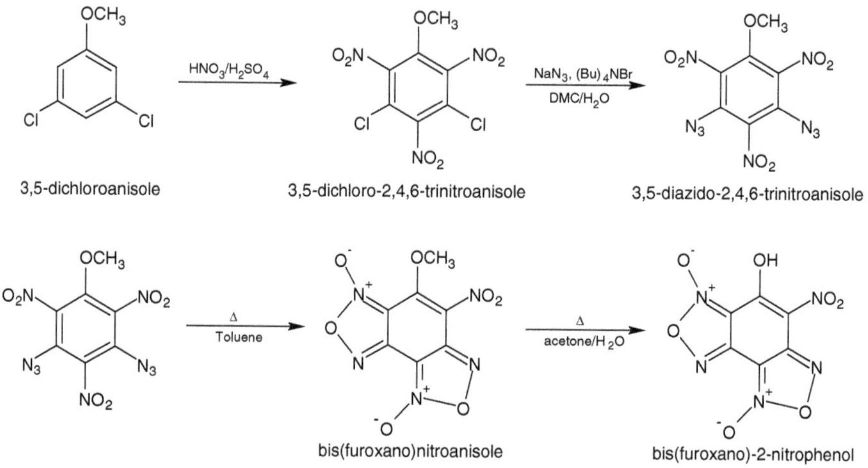

[Scheme showing synthesis: 3,5-dichloroanisole → (HNO₃/H₂SO₄) → 3,5-dichloro-2,4,6-trinitroanisole → (NaN₃, (Bu)₄NBr, DMC/H₂O) → 3,5-diazido-2,4,6-trinitroanisole → (Δ, Toluene) → bis(furoxano)nitroanisole → (Δ, acetone/H₂O) → bis(furoxano)-2-nitrophenol]

The salts of bis(furoxano)-2-nitrophenol can be prepared by reaction of bis(furoxano)nitrophenol with the relevant carbonate, acetate, or hydroxide in aqueous solution [54–56].

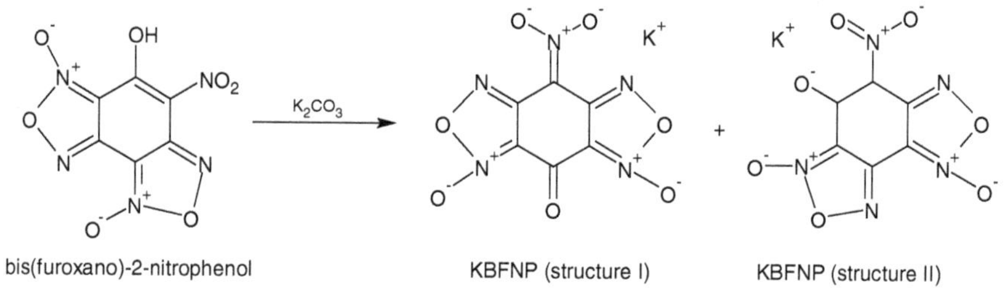

[Scheme: bis(furoxano)-2-nitrophenol + K₂CO₃ → KBFNP (structure I) + KBFNP (structure II)]

The salts can be also prepared by metal exchange from the potassium salt (e.g., cesium salt from cesium iodide and KBFNP in water) [2].

7.5.4 Uses

Alkaline salts of bis(furoxano)-2-nitrophenol (primarily KBFNP) were proposed as heavy metal-free primary explosives. Even the most promising candidate—KBFNP—exhibits low thermal stability at temperatures over 100 °C and even more importantly it has a very low performance in comparison with LA. As an unsuitable LA replacement, it may find some use in igniting compositions replacing LS. KBFNP has been also investigated for use in actuators or micro-propulsion systems [54].

References

1. McGuire, R.R.: The Properties of Benzotrifuroxan. Report UCRL-52353. California University and Lawrence Livermore Laboratory, Livermore (1978)
2. Sitzmann, M.E., Bichay, M., Fronabarger, J.W., Williams, M.D., Sanborn, W.B., Gilardi, R.: Hydroxinitrobenzodifuroxan and its salts. J. Heterocyc. Chem. **42**, 1117–1125 (2005)
3. Doherty, R.M., Simpson, R.L.: A comparative evaluation of several insensitive high explosives. In: Proceedings of 32nd International Annual Conference of ICT, vol. 28, pp. 32-31–32-33, Karlsruhe, Germany (1997)
4. Norris, W.P.: Synthesis of 7-amino-4,6-dinitrobenzofuroxan. US Patent H476, 1988
5. Mehilal, S.N., Chougule, S.K., Sikder, A.K., Gandhe, B.R.: Synthesis, characterization, and thermal and explosive properties of alkali metal salts of 5,7-diamino-4,6-dinitrobenzofuroxan (CL-14). J. Energetic Mater. **22**, 117–126 (2004)
6. Millar, R.W.: Lead Free Initiator Materials for Small Electro explosive Devices for Medium Caliber Munitions: Final Report 04 June 2003. Report QinetiQ/FST/CR032702/1.1, QuinetiQ, Farnborough, 2003
7. Norris, W., Spear, R.J., Read, R.W.: Explosive Meisenheimer complexes formed by addition of nucleophilic reagents to 4,6-dinitrobenzofurazan 1-oxide. Aust. J. Chem. **36**, 297–309 (1983)
8. Prout, C.K., Hodder, O.J.R., Viterbo, D.: The crystal and molecular structure of 4,6-dinitrobenzfuroxan. Acta Crystallogr. B **28**, 1523–1526 (1972)
9. Green, A.G., Rowe, F.M.: Nitro-derivates of isoOxadizole oxides and of isoOxadozoles. J. Chem. Soc. **113**, 67–74 (1918)
10. Boulton, A.J., Clifford, D.P.: Two explosive compounds: the potassium salt of 4,6-dinitrobenxofuroxan, and 3,4-dimethyl-4-(3,4-dimethyl-5-isoxazolylaxo)isoxazolin-5-one. J. Chem. Soc. 5414–5416 (1965)
11. Norris, W.P., Osmundsen, J.: 4,6-Dinitrobenzofuroxan. I. Covalent hydration. J. Org. Chem. **30**, 2407–2409 (1965)
12. Sinditskii, V.P., Egorshev, V.Y., Serushkin, V.V., Margolin, A.V., Dong, H.W.: Study on combustion of metal-derivatives of 4,6-dinitrobenzofuroxan. In: Proceedings of 4th International Autumn Seminar on Propellants, Explosives and Pyrotechnics, pp. 69–77, Shaoxing, 2001
13. Shinde, P.D., Mehilal, R.B.S., Agrawal, J.P.: Some transition metal salts of 4,6-dinitrobenzofuroxan: Synthesis, characterization and evaluation of their properties. Propellants Explosives Pyrotechnics **28**, 77–82 (2003)
14. Spear, R.J., Norris, W.P.: Structure and properties of the potassium hydroxide-dinitrobenzofuroxan adduct (KDNBF) and related explosive salts. Propellants Explosives Pyrotechnics **8**, 85–88 (1983)
15. Mehilal, S.A.K., Pawar, S., Sikder, N.: Synthesis, characterization, thermal and explosive properties of 4,6-dinitrobenzofuroxan salts. J. Hazard. Mater. **90**, 221–227 (2002)
16. Danilov, J.N., Ilyusin, M.A., Tselinskii, I.V.: Promyshlennye vzryvchatye veshchestva; chast I. Iniciiruyushchie vzryvshchatye veshchestva. Sankt-Peterburgskii gosudarstvennyi tekhnologicheskii institut, Sankt-Peterburg (2001)
17. Tomlinson, W.R., Sheffield, O.E.: Engineering Design Handbook, Explosive Series of Properties Explosives of Military Interest. Report AMCP 706 177, 1971
18. Drost, P.: Ueber Nitroderivates des o-Dinitrosobenzols. Justus Liebigs Annalen der Chemie **307**, 49–69 (1899)
19. Jones, D.E.G., Lightfoot, P.D., Fouchard, R.C., Kwok, Q., Turcotte, A.-M., Ridley, W.: Hazard characterization of KDNBF using a variety of different techniques. Thermochim. Acta **384**, 57–69 (2002)
20. Zhang, Y., Sheng, D., Ma, F., Zhu, Y., Chen, L., Yang, B.: New primary explosive bis-furoxano-nitrophenol potassium salt. Hanneng Cailiao **15**, 600–603 (2007)

21. Fedoroff, B.T., Sheffield, O.E., Kaye, S.M.: Encyclopedia of Explosives and Related Items. Picatinny Arsenal, New Jersey (1960–1983)
22. Li, Y., Zhang, T., Miao, Y., Zhang, J.: A new way to synthesize spherical KDNBF. Huozhayao Xuebao **26**, 53–56 (2003)
23. Matyáš, R., Šelešovský, J., Musil, T.: Sensitivity to friction for primary explosives. J. Hazard. Mater. **213–214**, 236–241 (2012)
24. Whelan, D.J., Spear, R.J., Read, R.W.: The thermal decomposition of some primary explosives as studied by differential scanning calorimetry. Thermochim. Acta **80**, 149–163 (1984)
25. Jones, D.E.G., Feng, H.T., Fouchard, R.C.: In: 26th International Pyrotechnics Seminar, pp. 195–202, Nanjing, 1999
26. Jones, D.E.G., Feng, H.T., Fouchard, R.C.: Kinetic studies of the thermal decomposition of KDNBF, a primer for explosives. J. Therm. Anal. Calorim. **60**, 917–926 (2000)
27. Jones, D.E.G., Lightfoot, P.D., Fouchard, R.C., Kwok, Q., Turcotte, A.-M.: In: Proceedings of the 28th NATAS Annual Conference on Thermal Analysis and Applications, pp. 308–313, Orlando, 2000
28. Li, Y., Zhang, T., Zhang, J.: Thermal decomposition processes and non-isothermal kinetics of KDNBF. Hanneng Cailiao **12**, 203–206 (2004)
29. Wang, S., Zhang, T., Sun, Y., Zhang, J.: Synthesis, structural analysis and sensitivity properties of RbDNBF. Hanneng Cailiao **13**, 371–374 (2005)
30. Wang, S., Zhang, T., Yang, L., Zhang, J., Sun, Y.: Synthesis, thermal decomposition and sensitivity study of CsDNBF. Propellants Explosives Pyrotechnics **32**, 16–19 (2007)
31. Spear, R.J., Norris, W.P., Read, R.W.: 4,6-Dinitrobenzofuroxan: An Important Explosive Intermediate. Report MRL-TN-470. Department of Defence Material Research Laboratories, Melbourne (1983)
32. Green, A.G., Rowe, F.M.: The conversion of orthonitroamines into isooxadiazole oxides (furoxans). J. Chem. Soc. **103**, 2023–2029 (1913)
33. Gaughran, R.J., Picard, J.P., Kaufman, J.V.R.: Contribution to the chemistry of benzofuroxan and benzfurazan derivates. J. Am. Chem. Soc. **76**, 2233–2236 (1954)
34. Piechowicz, T.: Procédé de préparation d'un explosif d'amorçage. FR Patent 1,519,799, 1968
35. Brown, N.E., Keyes, R.K.: Structure of salts of 4,6-dinitrobenzofuroxan. J. Org. Chem. **30**, 2452–2454 (1965)
36. McGirr, R.: Ignition composition for detonators and squibs. US Patent 3,135,636, 1964
37. McGuchan, R.: Improvements in primary explosive compositions and their manufacture. In: Proceedings of 10th Symposium on Explosives and Pyrotechnics, San Francisco, 1979
38. Piechowicz, T.: 4,6-Dinitrobenzofurazan 1-oxide barium salt explosive primer. FR Patent 1,522,297, 1969
39. Meyer, R., Köhler, J., Homburg, A.: Explosives. Wiley-VCH, Weinheim (2002)
40. Bjerke, R.K., Ward, J.P., Ells, D.O., Kees, K.P.: Primer composition. US Patent 4,963,201, 1990
41. Duguet, J.R.: Priming charge with annular percussion and process for its manufacture. US Patent 5,353,707, 1994
42. Scott, H.A.: Priming composition. WO Patent 9,914,171, 1999
43. Pile, D.A., Jonn Jr. Henry J.: Bismuth oxide primer composition. US Patent 2005 189053, 2005
44. Mei, G.C., Pickett, J.W.: Lead-free nontoxic priming mix. EP Patent 1,440,958, 2004
45. Fronabarger, J.W., Williams, M.D., Sanborn, W.B., Sitzmann, M.E., Bichay, M.: Preparation, characterization and output testing of salts of 7-hydroxy-4,6-dinitrobenzofuroxan. SAFE J. **35**, 14–18 (2007)
46. Carter, G.B.: Composition pour appret primer composition. CA Patent 2,156,974, 1996
47. Norris, W.P.: Primary explosive. US Patent 4,529,801, 1985
48. Fronabarger, J., Sanborn, W.B., Bichay, M.: An investigation of some alternatives to lead based primary explosive. In: Proceedings of 37th AIAA/ASME/SAE/ASEE (2001-3633); Joint Propulsion Conference and Exhibit, pp. 1–9, Salt Lake City, Utah, 2001
49. Norris, W.P., Chafin, A., Spear, R.J., Read, R.W.: Synthesis and thermal rearrangement of 5-chloro-4,6-dinitrobenzofuroxan. Heterocycles **22**, 271–274 (1984)

50. Boulton, A.J., Katritzky, A.R.: N-oxide and related compounds. Part XXII. The rearrangement of 4-nitrobenzofuroxans to 7-nitrobenzofuroxans. Revue de chimie **7**, 691–697 (1962)
51. Hein, D.W., Radkowski, S.J.: Process for preparing 3,3'-diamino-2,2',4,4',6,6'-hexanitrobiphenyl. US Patent 3,402,202, 1968
52. Bichay, M., Hirlinger, J.: New Primary Explosives Development for Medium Caliber Stab Detonators. Report SERDP PP-1364, p. 18, US Army ARDEC, 2004
53. Fronabarger, J., Williams, M., Sanborn, W.B.: Characterization and output testing of the novel primary explosive, bis(furoxano)nitrophenol, potassium salt. In: Proceedings of 41st AIAA/ASME/SAE/ASEE Joint Propulsion Conference and Exhibit, pp. 1–6, Tucson, AZ, 2005
54. Sitzmann, M., Bichay, M., Fronabarger, J., Williams, M., Sanborn, W., Gilardi, R.: Preparation, characterization and output testing of the novel primary explosive, bis(furoxano)nitrophenol, potassium salt. In: Proceedings of 31st International Pyrotechnics Seminar, pp. 729–734, Fort Collins, CO, 2004
55. Fronabarger, J., Sitzmann, M.: Nitrobenzodifuroxan compounds, including their salts, and methods thereof. US Patent 7,271,267, 2007
56. Ott, D.G., Benziger, T.M.: Preparation of 1,3,5-triamino-2,4,6-trinitrobenzene from 3,5-dichloranisole. J. Energetic Mater. **5**, 343–354 (1987)
57. Ott, D.G., Benziger, T.M.: Preparation of 1,3,5-triamino-2,4,6-trinitrobenzene from 3,5-dichloranisole. US Patent 4,952,733, 1990

Chapter 8
Tetrazoles

Tetrazoles are chemical compounds characterized by a doubly unsaturated five-membered ring containing four nitrogen atoms and one carbon atom. The tetrazole ring usually exists in two tautomeric forms, 1*H* and 2*H*:

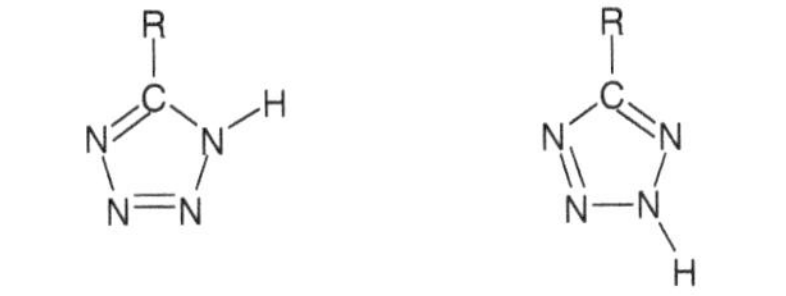

5-R-1*H*-Tetrazole 5-R-2*H*-Tetrazole

where *R* represents any substituent. The nitrogen atoms of its structure make tetrazole a strong acid. Many tetrazole derivatives are also acids and often yield explosive salts.

Unsubstituted tetrazole (1*H*- or 2*H*-tetrazole sometimes referred as "free tetrazole") forms colorless crystals with density 1.632 g cm^{-3} (X-ray) which melt at 155.5 °C without decomposition according to Fedoroff, Shefield, and Kaye's encyclopedia [1] or at 157–158 °C according to Bagal [2]. Tetrazole is easily soluble in water, ethanol, acetone, and acetic acid [2]. Heat of formation is 236 kJ mol^{-1} [3]. Free tetrazole does not have the characteristics of a primary explosive. Tetrazole easily forms metallic salts owing to the acidic nature of its hydrogen atom [2]. A large number of its derivatives or derivative's salts do have explosive properties and many fall into the category of primary explosives.

Explosive properties of tetrazoles and their salts vary considerably, ranging from nonexplosive behavior to being extremely sensitive and powerful explosives. In general terms, the 5-substituted tetrazoles and their salts may be ranked in terms of their explosive behavior in the following order [4]:

R. Matyáš and J. Pachman, *Primary Explosives*,
DOI 10.1007/978-3-642-28436-6_8, © Springer-Verlag Berlin Heidelberg 2013

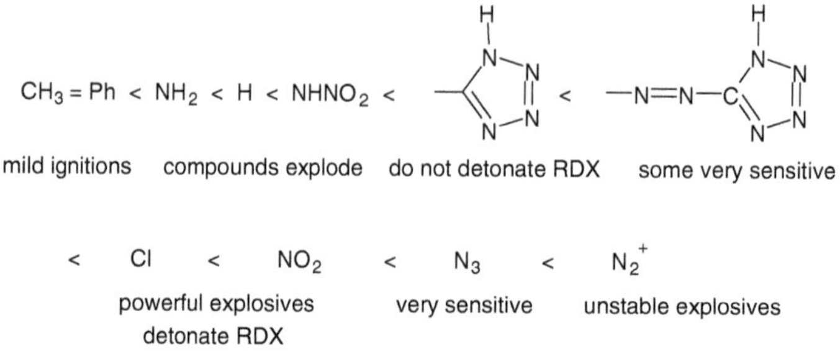

CH₃ = Ph < NH₂ < H < NHNO₂ < [tetrazole structure] < —N=N—C[tetrazole structure]

mild ignitions   compounds explode   do not detonate RDX   some very sensitive

< Cl < NO₂ < N₃ < N₂⁺

powerful explosives   very sensitive   unstable explosives
detonate RDX

This behavior is probably related to the electron-withdrawing power of the substituent in position 5 of the ring. Tetrazoles with substituents having higher electron-withdrawing power exhibit more explosive behavior [4] and also seem to be more sensitive [5]. In the case of metal salts, the nature of the metal ion also influences the explosive behavior. Heavy metals, including silver, lead and mercury, give more sensitive substances with greater initiating efficiency than alkali metal salts [4]. 5-Substituted tetrazoles are more stable than 1- and 2-substituted ones [5].

Haskins [6] also compared the empirical ranking of explosive behavior proposed by Bates and Jenkins [4] with one devised according to the total charge in the substituent group. Extended Hückel molecular orbital (EHMO) calculations provided a quantitative measure of electron-withdrawing power of the substituents in the following order:

$$CH_3 < H < NH_2 < NHNO_2 < N_3 < NO_2$$

Haskins [6] suggested that the EHMO calculations, although not in agreement with the qualitative ordering, were in good enough agreement to indicate a relationship between explosive behavior and electron withdrawal power from the ring.

Chen et al. [5] performed theoretical calculations (PM3 MO) on the thermolysis of 5-substituted tetrazole derivatives and their metal salts. They compared the observed sensitivity and the activation energies of the thermolysis and found that the two follow a similar order. The order of values of activation energies is as follows:

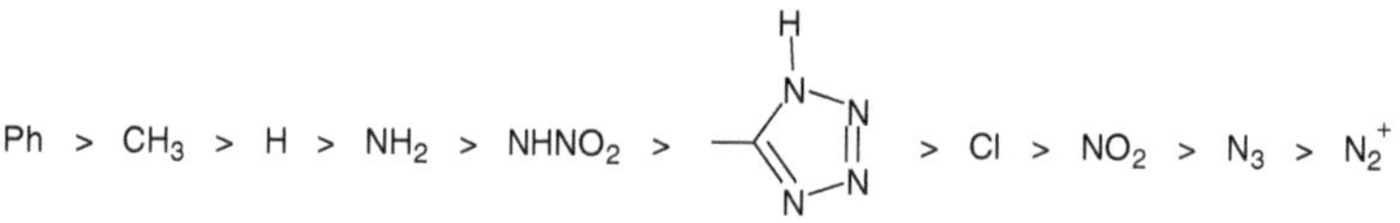

Ph > CH₃ > H > NH₂ > NHNO₂ > [tetrazole structure] > Cl > NO₂ > N₃ > N₂⁺

The sensitivities of these derivatives are in the reverse order [5].

A number of metal salts of 5-(*N*-nitramino)tetrazoles have a short predetonation zone (fast DDT). Silver salts of 1-(*N*-nitramino)- (I), 2-(*N*-nitramino)- (II), 5-(*N*-nitramino)- (III), and 1-methyl-5-(*N*-nitramino)-tetrazoles (IV) were prepared to study the effect of the position of the substituting group on the initiating efficiency.

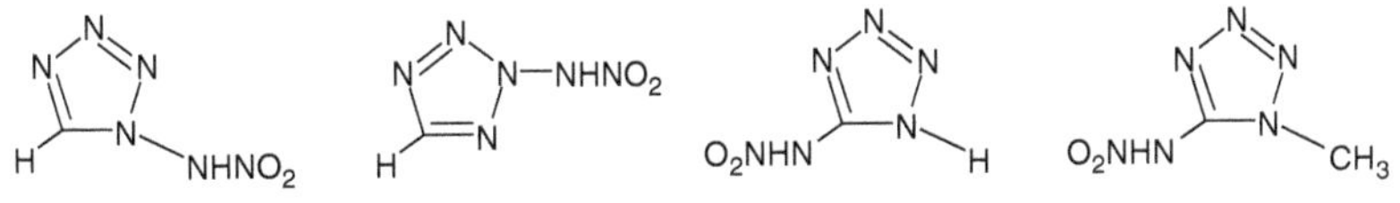

8.1 Tetrazene

[Structures I, II, III, IV of tetrazole derivatives with NHNO₂ substituents]

I II III IV

The silver salts of (I) to (IV) are white crystalline substances, sparingly soluble in water and organic solvents, but soluble in aqueous ammonia. The initiating efficiency was determined as the smallest amount of the salt needed for initiation of RDX in tubing of a number 8 detonator. The details of the procedure are reported by Avanesov [7]. The experimental values of the minimal amount increases in the following order:

$$\text{Ag(II):Ag(I):Ag(III):Ag(IV)} = 1:5:60:120$$

The salts of 1- and 2-(*N*-nitramino)tetrazole have much better initiating efficiency than their 5-(*N*-nitramino)tetrazole analogs. The silver salt of 2-(*N*-nitramino)tetrazole (II) is a powerful initiating explosive with efficiency higher than LA [8].

Tetrazoles, their salts, and their complexes have been studied as replacements for lead-containing substances such as lead azide or lead styphnate in detonators and priming mixtures. Despite the existence of a vast variety of such compounds, only a few have found practical application in primer mixes or as a primary explosive (tetrazene, the mercury salt of 5-nitrotetrazole, CP, BNCP). Many tetrazole derivatives often suggested for use in gas generators with application in automobile safety devices (airbags and seat belt tensioners) are subject to patent protection. Due to the large number of possible tetrazole compounds, we have decided to cover only a small group of selected candidates. Our intention has been to cover those that have already been applied as primary explosives or, to a lesser extent, those most often discussed.

8.1 Tetrazene

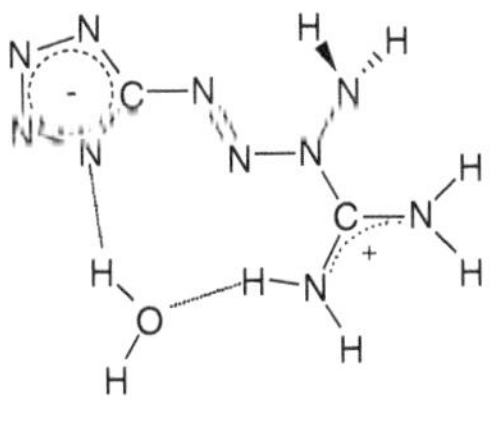

[Structure of tetrazene monohydrate]

1-Amino-1-(tetrazol-5-yldiazenyl)guanidin monohydrate known as tetrazene (GNGT, in earlier literature referred to as tetracene) is without question the most

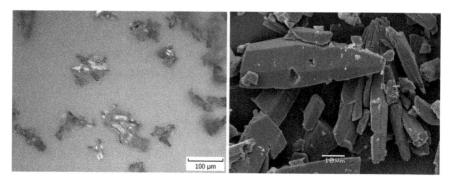

Fig. 8.1 Crystals of GNGT—*left* optical microscopy (by kind permission of Dr. Šelešovský), *right* SEM

common tetrazole substance used as a primary explosive. In fact it was the first tetrazole compound which found a use as an explosive. GNGT was first prepared by Hoffman and Roth in 1910 by reaction of aminoguanidine with nitrous acid [9]. Hoffman and his co-workers assigned tetrazene a linear structure on the basis of its chemical reactions (historical formula I). Their suggested structure was later disproved and substituted by a structure with the tetrazole ring (historical formula II) [10]. On the basis of X-ray analysis, it was, however, determined that the molecular structure of tetrazene is represented as the zwitterion with the hydrogen-bonded water molecule as shown above [11].

8.1.1 Physical and Chemical Properties

Tetrazene crystallizes as a monohydrate in the form of a fluffy solid made up of colorless or pale yellow crystals [12–15] (see Fig. 8.1) with crystal density 1.61–1.65 g cm^{-3} [15, 16] or 1.7 g cm^{-3} [13, 16]. The bulk density of GNGT is very low, only 0.45 g cm^{-3} according to Rinkenbach and Burton [15], Meyer's encyclopedia reports only 0.3 g cm^{-3} [16].

Tetrazene is practically insoluble in water and most common organic solvents (alcohol, acetone, ether, benzene, ethylene dichloride). It is slightly hygroscopic as it absorbs 0.77% water (at 30 °C; 90% relative humidity). In boiling water, GNGT decomposes [14, 16, 17]. GNGT does not react with concentrated ammonia, metals (steel, copper, aluminum) or high explosives (TNT, tetryl, PETN, RDX) at room

8.1 Tetrazene

temperature [2, 14]; sodium and potassium hydroxide provoke chemical decomposition producing ammonia [18], whilst carbon dioxide decomposes GNGT only in the presence of water [14].

GNGT does not react with diluted acids (10%) but in concentrated acids it dissolves, forming explosive salts (nitrates, perchlorates, sulfates, etc.) [2, 12, 13, 15]. Tetrazene may be precipitated from these salts by dilution [13] or by neutralization with ammonia or sodium acetate [11, 16]. One disadvantage of these salts is their tendency to undergo hydrolysis in a humid environment [14]. All of these salts also react with salts of heavy metals forming double salts. For example, $C_2H_7N_{10}OAg \cdot AgNO_3 \cdot 3H_2O$ (which does not undergo hydrolysis in humid air) may be precipitated from a slightly acidified solution of GNGT by addition of excess aqueous $AgNO_3$ [13, 14]. Tetrazene decomposes by the action of bases producing the relevant salt of 5-azidotetrazole [17].

It is stable at normal temperatures but its thermal stability decreases as the temperature increases. In presence of moisture, degradation occurs even at temperatures as low as 60 °C. The decomposition of GNGT in boiling water is practically quantitative leading to a variety of products including 5-azidotetrazole which is a very sensitive primary explosive. It is therefore not recommended to decompose larger amounts of GNGT in boiling water. Some explosions of decomposition solutions have been reported after cooling down the water with what was believed to be innocent decomposition products [14].

The poor thermal stability (it explodes at 135–140 °C [15]) along with its tendency to become easily dead-pressed are some of its drawbacks.

8.1.2 Explosive Properties

Impact sensitivity of GNGT is reported to be (a) about the same as MF [19]; (b) slightly more sensitive than MF [14, 15, 17, 18]; or (c) slightly less sensitive than MF [1, 20]. Values of impact sensitivity reported by various authors are summarized in Fig. 2.15. Sensitivity to friction is about the same as that of MF according to our own measurement (see Fig. 2.19) [21]. The sensitivity to electrostatic discharge was found to be 2.7 mJ and, as can be seen from Fig. 2.21, is comparable to that of DDNP. GNGT is often mixed with hard inert particles to increase its sensitivity especially when used in percussion and stab priming mixtures.

GNGT detonates more easily when uncompressed and undergoes "gradual dead-pressing" when subjected to higher compacting pressures. This is clearly demonstrated by the gradual decrease of apparent brisance (measured by the amount of crushed sand) with increasing compacting pressure (see Table 8.1). Pressing GNGT with 20.7 MPa results in material with a density of 1.05 g cm^{-3}. Increasing the compacting pressure above 30 MPa results in dead-pressed material [15, 17].

In addition, tetrazene shows a variation in the "order of detonation" depending on the method of initiation. This is reflected in a lower brisance when initiated by

Table 8.1 Brisance of GNGT as a function of compacting pressure [15]

Amount of GNGT (g)	Loading pressure (MPa)	Crushed sand (g)
0.4	0	13.1
0.4	250	9.2
0.4	500	7.5
0.4	3,000	2.0

flame or small amounts of other primary explosives, and a higher brisance when initiated by the severe shock of a large priming charge. A wide range of detonation parameters is therefore obtainable for GNGT by varying the compacting pressure and the type of initiation [15].

GNGT is easily ignited by flame and when small enough amount is used it deflagrates without observable flame.

The initiating efficiency of GNGT is rather poor even when the conditions are combined in such way that gives maximum brisance. It is capable of initiating tetryl but not TNT [15].

8.1.3 Preparation

Tetrazene is produced by the reaction of an aminoguanidine salt (aminoguanidine is not soluble in water; the most frequently used are sulfate, carbonate, or nitrate) with sodium nitrite [2, 13, 14, 17, 22]. The reaction of aminoguanidine with sodium nitrite can produce three different compounds depending on the environment [23]:

- Azidoformamidine (formerly guanyl azide) in a solution of a strong mineral acid
- Tetrazene in a slightly acidic environment
- 1,3-Di-1H-(tetrazol-5-yl)triazene in a solution of acetic acid

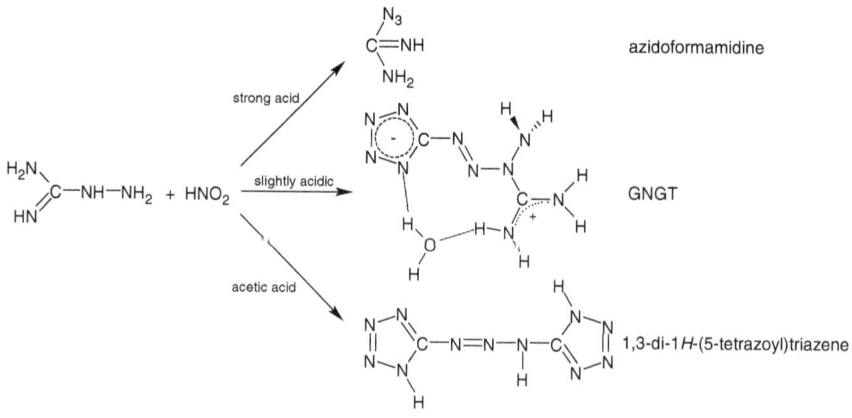

The preparation of GNGT is carried out in a slightly acidic environment; the solution of aminoguanidine is acidified before the reaction (pH ~4–5; nitric acid or

8.1 Tetrazene

acetic acid are often used). The solutions of aminoguanidine salt and sodium nitrite are preheated to 50–60 °C and the reaction temperature is kept at 50–55 °C. Tetrazene gradually precipitates from the solution. The dosage rate affects the crystal size of the product. Preparation of GNGT can be done using the same technology as that used for LA or LS production [2, 13, 14, 17, 22]. Dextrin is sometimes used for obtaining more uniform crystals [22]. A pure white product may be obtained by recrystallization from nitric acid [14].

If the reaction temperature is high (about 65 °C or more) and the pH of the reaction mixture too low, azidoformamidine forms instead of GNGT. By neutralization of the reaction mixture, azidoformamidine undergoes transformation to 5-aminotetrazole [17].

GNGT residues or any product of inappropriate crystal size can be chemically decomposed by boiling in water [17]. The risk associated with such operation was described earlier in this chapter (physical and chemical properties of GNGT).

8.1.4 Uses

Tetrazene is classified as an initiating explosive but its own initiating efficiency is low. It is capable of initiating only unpressed PETN and tetryl [12, 15]. It can be used in detonators when initiated by another primary explosive or in a mixture with another primary explosive. Addition of even small amounts (ca. 0.5–5%) of GNGT to other explosives dramatically increases the sensitivity to mechanical stimuli and flame of the resulting mixture. GNGT therefore more often acts as an energetic sensitizer in many compositions—particularly in primers [24] and various fuse mixtures. It is used in applications where high sensitivity to friction, impact, stab, flame, and spark is demanded. Mixtures of tetrazene with lead styphnate have replaced earlier mixtures based on MF in primers. These mixtures are less toxic (do not contain mercury), generally noncorrosive and, in particular, do not corrode firearm barrels [1, 18]. Tetrazene is also used in nontoxic lead-free primer compositions (alone or in combination with DDNP and other substances) [24–26]. The content of GNGT in explosive mixtures generally does not exceed 10% [17].

Tetrazene is a heavy metal-free primary explosive that could be used in many applications. Its main drawbacks are its poor thermal stability, low initiating efficiency, and the tendency toward dead-pressing [27]. The use of GNGT is further limited by its low flame temperature which is a problem if it is used in priming mixtures. One possible solution to this problem (employed in NONTOX priming mixture) is the addition of pyrotechnic components that increase the flame temperature sufficiently to reliably ignite the gun powder [26].

8.2 5-Aminotetrazole Salts

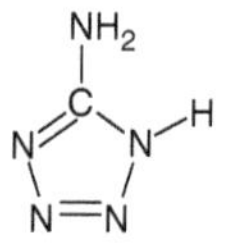

5-ATZ

5-Amino-1H-tetrazole (subsequently just 5-ATZ) was first synthesized by Johannes Thiele in 1892. He obtained this compound by cyclization of azidoformamidine in a boiling aqueous solution [28, 29].

8.2.1 Physical and Chemical Properties

Free 5-ATZ contains the acidic tetrazole ring and the basic amino group. Its chemical behavior is similar to that of amino acids. Free 5-ATZ forms a monohydrate that loses water of crystallization above 100 °C. The melting temperature of 5-ATZ is 203 °C [2]. This compound is very hygroscopic, soluble in hot water, acetone, and ethanol [1, 2].

Most metallic salts of 5-ATZ are soluble in water as well (except the silver, cupric, and mercurous salts). Many metallic salts form various hydrates [1]. The sodium salt forms yellow crystals, crystallizes with three molecules of water of crystallization and is soluble in water [2]. Some physical properties of heavy metallic salts are summarized in Table 8.2 [30].

8.2 5-Aminotetrazole Salts

Table 8.2 Physical properties of some metallic salts of 5-aminotetrazole [30]

	Color	Comment
AgATZ	Colorless	Microcrystalline precipitate
Cu(ATZ)$_2$·H$_2$O	Green	Precipitated from acetone
Pb(ATZ)$_2$	Colorless	Precipitated with acetone from water
Co(ATZ)$_2$·xH$_2$O	Pink	Precipitated from acetone
Ni(ATZ)$_2$·H$_2$O	Blue	Precipitated from acetone

Table 8.3 Sensitivity and ignition properties of some metallic salts of 5-aminotetrazole

	Impact sensitivity; 2.5 kg [31] (cm)	Ignition temperature; hot bar [31] (°C)	Ignition temperature; heating rate 5 °C min^{-1} [30](°C)
AgATZ	22	366	352 (smoke)
HgATZ	38	256 (expl)	–
Cu(ATZ)$_2$	68	256	164 (flame)
Pb(ATZ)$_2$	–	–	303 (flame)
Co(ATZ)$_2$·xH$_2$O	–	–	228 (expl.)
Ni(ATZ)$_2$·H$_2$O	–	–	290 (expl.)

8.2.2 Explosive Properties

Most metallic salts of 5-ATZ are not typical primary explosives. Similarly to free 5-ATZ, the sodium salt is not sensitive to impact [2] and even the sensitivity to impact of heavy metal salts is relatively low in comparison with common primary explosives (see Table 8.3).

Bates and Jenkins reported that all salts of 5-ATZ mentioned in Table 8.3 show irregular burning behavior with occasional explosions. They are considered as "substances with little promise of being technically useful" [30].

8.2.3 Preparation

5-ATZ was first successfully prepared by Thiele by diazotization of aminoguanidine nitrate. Aminoguanidine salts (nitrate, carbonate, sulfate) are used for synthesis because free aminoguanidine has low stability. The intermediate azidoformamidine is sufficiently stable to be isolated. The cyclization of azidoformamidine by heating with sodium acetate, sodium carbonate, diluted acid, or an aqueous solution of ammonia yields 5-ATZ [28, 29, 32–34].

Another method was published by Stollé [35, 36]. 5-ATZ is prepared by treating cyanamide, or the more readily available dicyandiamide, with hydrazoic acid [32, 36]. Alternatively, hydrazoic acid can be substituted by its sodium salt in presence of hydrochloric acid (hydrazoic acid forms in situ) [28].

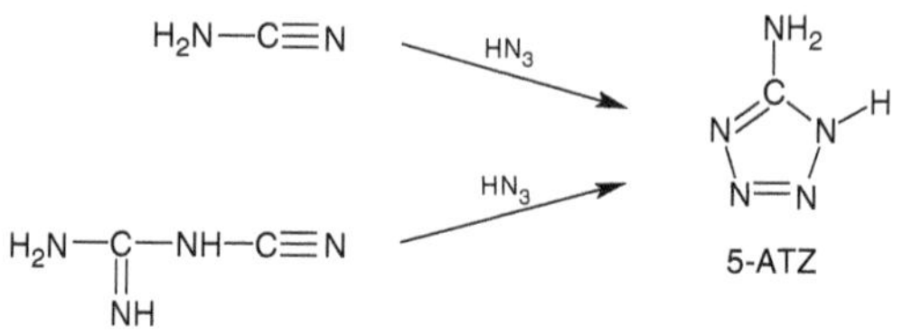

Metallic salts of 5-ATZ can be prepared by treating an aqueous solution of 5-ATZ (or its sodium salt) with the relevant soluble salt [29, 37]. In case of salts of first row transition elements, it is recommended to carry out the precipitation in acetone [30].

The type of anion of the metallic salt influences the purity of the resulting 5-ATZ salt. Using nitrate, sulfate, or carbonate results in a product contaminated by relevant anion. These ions cannot be removed from the product by washing with water [29, 37]. Daugherty, however, reported that contaminated product forms when using sulfate and chloride, but that nitrate and perchlorate yield a pure product. In the case of the latter two, the product precipitates only after standing for a week or more (in case of $Cu(ATZ)_2$) [37]. Bates and Jenkins reported that salts of 5-ATZ free from this unwanted anion can be prepared by repeated precipitation (precipitation of aqueous solution of salt of 5-ATZ with acetone) and centrifuging out the precipitate [30].

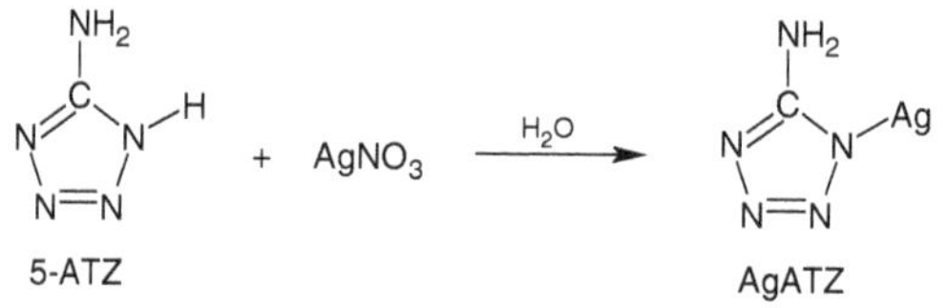

The reaction of 5-ATZ with silver salts may, in certain conditions, lead to complex rather than normal salts. The only metallic complex salt of 5-amino-1H-tetrazole that has been reported as suitable as a primary explosive is the complex disilver 5-amino-1H-tetrazolium perchlorate $Ag_2(ATZ)ClO_4$. It was first reported by Charles Rittenhouse in 1972 [38]. The analogous nitrate complex has been published recently [39]. Both complexes are insoluble in water and other common solvents. They are only soluble in concentrated acid solutions (e.g., hydrochloric, nitric, perchloric acid) forming the corresponding salts [39].

The disilver 5-amino-1H-tetrazolium perchlorate has comparable explosive properties to those of lead azide—it is highly thermally stable and easily initiated by standard bridge-wire initiators [38]. The impact sensitivity is 2 J and 15 J for its nitrate analogue. The friction sensitivity is reported below 5 N for the perchlorate and 18 N for the nitrate. Relatively high thermal stability has been observed for both

8.3 5-Nitrotetrazole Salts

substances. The perchlorate decomposed at 319 °C and the nitrate at 298 °C when heated in DSC with heating rate 5 °C min^{-1} [39].

The complex disilver 5-amino-1H-tetrazolium perchlorate and nitrate are formed by the reaction of an aqueous solution of silver perchlorate with 5-ATZ in presence of perchloric acid [38].

$$5\text{-ATZ} + \text{AgClO}_4 \xrightarrow{\text{HClO}_4} \text{Ag}_2(5\text{-ATZ})\text{ClO}_4$$

Preparation of a nitrate derivative may be carried out in a similar way [39].

8.2.4 Uses

Free 5-ATZ, its simple or complex salts and its other derivatives are not used as standalone primary explosives. They are however often mentioned in patent literature as suitable replacements for azide-containing pyrotechnic mixtures (due to the high nitrogen content in the molecule). These mixtures are suggested in gas-generating compositions for inflation of vehicle airbags, seat belt pretensioners, and for other similar devices. Free 5-ATZ is also used as a starting material for the preparation of other tetrazole derivatives.

8.3 5-Nitrotetrazole Salts

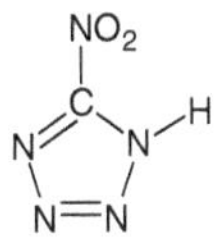

HNT

Free 5-nitro-1H-tetrazole (subsequently just 5-nitrotetrazole; HNT) is an extremely sensitive substance that explodes with the slightest stimulus or even spontaneously [40–42]. However, its metallic salts are stable and heavy metallic salts have the characteristics of primary explosives.

8.3.1 Physical and Chemical Properties

Free 5-nitrotetrazole forms colorless, extremely hygroscopic, crystals which, if exposed to humid air, dissolve into a simple solution (the material shows strong deliquescence). Melting point of HNT is 101 °C; enthalpy of formation is

Table 8.4 The properties of metallic salts of 5-nitrotetrazole

Compound	Physical appearance	Solubility
$Cu(NT)_2 \cdot HNT \cdot 4H_2O$	Light blue needles [1]	Slightly soluble in water [44]
$Cu(en)_2(NT)_2$[a]	Violet crystals [4]	Slightly soluble in cold water, readily soluble in hot water [4]
$Hg(NT)_2$	Heavy granular crystals [45]	Poor solubility in water (0.09 g in 100 ml at 30 °C [46]), soluble in ammonium acetate solution [1]
AgNT	White fluffy needles [45]	Practically insoluble in water [1]
$Cu(NT)_2$	Blue crystals [47]	Slightly soluble in water [48]
CuNT	Brown crystals, density 2.81 g cm^{-3} [49, 50]	–
$Co(NT)_2$, $Ni(NT)_2$	Reddish-white, microcrystalline [45]	Practically insoluble in water [45], $Ni(NT)_2$ forms intractable gelatinous precipitate [4]
$Pb(NT)_2 \cdot Pb(OH)_2$	Pale yellow crystals [45]	Soluble in hot water, slightly in cold water [1]

[a]en - Ethylenediamine.

Fig. 8.2 Crystals of the cupric salt of 5-nitrotetrazole [47]. Reprinted by permission of the Department of the Navy employees, M. Bichay, K. Armstrong, R. J. Cramer and the Pacific Scientific employess, J. Fronabarger and M. D. Williams

261 kJ mol^{-1}. It is strongly acidic ($pK_a = -0.8$) [43]. Some properties of heavy metallic salts of 5-nitrotetrazole are summarized in Table 8.4.

Metallic salts are mostly soluble in water (e.g., alkaline, salts of alkaline earths, cadmium, and thallium). The silver, mercury, cobalt, and nickel salts are insoluble in cold water; cobalt and cupric salts give a dihydrate [4]. Nitrotetrazole salts have good chemical stability and are not sensitive to carbon dioxide nor influenced by moisture. They may be stored without deterioration in hot and humid climates [1]. The cupric salt of 5-nitrotetrazole is not hygroscopic [48]. The crystals of this salt are shown in Fig. 8.2.

Table 8.5 Properties of alkali salts of 5-nitrotetrazole [51]

	Li	Na	K	Rb	Cs
Water molecules in crystal	5	2	0	0	0
Crystal system	Monoclinic	Triclinic	Monoclinic	Monoclinic	Monoclinic
Density (g cm^{-3})	1.609	1.731	2.027	2.489	2.986

Recently reported properties of alkali salts of 5-nitrotetrazole are summarized in Table 8.5 [51]. The sodium salt is not stable; it forms various hydrates with one to four molecules of water of crystallization [52] in the form of colorless crystals highly soluble in water. Ferrous 5-nitrotetrazole forms a dihydrate (even after drying in vacuo over P_2O_5) [53].

Silver salt of 5-nitrotetrazole is known to exist in six polymorphic forms which is the major disadvantage of this compound [4], whereas the mercuric salt does not exhibit polymorphism [54]. The very wide range of crystal structures of $Hg(NT)_2$ has been reported [4]. The big drawback of mercury 5-nitrotetrazole is its incompatibility with aluminum [42].

Compatibility of the silver salt of 5-nitrotetrazole has been investigated by Blay and Rapley [55]. They exposed a thin layer of AgNT on a metal surface at 50–60 °C to humid conditions for up to 8 weeks. They observed that it is incompatible with aluminum, the metal being corroded at humidity levels as low as 75% (significantly more than for SA). Copper and brass were less affected, and only at 100% relative humidity, and no corrosion of stainless steel was observed under the conditions of the test. The silver salt of 5-nitrotetrazole is compatible with HMX, RDX, and the lead salt of dinitroresorcine. However, when a mixture of AgNT with tetrazene was stored at high humidity, both tetrazene and AgNT rapidly decomposed, but a similar mixture stored at low humidity showed no decomposition. The silver salt of 5-nitrotetrazole is more resistant to rubbers and plastics than SA. Spectrophotometric determination of the silver salt of 5-nitrotetrazole has also been published by Blay and Rapley [55].

8.3.2 Explosive Properties

Free 5-nitrotetrazole is mostly reported as an extremely sensitive substance. Despite this fact, Koldobskii et al. experimentally determined its detonation velocity as 8,900 m s^{-1} at density 1.73 g cm^{-3} [43].

Sensitivity of the alkaline salts basically depends on the amount of water of crystallization in the molecule. Lithium and sodium salts form hydrates and the sensitivity of both to impact and friction is therefore low. However, the potassium, rubidium and cesium salts form anhydrous compounds with sensitivity significantly higher, approaching that for standard primary explosives [51].

The sensitivity to impact of the heavy metallic salts of 5-nitrotetrazole, and their initiating efficiency to tetryl are summarized in Table 8.6. The ammonium salt has

Table 8.6 Impact sensitivity, ignition temperature and initiating efficiency against tetryl for certain tetrazole salts

Compound	Impact sensitivity		Initiating efficiency (against tetryl)	Ignition temperature	
	As reported	Converted to energy (J)	(g)	(°C)	References
NaNT·2–4H$_2$O	insens. (compl. dry sensitive)	–	–	–	[1]
Cu(NT)$_2$·HNT·4H$_2$O	5 cm/5 kg^a	2.5	–	–	[1]
	–	–	–	237	[52]
[Cu(en)$_2$](NT)$_2$	20 cm/5 kg	9.8	–	–	[1]
	50 cm/2.5 kg (burns)	12	–	–	[46]
	–	–	–	204	[52]
Hg(NT)$_2$	7 cm/2.5 kg	1.7	–	–	[46]
	Same as MF		0.006	~215	[45]
	Significantly more than LA	–	–	202–235^b	[56]
	7 cm/2.5 kg	1.7		210, 215	[1]
	5 cm/5 kg^c	2.5			
	–		–	268	[2]
AgNT	Little more than MF		0.005	230	[45]
	22.7 cm^f	–	–	259, 262	[52]
	~LA		0.004^d	–	[2]
	More than LA		–	273^e	[57]
Co(NT)$_2$, Ni(NT)$_2$	~MF		–	220	[45]
Pb(NT)$_2$·Pb(OH)	Little more than MF		0.02	220	[45]
LA (for comparison)	–		0.02	–	[45]
	19 cm (F of I 29)^f		–	319	[52]

^{a}>60 cm/5 kg for wet product.
^{b}Range for various batches.
^{c}Same value for wet substance.
^{d}Pressed at 15.7 MPa.
^{e}Decomposition temperature (DSC, heating rate 5 °C min^{-1}).
^{f}(h_{50}, Ball & Disc test).

the characteristics of a secondary explosive; it is less sensitive to impact than RDX [58]. The dihydrate of the sodium salt is reported to have relatively low sensitivity (Table 8.6). On dehydration, however, it forms the very sensitive anhydrous sodium salt. An initiation attributed to dehydration of this substance has been reported during manipulating NaNT·2H$_2$O with a wooden spatula [56].

Comparison of the sensitivity of AgNT and Cu(NT)$_2$ with their complexes with ethylenediamine (EDA) and the ammonium ligand has been published recently. According to this study, both AgNT and Cu(NT)$_2$ are more impact sensitive than lead azide; ethylenediamine complex compounds are significantly less sensitive [57].

Nitrotetrazole salts have good flammability that is not affected by moisture [45]. Most of them (even the alkaline salts) are powerful primary explosives, e.g., the cesium salt has a performance similar to that of lead azide [42].

Mercuric Salt of 5-Nitrotetrazole

The mercuric salt is sensitive to stab initiation. Larger particles are more sensitive than smaller ones. Just as in the case of LA, a small content of tetrazene significantly increases stab sensitivity of the resulting mixture. Initiation by hot wire is slightly more difficult than in the case of milled dextrinated LA and LS [59]. The variability of the explosive properties (impact and stab sensitivity, ESD, initiation temperature, plate dent test) of $Hg(NT)_2$ for different methods of preparation and batches have been examined by Redman and Spear [56]. Dead-pressing of the mercury salt depends on the way it is prepared. The pure mercury salt cannot be dead-pressed by pressure up to 276 MPa according to Bates and Jenkins [4]; or even up to 700 MPa according to Scott [59]. Mercuric salt of 5-nitrotetrazole is more powerful than LA (RD 1333—co-precipitated with carboxymethylcellulose) in a plate dent test when loaded into an electric detonator (modified MK71) [59].

Silver Salt of 5-Nitrotetrazole

The sensitivity to impact of AgNT, initiating efficiency, and ignition temperature are summarized in Table 8.6. Sensitivities of this substance have also been examined by Millar [52]. Sensitivity to friction was determined by the emery paper friction test as the pendulum velocity required to produce a 50% probability of ignition. The value for AgNT is 1.7 m s^{-1} (for LA 1.2 m s^{-1} and 1.8 m s^{-1} for LS). Electric discharge sensitivity of AgNT is presented in Table 8.7. The author tested sensitivity using two tests, the standard test and the test for sensitive explosives No. 7 (this test is used for highly sensitive explosives).

Thermal stability of AgNT has been studied by Blay and Rapley [55]. Temperature of ignition is 262 °C at a heating rate of 5 °C min^{-1}. The silver salt of 5-nitrotetrazole is thermally less stable than SA. Typically, a 20% mass loss occurs within 3 h at 240 °C (compared with 3% for SA) [55].

The silver salt is a powerful primary explosive with a performance similar to that of lead azide [42]. Unlike $Hg(NT)_2$, the silver salt can be dead-pressed even at relatively low pressures (~20 MPa) [2, 4].

Cupric Salt of 5-Nitrotetrazole

The cupric salt of 5-nitrotetrazole has been studied as a potential green primary explosive by Fronabarger, Sanborn and Bichay et al. [48]. The values of sensitivity and performance reported by these authors are summarized in Tables 8.8 and 8.9. Bagal reported an ignition temperature of 224 °C (heating rate 5 °C min^{-1}) [2].

Table 8.7 Electric discharge sensitivity of AgNT [52]

Compound	Standard test (µJ)	Test for sensitive explosives (No. 7) (µJ)
AgNT	45 ignition	2.6 ignition; not 1.9
LA[a]	–	2.5
LS[a]	–	7

[a]Service material RD 1343—LA (precipitated from sodium carboxymethylcellulose/sodium hydroxide) and RD 1303—LS.

Table 8.8 Sensitivity and DSC data of the cupric salt of 5-nitrotetrazole [48]

Compound	Impact sensitivity Ball drop (mJ)	Friction sensitivity Small BAM (g)		DSC onset (20 °C min^{-1}, argon atm.) (°C)
		No-fire level	Low fire level	
Cu(NT)$_2$[a]	57	55	65	245
Cu(NT)$_2$[b]	79	1,000	1,030	206
LA (RD 1333)	50	Not determined	10[c]	332

[a]Prepared from NaNT and the cupric salt.
[b]Prepared by hydrolysis of acidic cupric salt.
[c]Minimum weight setting; samples discharged at this level.

Table 8.9 Results of plate dent test of the cupric salt of 5-nitrotetrazole compared with LA (aluminum plate)

Loading pressure (MPa)	Cu(NT)$_2$[a] [48, 49, 60, 61]		LA[b] [47, 48]		LA[b] [60, 62]	
	Density (g cm^{-3})	Average dent in Al block (mm)	Density (g cm^{-3})	Average dent in Al block (mm)	Density (g cm^{-3})	Average dent in Al block (mm)
34.5	1.26	9.0	2.62	27.7	2.87	30.5
69	1.38	22.6	2.99	33.7	3.17	34.0
138	1.53	31.7	3.18	39.0	3.60	36.9
276	1.77	29.0	3.55	46.3	3.98	39.4

[a]Prepared from NaNT and the cupric salt.
[b]Different values for LA (RD 1333) in the literature.

Brisance of the cupric salt of 5-nitrotetrazole at various densities has been measured by the plate dent test (see results in Table 8.9). Cu(NT)$_2$ detonates even at low density but it did not match the performance achieved by LA [47, 48]. This may be because of its longer DDT compared to LA. It is interesting that the method used for synthesizing has a significant impact on the detonation parameters of Cu(NT)$_2$. The cupric salt of 5-nitrotetrazole prepared from NaNT and a cupric salt detonates in witness plate test while the same compound prepared directly by hydrolysis of an acidic cupric salt of 5-nitrotetrazole does not detonate [48].

The dibasic salt Cu(NT)$_2$·2Cu(OH)$_2$ does not have the characteristics of a primary explosive, and the ethylenediamino and propanediamino complex salts behave more like pyrotechnic compositions [4].

Table 8.10 Sensitivity and brisance of CuNT compared to LA [49, 50]

Compound	Sensitivity to impact (mJ)	Sensitivity to friction Low fire level/no fire level (g)[a]	Dent block testing Average dent (mm)[b]
CuNT	40	10/0	0.94
LA (RD 1333)	50	10/0	0.84

[a]Small-scale Julius Peters BAM tester.
[b]Aluminum block; materials pressed at 69 MPa.

Cuprous Salt of 5-Nitrotetrazole

The cuprous salt of 5-nitrotetrazole (CuNT) has been patented as a potential green primary explosive by Fronabarger et al. Sensitivity to impact is higher than that of LA and sensitivity to friction is about the same. The brisance of CuNT measured by PDT is superior to that of LA. Sensitivity and performance data are summarized in Table 8.10 [49, 50].

Thermal stability is superior to that of LA (by TGA); onset of thermal decomposition is at 322–331 °C (DSC, heating rate not reported) [50].

8.3.3 Preparation

Free 5-nitrotetrazole can be prepared by treating a warm aqueous suspension of its acid copper salt with hydrogen sulfide [1, 45, 46]. If hydrochloric acid is used instead of hydrogen sulfide, the 5-nitrotetrazole becomes unstable and explodes without any external stimulus after a period of 2–3 weeks in storage [1, 4].

$Cu(NT)_2 \cdot HNT \xrightarrow{H_2S} 3\ HNT$

Free 5-nitrotetrazole can be also synthesized by acidification of its sodium salt followed by its extraction by diethyl ether [43]. The ion-exchange column can be used as a safe and convenient method for preparing a dilute solution of free 5-nitrotetrazole from its sodium salt. This solution can be used for preparation of other metallic salts (e.g., Tl^+, Rb^+, Cs^+, K^+, Ba^{2+}, Pb^{2+}). The product is obtained after crystallization from the reaction solution. Spontaneous explosions occur during isolation of the thallium and lead salts [42].

The monobasic lead salt of 5-nitrotetrazole can be easily prepared by dripping the 5-nitrotetrazole solution from the ion-exchange column directly into a warm suspension of lead oxide or hydroxide [42].

The preparation of 5-nitrotetrazole by direct nitration of tetrazole is not possible due to the acidic nature of hydrogen bound on the tetrazole ring. 5-Nitrotetrazole is therefore synthesized by the Sandmeyer reaction of 5-ATZ (treating an acidic solution of 5-ATZ with an excess of sodium nitrite in the presence of finely divided copper or copper compounds) when the acidic copper salt forms [1, 42, 45, 46, 52, 61, 63].

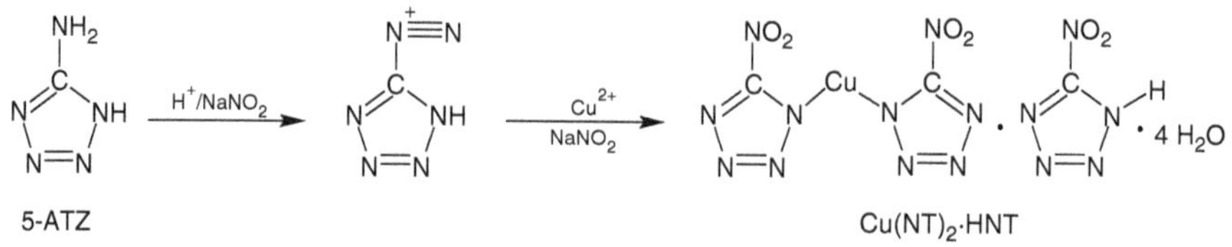

The other metallic salts of 5-nitrotetrazole can be prepared via the sodium salt according to the original von Herz procedure [45]. The sodium salt is prepared by the reaction of sodium hydroxide with acid copper 5-nitrotetrazole $Cu(NT)_2 \cdot HNT$ in an aqueous solution. The sodium 5-nitrotetrazole is isolated by crystallization [1, 46, 57, 63, 64]. The product can by purified by recrystallization from acetone [42] followed by filtration and precipitation by addition of hexane [64] or crystallized by evaporation of most of the acetone [56].

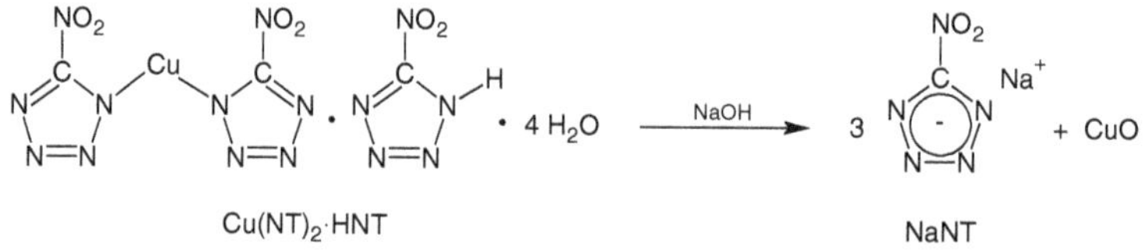

Spontaneous explosion of $NaNT \cdot 2H_2O$ during drying has been reported. The explosion was attributed to acid spray from the nearby hydrochloric solution when it is probable that the extremely unstable free 5-nitrotetrazole formed [56].

The sodium salt can also be prepared in one step from 5-ATZ without isolation of the acid copper salt [58, 65].

The original von Herz procedure (via sodium salt) was found to have several drawbacks, which are described below [46, 66]:

- Minor explosions occur during the diazotization stage (increasing frequency with increasing temperature).
- Acid copper salt is gelatinous, long periods to separate by filtration, problematic purification.
- Sodium salt forms variable hydrates; sensitive if allowed to dehydrate.
- Yield and quality of $Hg(NT)_2$ is very dependent on purity of sodium salt.
- Sodium salt must be purified by recrystallization.
- Overall yield on 5-ATZ is below 50%.

The $Hg(NT)_2$ obtained by the von Herz procedure further exhibits a tendency to be dead-pressed unlike the product obtained from recrystallization. The dead-pressing of $Hg(NT)_2$ is closely related to the quality of the sodium salt used as a starting material in the synthesis. The pure salt gives material which is more

Table 8.11 The effect of recrystallization of NaNT on properties of the Hg salt [4]

Treatment of sodium salt intermediate	Ignition temperature (at 5 °C min^{-1})	Bulk density (g cm^{-3})	Dead-pressing level (MPa)
Crude—no purification	190	0.75	103–138
Partially purified	234	0.61	241–276
Once recrystallized from acetone	242	0.88	>276
Twice recrystallized from acetone	246	0.86	>276

resistant to dead-pressing than the unpurified material as can be seen from Table 8.11 [46, 66]. Technologically it is not suitable, as the recrystallization of sodium salt is quite a time-consuming process.

Careful optimization of the process leads to elimination of the minor explosions and improved yield. The method itself has however been superseded by the diamine complex route. These complexes are synthesized by the reaction of acidic copper salt of 5-nitrotetrazole with the relevant diamine (mostly 1,2-ethylenediamine or less frequently 1,3-diaminopropane) in the presence of copper sulfate [4, 40, 41, 46, 56, 66]. These coordination compounds could be readily purified by recrystallization from water and they are safer and easier to handle in the dry state than the sodium salt (1,2-ethylenediamine complex only burns when impacted by 2.5 kg hammer from 50 cm) [42, 46].

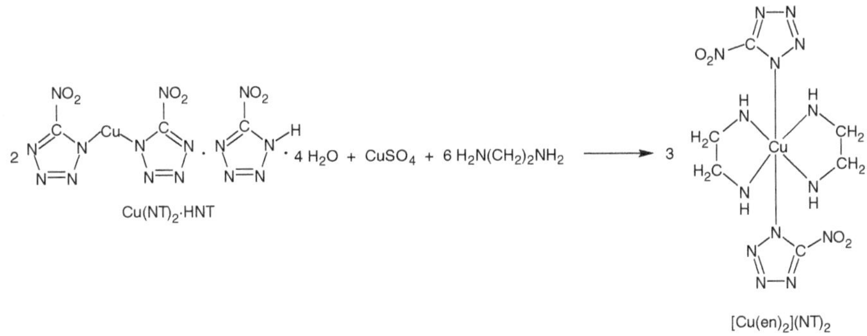

$$2 \text{ Cu(NT)}_2 \cdot \text{HNT} \cdot 4\text{H}_2\text{O} + \text{CuSO}_4 + 6\text{ H}_2\text{N(CH}_2)_2\text{NH}_2 \longrightarrow 3 \text{ [Cu(en)}_2\text{](NT)}_2$$

Other metallic salts of 5-nitrotetrazole (e.g., silver) are prepared by treating aqueous solutions of NaNT, [Cu(en)$_2$](NT)$_2$, or [Cu(pn)$_2$](NT)$_2$ with the soluble version of the relevant metallic salt [4, 41, 42, 46, 47, 53, 54, 59, 65, 67]. In the case of [Cu(en)$_2$](NT)$_2$ or [Cu(pn)$_2$](NT)$_2$ diamine must first be broken down and destroyed or made inactive to prevent complex formation with the metal. This is easily achieved by reaction with nitric acid (to tie up the diamine) followed by addition of the mercuric nitrate solution (mercuric salt forms) [4, 42, 46, 56, 66]. The several methods of Hg(NT)$_2$ analysis are summarized in [67].

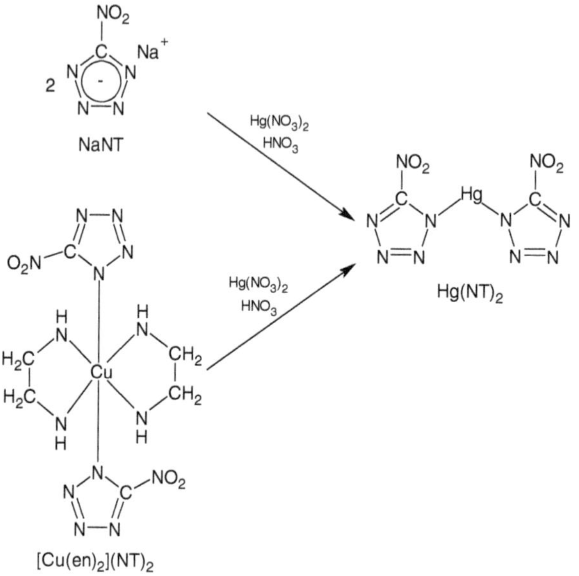

[Structural diagrams showing: 2 NaNT reacting with Hg(NO₃)₂/HNO₃ to form Hg(NT)₂; and [Cu(en)₂](NT)₂ reacting with Hg(NO₃)₂/HNO₃ to form Hg(NT)₂]

The cuprous salt of 5-nitrotetrazole is prepared by the reaction of the sodium salt of 5-nitrotetrazole with cuprous chloride suspended in boiling water (reflux) [49, 50]. The addition of acid to the reaction during preparation improves its thermal stability [49]. Recently, Fronabarger et al. improved the preparation procedure by using the more accessible cupric salt as a starting reactant which is reduced to the cuprous salt directly in the reaction mixture. Sodium ascorbate or ascorbic acid is recommended as the reducing agent [68].

The preparation of various alkaline salts via the ammonium salt of 5-nitrotetrazole has been reported recently. They are prepared by the action of the relevant hydroxide or carbonate on the ammonium salt in a methanol or ethanol solution upon reflux [51].

The preparation and analytical characterization of a number of organic salts of 5-nitrotetrazole have recently been published by Klapötke et al. However, these salts have the characteristics of secondary rather than primary explosives and are therefore not discussed in this book [57].

8.3.4 Uses

The salts of 5-nitrotetrazole have been studied as possible lead azide replacements. They were however for various reasons rejected for practical purposes [4].

The mercury salt of 5-nitrotetrazole has been developed as a less hazardous replacement for lead azide [61, 70] and mercury fulminate [18]. This substance is superior to dextrinated lead azide with lower sensitivity to impact and electrostatic discharge, and higher initiating efficiency and, in addition, it does not react with carbon dioxide and metals [62, 63, 71, 72]. The mercury salt has also been suggested as a primary explosive applied in electric detonators [69], in stab-initiated detonators with a small amount of tetrazene (5–10%) [71] or in a composition applied to the head of percussion fuses [72]. In today's environment, it has been however rejected on toxicological grounds.

The silver salt has also been suggested for practical use in percussion-initiated devices such as heads of percussion fuses [72]. This salt has recently been proposed as a potential green primary explosive that could replace LA. A potential drawback, however, of AgNT is its high sensitivity to electrostatic discharge [52].

The cupric salt of 5-nitrotetrazole has also been recently studied as a promising alternative "green" LA replacement [47, 48, 60, 62].

The cesium salt of 5-nitrotetrazole has been reported by Hagel and Bley [73] for use in electrical igniters in the automotive industry. Most other salts (e.g., barium and thallium) have been found unstable due to excessive hygroscopicity or high solubility in water [40, 41].

8.4 5-Chlorotetrazole Salts

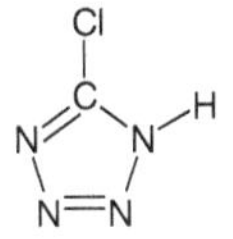

8.4.1 Physical and Chemical Properties

5-Halogentetrazoles form a variety of metallic salts including the most common, the cupric salt of 5-chlorotetrazole, which forms light blue crystals with density 2.04 g cm^{-3} [1, 2]. Its bromo counterpart exists as a green compound. Both of them are soluble in hot diluted mineral acids. Melting point of 5-chlorotetrazole is 305 °C (with decomposition); it is slightly hydroscopic as it absorbs 3.11% water (at 30 °C; 90% relative humidity) [1].

8.4.2 Explosive Properties

The cupric salt of 5-chlorotetrazole is as sensitive as its 5-bromo derivate (2.54 cm for 2 kg hammer). Brisance of the chloro-derivative is about the same as bromo-derivative. They both also have about the same initiating efficiency (0.3 g for RDX) [1]. Bagal [2] reported ignition efficiency of the chloro-derivative to be 0.1 g for tetryl and an ignition temperature of 305 °C (heating rate 5 °C min^{-1}). The substance is also sensitive to stabbing [2]. Silver 5-chlorotetrazole was found to be an effective initiating substance at low pressing loads but it gets readily dead-pressed at higher loads [42].

8.4.3 Preparation

Both 5-chloro and 5-bromotetrazoles are prepared from 5-ATZ by the Sandmeyer reaction [32]. The reaction is carried out in sufficiently diluted solutions in presence of the cupric halide and produces the cupric salt of 5-halogentetrazole. The reaction must be cooled to low temperatures, as is usual for most diazotizations. Preparation of the cupric salt of 5-chlorotetrazole has been described by several authors, e.g., Stollé [35]. If the solutions of the reacting substances are diluted too much, blue crystals of $CuCl_2 \cdot 4(5\text{-ATZ})$ are formed [74].

Optimizing the reaction conditions by inverting the order of chemical addition tends to eliminate otherwise frequent explosions. The idea of the optimization is based on the addition of nitrite to the 5-ATZ hydrochloride solution already containing $CuCl_2$. The product is formed immediately after addition of the nitrite. A carefully controlled rate of addition of the nitrite can therefore lower the amount of the diazonium salt in the mixture [74].

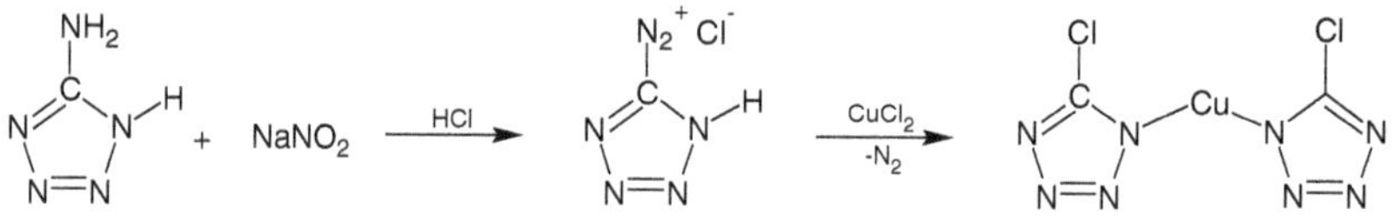

The other metallic salts of 5-halotetrazoles can be prepared via the sodium salt, which is obtained by reaction of the cupric salt of 5-halotetrazole with sodium hydroxide to give the sodium salts [42, 53]. The sodium salt can be also obtained directly from 5-ATZ [75].

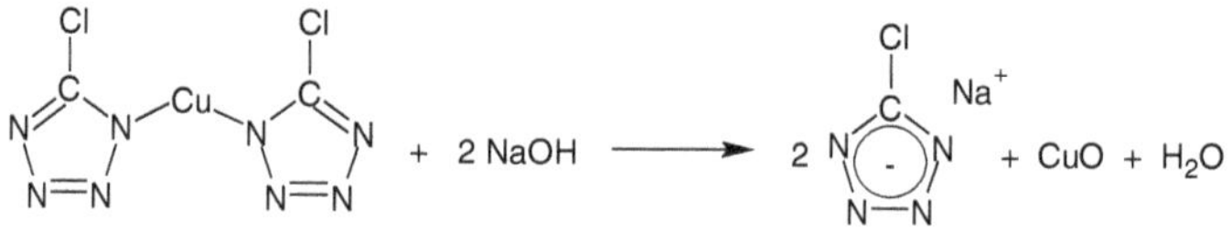

In earlier work, the cupric salt of 5-chlorotetrazole was rejected for practical applications due to its corrosive nature toward metals. However, it was later found that the corrosive behavior is not caused by the cupric salt itself but by chloride ions absorbed in the product from the hydrochloric acid medium in which it was prepared. This impurity is very difficult to remove by washing. It is therefore not recommended to prepare the cupric salt of 5-chlorotetrazole directly by the Sandmeyer reaction from 5-ATZ, but rather by precipitation from the sodium salt of 5-chlorotetrazole [42].

8.4.4 Uses

Out of the whole range of 5-halotetrazole salts, only the cupric salt of 5-chlorotetrazole has found a use—e.g., in primers for space applications [1]. Many other metallic salts of 5-chlorotetrazole (e.g., silver salt) have been disqualified from application mainly due to their high mechanical sensitivity [76].

8.5 5-Azidotetrazole Salts

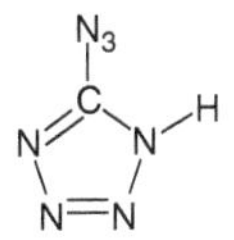

8.5.1 Physical and Chemical Properties

5-Azido-1H-tetrazole (subsequently just 5-azidotetrazole) forms monoclinic colorless needles with a theoretical maximum density 1.72 g cm^{-3}. Melting point is 75 °C [77]. It is, just like its alkali metal salts, soluble in water, acetone, ether, and ethanol. Lead, mercury, and silver salts are insoluble in these solvents [1, 78]. The potassium salt of 5-azidotetrazole is stable in aqueous and alcoholic solutions, but it explodes spontaneously if traces of acetic acid are present, even in acetone solution [78]. The 5-azidotetrazoles are fairly stable in storage when in the pure state [1].

8.5.2 Explosive Properties

Free 5-azidotetrazole is reported in Fedoroff, Sheffield, and Kaye's encyclopedia as less sensitive to impact than its alkaline salts [1]. Other authors reported 5-azidotetrazole as a highly sensitive compound [2, 77]. Stierstorfer et al. [77] reported sensitivity to impact below 1 J, sensitivity to friction below 5 N, and sensitivity to electrostatic discharge 2.2 mJ. 5-Azidotetrazole melts and decomposes at 165 °C when heated with heating rate 5 °C min^{-1} [77]. Pure 5-azidotetrazole is fairly stable in storage. However, it may spontaneously explode in presence of some impurities especially in presence of traces of acetic acid (it can explode even in acetone solution; explosions do not occur in aqueous or ethanol solutions) [1].

Sensitivities of the silver and potassium salts are extremely high. The potassium salt can even be detonated by touching it with a spatula [1]. The cesium salt is another very sensitive compound which is, together with the potassium salt, reported to be a substance which "cannot be handled without violent explosion" [77]. On the other hand, the published value for impact sensitivity of the barium salt is 35 cm/0.5 kg (h_{50}) [1]. Friedrich published an impact sensitivity for 1 kg hammer of 3 cm for the copper salt and 3–5 cm for the cadmium salt [79]. The lead salt detonates when touched with a flame [1].

Metallic salts of 5-azidotetrazole are brisant primary explosives (even amounts as small as 10 mg of the potassium salt can cause damage) [1]. Initiating efficiency of these substances is high; an amount of only 1–2 mg is sufficient to initiate PETN [80].

8.5.3 Preparation

5-Azidotetrazole can be prepared by several methods. The simplest method is the direct cycloaddition reaction between alkaline azides and cyanogen halogen. An alkaline salt of 5-azidotetrazole forms according to the following reaction [81, 82]:

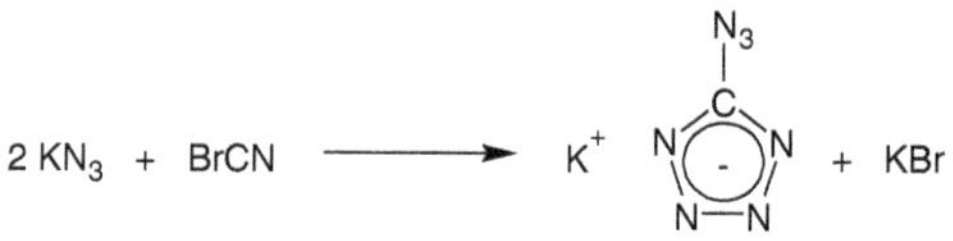

$$2\ KN_3 + BrCN \longrightarrow K^+ \ [\text{tetrazolate-}N_3] + KBr$$

The reaction is carried out in an aqueous solution and at low temperatures; sodium azide may be used as well [77]. The resulting salt can be converted to free 5-azidotetrazole by hydrochloric acid [83].

Pure 5-azidotetrazole can be prepared by the same reaction in presence of azoimide [81].

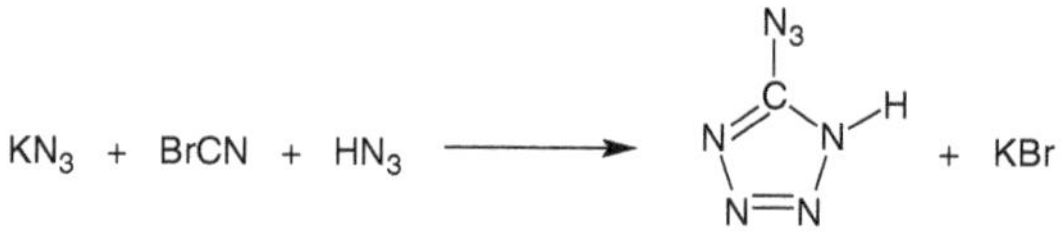

$$KN_3 + BrCN + HN_3 \longrightarrow [\text{5-azidotetrazole}] + KBr$$

Another way of preparing 5-azidotetrazole is based on the diazotization reaction. 5-ATZ is mostly used as the starting material. Tetrazole diazonium salt forms as an intermediate which is converted into the desired tetrazole azide by treatment with sodium azide [84].

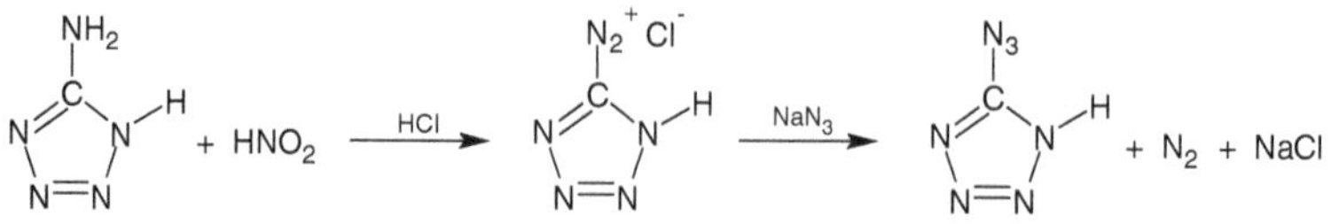

$$[\text{5-amino-tetrazole}] + HNO_2 \xrightarrow{HCl} [\text{tetrazole-}N_2^+ Cl^-] \xrightarrow{NaN_3} [\text{5-azidotetrazole}] + N_2 + NaCl$$

8.5 5-Azidotetrazole Salts

The 5-azidotetrazole can be purified by recrystallization from anhydrous diethylether [84].

5-Hydrazino-1*H*-tetrazole can also be used as a starting material for 5-azidotetrazole preparation. Preparation may be done by treating sodium nitrite and hydrochloric acid with 5-hydrazino-1*H*-tetrazole in an aqueous environment [77, 85, 86]:

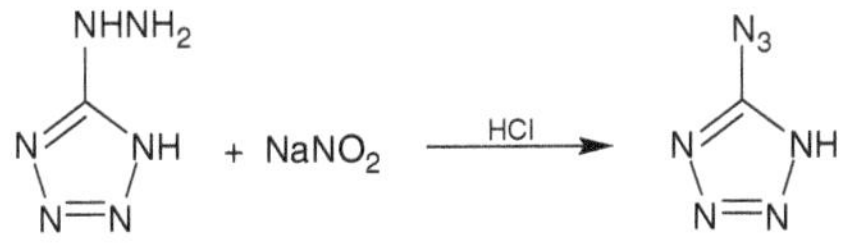

Diaminoguanidine is another possible starting material which can be used for preparation of 5-azidotetrazole and its salts. It forms when diaminoguanidine nitrate is treated with two moles of nitrous acid in an acetic acid solution [78].

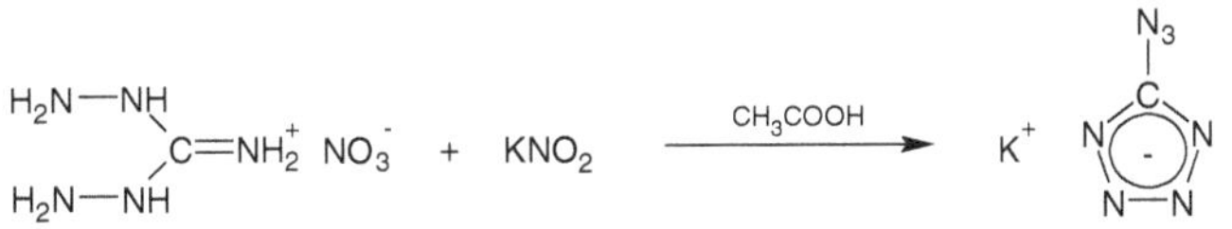

Another, rather unusual, way to 5-azidotetrazole referred to in [77] is partial degradation of tetrazene with barium hydroxide [10]:

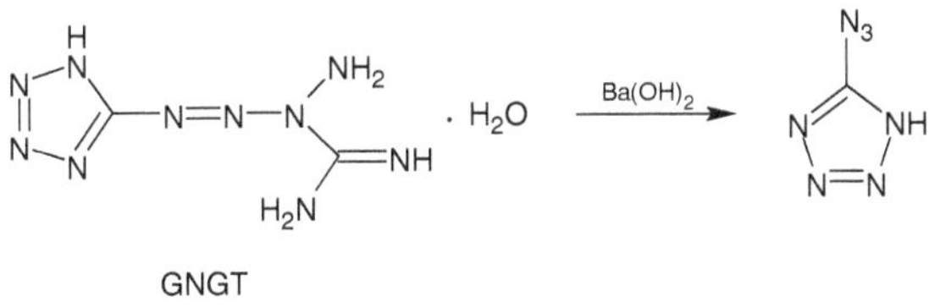

GNGT

Various salts of 5-azidotetrazole can be prepared by treating it with relevant salts [78]. Preparation, crystallographic studies, and characterization of hydrazinium, ammonium, aminobiguanidinium, guanidinium, lithium, sodium, potassium, cesium and calcium salts of 5-azidotetrazole were described by Klapötke and Stierstorfer [83]. Some experimental data from this paper are summarized in Table 8.12. The guanidinium salt forms as a semihydrate, whereas the lithium and sodium salts of 5-azidotetrazole form hydrates. The structure of the calcium salt is more complicated as it forms a complex with the following formula: $[Ca(CN_7)_2(H_2O)_{10}][Ca(H_2O)_6](CN_7)_2$.

8.5.4 Uses

Despite the high initiating efficiency of many metallic salts of 5-azidotetrazole, most of them are too sensitive for practical use.

Table 8.12 Selected experimental properties of 5-azidotetrazole salts [83]

	TMD (g cm^{-3})	Crystal system	Color	Impact (J)	Friction (N)	ESD (mJ)	T_{dec} (°C)
Hydrazinium	1.57	Monoclinic	Colorless needles	<1	5	5	136
Ammonium	1.61	Monoclinic	Colorless needles	<1	5	10	157
Aminoguanidinium	1.52	Triclinic	Colorless blocks	1	7	40	159
Guanidinium semihydrate	1.55	Monoclinic	Colorless rods			70	
Lithium hydrate	1.68	Monoclinic	Colorless needles			70	
Sodium hydrate	1.74	Monoclinic	Colorless needles	<1	<5	20	
Potassium	1.92	Monoclinic	Colorless rods				148
Cesium	2.81	Orthorhombic	Colorless plates				
Calcium complex	1.66	Monoclinic	Colorless needles	2	25	160	
RDX	1.82			7	120	>150	213

8.6 5,5′-Azotetrazole Salts

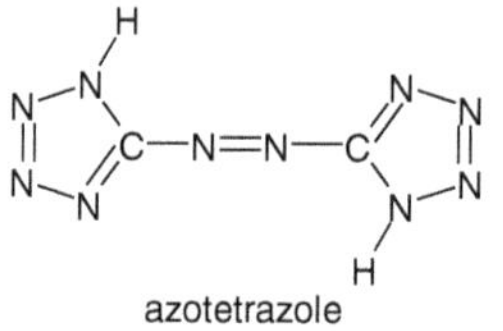

azotetrazole

Free 5,5′-azo-1H-tetrazole (subsequently just 5-azidotetrazole; AzTZ) is an unstable compound that immediately decomposes at room temperature. However it forms stable metallic salts, many of which have the characteristics of primary explosives. The first salts of AzTZ were prepared by Johannes Thiele in 1893. He described their properties and determined that they form hydrates [2, 29, 87, 88]. Salts of 5,5′-azotetrazole with protonated nitrogen bases (e.g., ammonium, guanidinium, triaminoguanidinium) are suggested as nitrogen rich gas generating agents for gas generators.

8.6 5,5′-Azotetrazole Salts

Table 8.13 Density and water of crystallization contents in one molecule of relevant alkaline and alkaline earth metal salts of AzTZ [89, 91, 92]

Salt of AzTZ	Li	Na	K	Rb	Cs	Mg	Ca	Sr	Ba
Moles of water of crystallization	6	5	5	1	2	8	8	6	5
Density (g cm^{-3})	1.528	1.684	–	2.396	–	–	1.677	–	2.298

8.6.1 Physical and Chemical Properties

Free 5,5′-azotetrazole forms a yellow solid that is unstable and decomposes within seconds at room temperature. However, it can be handled at −30 °C and stored for several months at −80 °C [89].

Most metal salts are yellow (K, Ca, Zn, Pb), while others range from orange (basic lead salt) to brown-colored solids (mercuric, iron) [2, 90, 91]. The cupric salt forms dark green crystals [2].

Most metallic salts of 5,5′-azotetrazole form hydrates. The water of crystallization content in these salts is summarized in Table 8.13. During storage of alkaline, alkaline earth metal salts and some trivalent metal salts (Al, La, Y, Ce, Nd, and Gd) all compounds lose water; this takes place over a period of hours for lithium, potassium, and the trivalent salts studied, and for other salts it takes between days and weeks. Upon loss of water of crystallization, magnesium and the trivalent salts of AzTZ decompose to colorless substances [89, 91, 92]. The cupric and mercuric salts form anhydrous salts, and the monobasic lead salt (PbAzTZ·PbO·5H$_2$O) forms a pentahydrate [2].

According to Bagal [2], the sodium salt forms a pentahydrate at room temperature that loses two molecules of crystalline water above 30 °C and forms an anhydride salt above 75 °C. Contrary to other authors, Reddy and Chatterjee [90] have reported the formation of a monohydrate for the sodium and barium salts (on the basis of TGA results). Lead and mercury salts form as anhydrides [90].

Density and crystal structure of alkali metal, alkaline earth metal, and several trivalent cations have been described by Hammerl et al. using X-ray diffraction [89]; the basic lead salt (density 4.62 g cm^{-3}) has been described by Pierce-Butler [93].

The silver and mercuric salts are highly hygroscopic [2] unlike the basic lead salt of 5,5′-azotetrazole, which is not, and further does not react with common metals (including copper) [80].

Alkali metal and alkaline earth metal salts of AzTZ are mostly soluble in water with the exception of the barium and strontium salts that are almost insoluble [89]. The potassium salt is hardly soluble in cold water, but more soluble in hot water [2]. The mercury and lead salts are also almost insoluble [90]. The lead salt is insoluble in inorganic solvents but soluble in weak acids and bases. The monobasic salt is insoluble in water and slightly soluble in most organic solvents. The best solvent for this salt is diluted nitric acid or a solution of ammonium acetate [2].

The sodium salt is not stable in acids. It decomposes to the unstable intermediate free AzTZ which decomposes to diazomethyl(tetrazol-5-yl)diazene and nitrogen. Diazomethyl(tetrazol-5-yl)diazene is the final product of decomposition in less acidic environments (e.g., in 2% HCl). In the more acidic environment, decomposition of diazomethyl(tetrazol-5-yl)diazene continues to salts of 5-hydrazino-1H-tetrazole, nitrogen, and formic acid [94].

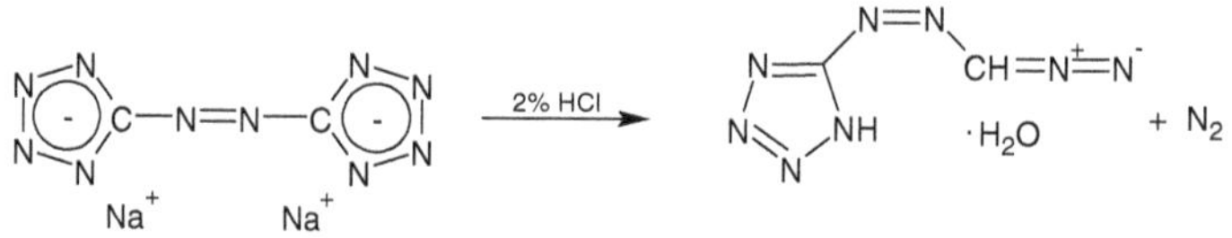

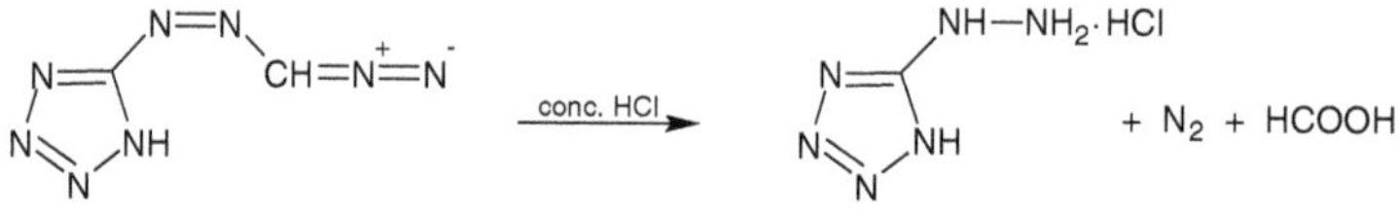

8.6.2 Explosive Properties

Sensitivity of metallic salts of 5,5′-azotetrazole noticeably depends on the presence of water of crystallization. Upon loss of water, the sensitivity of the compounds to impact and friction increases dramatically [89, 92, 95]. For example, the anhydrous sodium salt is highly sensitive to friction and impact while its hydrates are significantly less sensitive. Pentahydrate and trihydrate of sodium salt only crackle after ignition by a Bickford fuse while the anhydrous salt explodes. It is possible to grind the pentahydrate in a mortar while the anhydrous salt explodes with such treatment [2]. Many explosions of 5,5′-azotetrazole salts have occurred during drying under vacuum [89, 92].

According to Reddy and Chatterjee, sodium and barium monohydrates are not sensitive to friction and impact under the conditions generally used for primary explosives, whereas lead and mercury salts are highly sensitive to both [90]. In the Fedoroff, Sheffield, and Kaye's encyclopedia, it is however mentioned that sensitivity of the sodium, potassium, and barium salts is only about twice as high as for lead azide (not specified if the values concern anhydride or hydrates) [1].

Heavy metal salts of 5,5′-azotetrazole are highly sensitive to impact and friction [2, 14]. The lead salt of 5,5′-azotetrazole coated with dextrine showed reduced sensitivity [90]. The values of impact and friction sensitivity, ignition temperature, and sensitivity to electric discharge given by various authors are summarized in Tables 8.14 and 8.15.

As can be seen in Table 8.14, the mercury salt of 5,5′-azotetrazole is more sensitive to electric discharge than lead styphnate [90].

Sensitivity of several salts to impact is also reported by Bagal [2]; sensitivity of AgAzTZ to impact is about the same as for MF; mercuric and cupric salts are reported as very sensitive. The monobasic lead salt exceeds MF in sensitivity to impact and friction. On the contrary, the iron salt is reported as not sensitive to

8.6 5,5'-Azotetrazole Salts

Table 8.14 Initiation temperature, sensitivity to friction (weight used 4 kg), and sensitivity to electric discharge (gap between the needle and vial 2 mm) of metallic salts of 5,5'-azotetrazole

Compound	Ignition temperature (°C)			Friction sensitivity (cm s^{-1}) [90]		Electrostatic charge sensitivity E_{min} (mJ) [90]
	[90]	[14]	[2]	0 % Prob.	50 % Prob.	
Na$_2$AzTZ·H$_2$O	265	–	–	Not sensitive		Not sensitive
BaAzTZ·H$_2$O	180	–	–	Not sensitive		Not sensitive
PbAzTZ	180	173	–	5.1	7.6	30.3
PbAzTZ—dextr.	–	–	–	6.4	8.3	58
HgAzTZ	144	152	155	3.8	6.4	2.2 × 10^{-3}
PbAzTZ·PbO·5H$_2$O	–	–	194–196	–	–	–
Ag$_2$AzTZ	–	185	186	–	–	–
CuAzTZ	–	–	162	–	–	–
ZnAzTZ	–	–	161–176	–	–	–
LA—dextr.	–	–	–	10	11	157.5
LS	–	–	–	18	20	9 × 10^{-3}

Table 8.15 Impact sensitivity of metallic salts of 5,5'-azotetrazole [90]

Compound	Impact sensitivity		Not specified method, 0.5 kg hammer (cm) [14]	ATS apparatus, 2 kg (cm) [1]
	Not specified method, ball weight = 28 g (cm) [90]			
	0 % Prob.	50 % Prob.		
Na$_2$AzTZ·H$_2$O	Not sensitive		–	40
K$_2$AzTZ	–	–	–	40
BaAzTZ·H$_2$O	Not sensitive		–	40
PbAzTZ	5.0	7.0	12	12
PbAzTZ—dextr.	8.0	9.5	–	–
CdAzTZ	–	–	–	13
HgAzTZ	4.0	6.5	5	–
Ag$_2$AzTZ	–	–	8	–
LA—dextr.	12.0	13.5	–	22
LS	5.0	6.0	–	–

impact and friction [2]. Thermal stability of the basic lead salt of azotetrazole has been reported by Blay and Rapley [55]. This salt loses about 1% mass in 5 h at 172 °C under vacuum [55].

The initiating efficiency of the lead salt of 5,5'-azotetrazole exceeds that of other tested metallic salts, but it does not exceed that of standard composition lead azide/lead styphnate/aluminum (ASA compound). The mercury salt was not tested because of the "press fire problem" [90]. The values of initiating efficiency are summarized in Table 8.16.

Initiating efficiency of the silver and mercuric salts has been reported by Bagal [2]. Initiating efficiency of AgAzTZ is high, 0.03 g for PETN (AgAzTZ pressed under 15.7 MPa) and 0.13 g when AgAzTZ is pressed under 49 MPa. This implies that the silver salt is susceptible to dead-pressing. Initiating efficiency of the

Table 8.16 Initiating efficiency of metallic salts of 5,5'-azotetrazole compared with ASA composition in no. 6 detonator [90]

Compound	Initiating efficiency (g)
Na$_2$AzTZ·H$_2$O	0.35
BaAzTZ·H$_2$O	0.30
PbAzTZ	0.20
HgAzTZ	–
ASA comp.	0.060

PETN was used as secondary charge.

mercuric salt is 0.11 g for PETN (pressure is not reported). The monobasic lead salt lacks initiating ability [2].

The lead salt of 5,5'-azotetrazol is a powerful primary explosive that can be used as an initiator either alone or in a mixture. Its flame sensitivity is higher than that of lead azide and therefore PbAzTZ can be used in a mixture with lead azide to improve its flammability [90].

8.6.3 Preparation

The sodium salt of 5,5'-azotetrazole is prepared by oxidation of 5-ATZ. Potassium permanganate (or persulfate [85]) is used as an oxidizing agent in the presence of excess sodium hydroxide [18, 90, 91, 95]. The sodium salt of 5,5'-azotetrazole crystallizes from aqueous solution as a pentahydrate [95]. The alkalinity of the permanganate solution is an important parameter. Reaction of weakly alkaline permanganate and 5-ATZ results in the formation of hydrogen cyanide and carbon dioxide, while strongly alkaline permanganate leads to salts of 5-azotetrazole [29, 32].

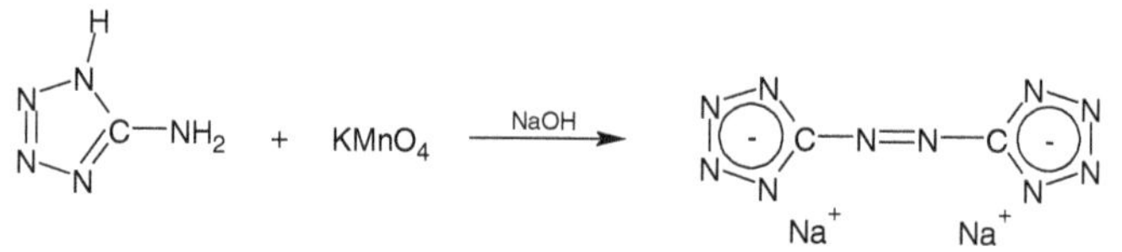

The heavy metallic salts of 5,5'-azotetrazole can be prepared by precipitation from a solution of the sodium salt by addition of a soluble metallic salt [14, 90–92].

Reaction of the barium salt of 5,5'-azotetrazole (prepared by treating barium chloride with the sodium salt of 5,5'-azotetrazole) with the relevant sulfate can be also used for preparation of soluble salts of 5,5'-azotetrazole (e.g., alkali salts). Synthesis of many salts (alkaline and alkaline earth), including analytical data, is described by Hammerl et al. [89].

The basic lead salt of 5,5′-azotetrazole is prepared by the same reaction in presence of sodium hydroxide, ammonia, etc. [2, 18, 96]. This salt is probably the only one which has been practically manufactured. The same technology equipment can be used for this as is used for production of LA. The production details are described in Bagal [2].

Free 5,5′-azotetrazole can be synthesized from the sodium salt and $HBF_4 \cdot Et_2O$ at −30 °C in methanol. A crystal structure analysis showed the presence of two methanol molecules [89].

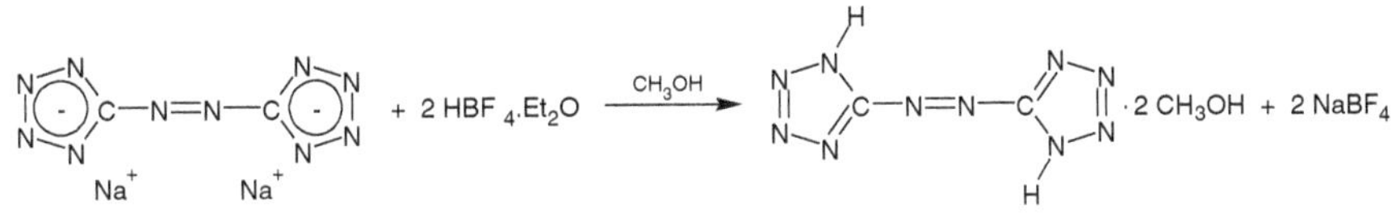

$$\text{Na}_2[\text{tetrazole-N=N-tetrazole}] + 2\ HBF_4 \cdot Et_2O \xrightarrow{CH_3OH} [\text{H-tetrazole-N=N-tetrazole-H}] \cdot 2\ CH_3OH + 2\ NaBF_4$$

8.6.4 Uses

Taylor and Jenkins have reported that the cadmium, cupric, and silver salts of 5,5′-azotetrazole are too sensitive for any technical use [76]. The lead salt of 5,5′-azotetrazole has been suggested for primers and detonators [80, 97] but it is hygroscopic at high relative humidity and has worse properties than its monobasic salt [76]. These disadvantages have not been observed in case of the lead monobasic salt which is a base compound for many patents for fuse heads [2]. This salt was also used with nitrocellulose varnish on the resistance bridge for electric igniters in Germany during World War II [18] and later in the British armed forces [42]. The ammonium, guanidinium, and triaminoguanidinium salts of 5,5′-azotetrazole have been investigated for application in gas generators.

8.7 Tetrazoles with Organic Substituent

Many tetrazoles with an organic substituent have been published in open literature. Some of them have the characteristics of primary explosives especially those containing azido, picryl, or other nitrogen-rich functional groups. Unfortunately, many of these substances are too sensitive for practical application. We have therefore decided to limit our selection to only those substances that are referred to as potentially useful primary explosives.

Table 8.17 Comparison of sensitivity of 5-PiATZ with LA and LS [52]

Compound	Impact sensitivity—h_{50}; Ball & Disc test (cm)	Friction sensitivity (m s^{-1})[a]	Sensitivity to electric discharge	
			Standard test (J)	Test for sensitive expl. (No.7) (µJ)
5-PiATZ	>50	>3.7	0.45 but not at 0.045	–
LA[b]	19	1.2	–	2.5
LS[b]	11	1.8	–	7

[a]Emery paper friction test; values are pendulum velocities required to produce a 50 % probability of ignition

[b]Service material RD 1343 for LA (precipitated from sodium carboxymethylcellulose/sodium hydroxide) and RD 1303 for LS

8.7.1 5-Picrylaminotetrazole

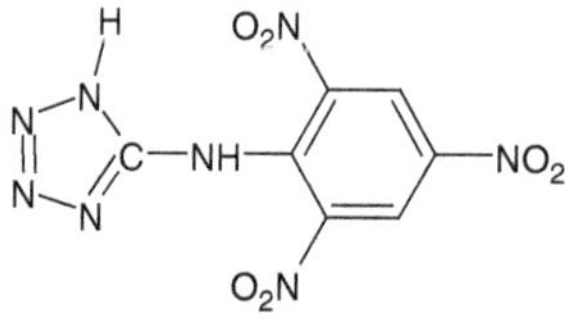

8.7.1.1 Physical and Chemical Properties

5-(2,4,6-Trinitrofenylamino)-1H-tetrazole (subsequently just picrylaminotetrazole; 5-PiATZ) exists as a yellow powder with density 1.91 g cm^{-3} and melting point 224 °C [98] or 232 °C [52]. It is soluble in acetone, and slightly soluble in ether and water [98].

5-PiATZ forms metallic salts. The silver salt forms yellow crystals that are insoluble in nitric acid and only slightly soluble in ammonia. The cupric salt exists as a green powder and the lead salt forms golden-colored crystals [98].

8.7.1.2 Explosive Properties

Sensitivity values of 5-picrylaminotetrazole compared with LA and LS are summarized in Table 8.17 [52]. It follows from the table that the sensitivity of 5-picrylaminotetrazole is relatively low, almost at a level of secondary explosives.

The silver, cupric, and lead salts are sensitive to impact. These salts explode by heating (ignition temperature is not specified) [98].

8.7.1.3 Preparation

5-Picrylaminotetrazole can be prepared by the reaction of 5-ATZ with 1-chloro-2,4,6-trinitrobenzene (picrylchloride) in acetic acid. The raw product can be purified by recrystallization from glacial acetic acid [52, 98].

Metallic salts can be prepared by reaction of 5-picrylaminotetrazole with soluble metallic salts (nitrates, acetates) [98].

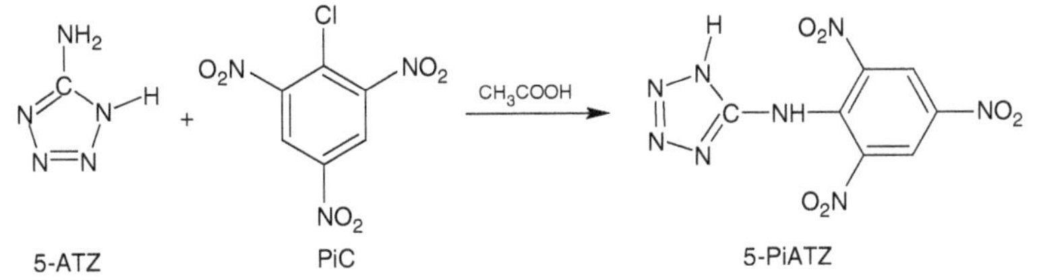

8.7.1.4 Uses

5-Picrylaminotetrazole has been studied as a potential green primary explosive replacement for LS, but its low sensitivity excluded it from this application [52].

8.7.2 *1-(1*H*-Tetrazol-5-yl)guanidinium Nitrate*

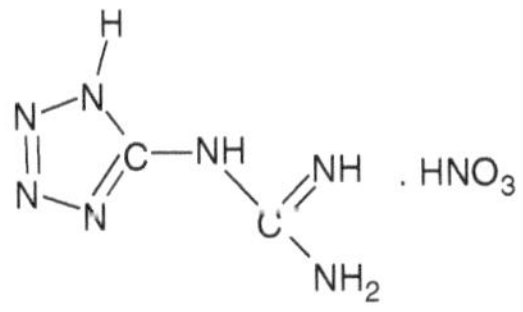

8.7.2.1 Physical and Chemical Properties

1-(1*H*-Tetrazol-5-yl)guanidinium nitrate (formerly 5-(guanylamino)-tetrazolium nitrate) forms white plate crystals [35]. The compound is referred to as an anhydride by Stollé [35] or as a monohydrate by Millar [52]. Crystal density is 1.58 g cm^{-3} [52] and melting point 183 or 185 °C (decomp. with crack) [35, 52]. It is soluble in water, hardly soluble in ethanol, and insoluble in ether. It forms a free base by repeated crystallization from water [35].

Table 8.18 Sensitivity of 1-(1H-tetrazol-5-yl)guanidinium nitrate compared with LA and LS [52]

Compound	Impact sensitivity h_{50}; Ball & Disc test (cm)	Friction sensitivity (m s^{-1})[a]	Sensitivity to electric discharge	
			Standard test (J)	Test for sensitive expl. (No.7) (µJ)
1-(1H-tetrazol-5-yl) guanidinium nitrate	>50	>3.7	Not at 4.5	–
LA[b]	19	1.2	–	2.5
LS[b]	11	1.8	–	7

[a]Emery paper friction test; values are the pendulum velocity required to produce a 50 % probability of ignition.
[b]Service material RD 1343 for LA (precipitated from sodium carboxymethylcellulose/sodium hydroxide) and RD 1303 for LS.

8.7.2.2 Explosive Properties

Sensitivities of 1-(1H-tetrazol-5-yl)guanidinium nitrate are compared with LA and LS in Table 8.18 [52].

It follows from Table 8.18 that the sensitivity of 1-(1H-tetrazol-5-yl) guanidinium nitrate is relatively low, similar to secondary explosives. Ignition of 1-(1H-tetrazol-5-yl)guanidinium nitrate takes place at 186 °C and the DSC onset is at 163 °C (10 °C min^{-1}) [52].

8.7.2.3 Preparation

1-(1H-Tetrazol-5-yl)guanidinium nitrate is easily formed by the reaction of 1-(1H-tetrazol-5-yl)guanidine with nitric acid, the nitrate crystallizing from the solution by cooling. 1-(1H-Tetrazol-5-yl)guanidine is prepared by the reaction of 5-ATZ with cyanamide in an aqueous environment [52]. The product can be recrystallized from diluted nitric acid [35].

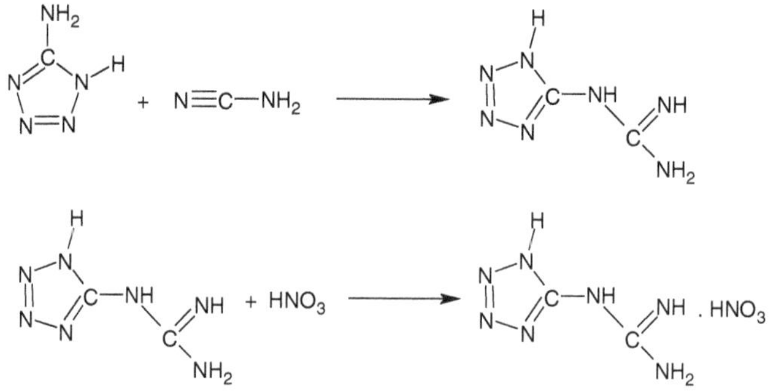

8.7.2.4 Uses

1-(1*H*-Tetrazol-5-yl)guanidinium nitrate has been studied as a green primary explosive potentially replacing LS, but the low sensitivity of the compound excluded it from this application [52].

8.8 Organic Derivatives of 5-Nitrotetrazole

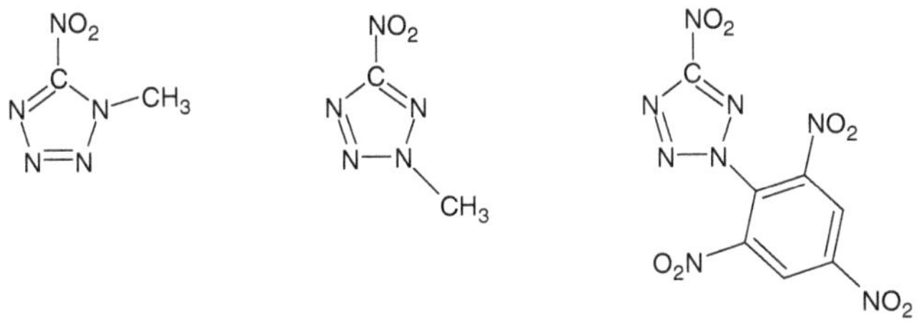

The organic derivatives of 5-nitrotetrazole form two isomers substituted either in position 1 or 2 of the ring. Some attention has been given to methyl-5-nitrotetrazoles and to 2-picryl-5-nitrotetrazole. The melting point of 1-methyl-5-nitrotetrazole is 56 °C and of 2-methyl-5-nitrotetrazole is 86 °C. However 2-methyl-5-nitrotetrazole was found not to initiate RDX in a commercial detonator setup. 5-Nitro-2-(2,4,6-trinitrofenyl)tetrazole (subsequently just 5-nitro-2-picryltetrazole) has the characteristics of primary explosives but unfortunately is very sensitive to friction. The *p*-nitrobenzyl derivate of 5-nitrotetrazole is not a primary explosive and behaves more like a secondary explosive [4].

Alkyl derivatives of 5-nitrotetrazole form by alkylation of the metallic salts of 5-nitrotetrazole (e.g., sodium, silver) by alkylhalogenide [4].

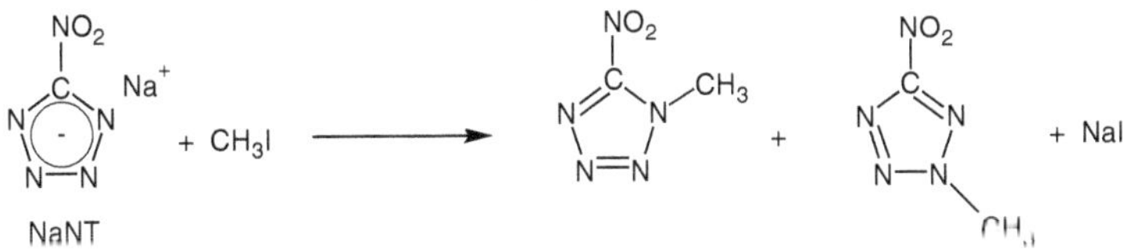

The alkylation leads to a mixture of isomers with the methyl group in positions 1 and 2. Electronegative substituent on carbon makes substitution in position 2 predominant. The methyl derivate is formed according to the reaction shown, but the ethyl, propyl, and isopropyl derivatives could not be isolated (silver salt of 5-nitro-1*H*-nitrotetrazole was used for synthesis) by Bates and Jenkins [4].

Table 8.19 Impact sensitivities of organic derivatives of 5-azidotetrazole in comparison with the cupric and cadmium salts of 5-azidotetrazole [79]

Compound	Impact sensitivity for 1 kg hammer (cm)
5-Azido-1-methyltetrazole	5–8
5-Azido-1-picryltetrazole	25–30
Cupric salt of 5-azidotetrazole	3
Cadmium salt of 5-azidotetrazole	3–5

5-Nitro-2-picryltetrazole can be prepared by the reaction of picrylchloride with sodium salt of nitrotetrazole in acetone. When water is present, picric acid is formed, casting doubts on the hydrolytic stability of the picryl derivative [4].

8.9 Organic Derivatives of 5-Azidotetrazole

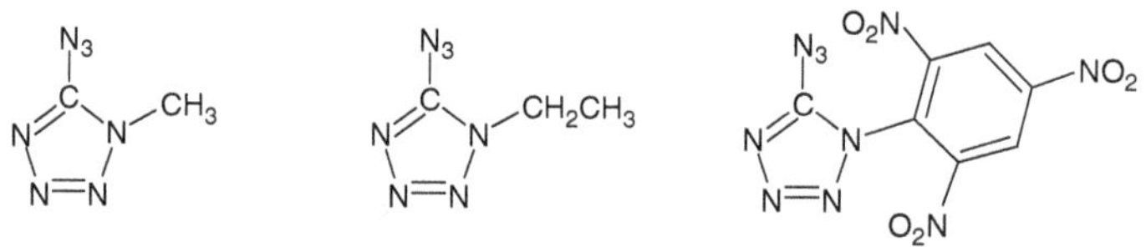

8.9.1 *Explosive Properties*

Organic derivatives of 5-azidotetrazole have been reported by Friedrich [99]. They are more stable and less sensitive than the metallic salts. The methyl and ethyl derivatives are very strong igniting explosives. Ethylene tetrazylazide, that forms an oily liquid, possesses extraordinary brisance and is easily ignited by spark and flames. It easily gelatinizes nitrocellulose [1, 99]. Impact sensitivities of 5-azido-1-methyltetrazole and 5-azido-1-picryltetrazole in comparison with the copper salt are summarized in Table 8.19 [79].

8.9.2 *Preparation*

Organic derivatives of 5-azidotetrazole can be prepared from the alkali salts of 5-azidotetrazole and organic chlorides of sulfates [99].

References

1. Fedoroff, B.T., Sheffield, O.E., Kaye, S.M.: Encyclopedia of Explosives and Related Items. Picatinny Arsenal, New Jersey (1960–1983)
2. Bagal, L.I.: Khimiya i tekhnologiya iniciiruyushchikh vzryvchatykh veshchestv. Mashinostrojenije, Moskva (1975)
3. Matushin, Y.N., Lebedev, V.P.: Thermochemical properties of mononitroderivates of azoles and oxadiazoles. In: Proceedings of 28th International Annual Conference of ICT, pp. 98/1–98/10, Karlsruhe, 1997
4. Bates, L.R., Jenkins, J.M.: Search for new detonators. In: Proceedings of International Conference on Research in Primary Explosives, pp. 14/1–14/18, Waltham Abbey, England, 1975
5. Chen, Z.X., Xiao, H., Yang, S.: Theoretical investigation on the impact sensitivity of tetrazole derivates and their metal salts. Chem. Phys. **250**, 243–248 (1999)
6. Haskins, P.J.: Electronic structure of some explosives and its relationship to sensitivity In: Jenkins, J.M., White, J.R. (eds.) International Conference on Research of Primary Explosives, vol. 1, pp. 6/1–6/28, Waltham Abbey (1975)
7. Avanesov, D.C.: Praktikum po fiziko-khimicheskim icpytaniyam vzryvchatykh veshchectv. Gosudarstbennoe izdatelstvo oboronnoi promyshlennosti, Moskva (1959)
8. Ilyushin, M.A., Tselinskii, I.V., Sudarikov, A.M.: Razrabotka komponentov vysokoenergicheskikh kompozitsii. SPB:LGU im. A. S. Pushkina – SPBGTI(TU), Sankt-Peterburg (2006)
9. Hoffman, K.A., Roth, R.: Aliphatische Diazosalze. Berichte der deutschen chemischen Gesellschaft **43**, 682–688 (1910)
10. Patinkin, S.H., Horwitz, J.P., Lieber, E.: The structure of tetracene. J. Am. Chem. Soc. **77**, 562–567 (1955)
11. Duke, J.R.C.: X-ray crystal and molecular structure of "tetrazene", ("tetracene"), $C_2H_8N_{10}O$. Chem. Commun. **1**, 2–3 (1971)
12. Davis, T.L.: The Chemistry of Powder and Explosives. Wiley, New York (1943)
13. Krauz, C.: Technologie výbušin. Vědecko-technické nakladatelství, Praha (1950)
14. Špičák, S., Šimeček, J.: Chemie a technologie třaskavin. Vojenská technická akademie Antonína Zápotockého, Brno (1957)
15. Rinkenbach, W.H., Burton, O.E.: Explosive characteristics of tetracene. Army Ordonance **12**, 120–123 (1931)
16. Meyer, R., Köhler, J., Homburg, A.: Explosives. Wiley-VCH, Weinheim (2002)
17. Danilov, J.N., Ilyushin, M.A., Tselinskii, I.V.: Promyshlennye vzryvchatye veshchestva; chast I. Iniciiruyushchie vzryvshchatye veshchestva. Sankt-Peterburgskii gosudarstvennyi tekhnologicheskii institut, Sankt-Peterburg (2001)
18. Urbański, T.: Chemistry and Technology of Explosives. Pergamon, Oxford (1984)
19. Wallbaum, R.: Sprengtechnische Eigenschaften und Lagerbeständigkei der wichtigsten Initialsprengstoffe. Zeitschrift für das gesamte Schiess- und Sprengstoffwesen **34**, 161–163 (1939)
20. Tomlinson, W.R., Sheffield, O.E.: Engineering Design Handbook, Explosive Series of Properties Explosives of Military Interest. Report AMCP 706-177, 1971
21. Matyáš, R., Šelešovský, J., Musil, T.: Sensitivity to friction for primary explosives. J. Hazard. Mater. **213–214**, 236–241 (2012)

22. Urbański, T.: Chemistry and Technology of Explosives. PWN—Polish Scientific Publisher, Warszawa (1967)
23. Lieber, E., Smith, G.B.L.: The chemistry of aminoguanidine and related substances. Chem. Rev. **25**, 213–271 (1939)
24. Hagel, R., Redecker, K.: Sintox—a new, non-toxic primer composition by Dynamit Nobel AG. Propellants Explosives Pyrotechnics **11**, 184–187 (1986)
25. Brede, U., Hagel, R., Redecker, K.H., Weuter, W.: Primer compositions in the course of time: From black powder and SINOXID to SINTOX compositions and SINCO booster. Propellants Explosives Pyrotechnics **21**, 113–117 (1996)
26. Nesveda, J., Brandejs, S., Jirásek K.: Non toxic and non-corrosive ignition mixtures. WO Patent 01/21558, 2001
27. Whelan, D.J., Spear, R.J., Read, R.W.: The thermal decomposition of some primary explosives as studied by differential scanning calorimetry. Thermochim. Acta **80**, 149–163 (1984)
28. Mihina, J.S., Herbst, R.M.: The reaction of nitriles with hydrazonic acid: Synthesis of monosubstituted tetrazoles. J. Org. Chem. **15**, 1082–1092 (1950)
29. Thiele, J.: Ueber Nitro- und Aminoguanidin. Justus Liebigs Annalen der Chemie **270**, 1–63 (1892)
30. Bates, L.R., Jenkins, J.M.: Salts of 5-Substituted Tetrazole: Part 2: Metallic Salts and Complexes of Tetrazole 5-Aminotetrazole, 5-Phenyltetrazole, and 5-Methyltetrazole. Report AD 727350. Explosives Research and Development Establishment, Waltham Abbey (1970).
31. Taylor, F.: Primary Explosive Research. Report NAVORD 2800, Naval Ordinance, 1953
32. Benson, F.R.: The chemistry of the tetrazoles. Chem. Rev. **41**, 1–61 (1947)
33. Hantzsch, A., Vagt, A.: Ueber das sogenannte Diazoguanidin. Justus Liebigs Annalen der Chemie **314**, 339–369 (1901)
34. Arient, J., Vobořil, I.: Způsob přípravy 5-aminotetrazolu. CS Patent 190,055, 1979
35. Stollé, R.: Zur Kenntnis des Amino-5-tetrazols. Berichte der deutschen chemischen Gesellschaft **62**, 1118–1126 (1929)
36. Stollé, R., Schick, E.: Verfahren zur Darstellung von Aminotetrazol. DE Patent 426,343, 1926
37. Daugherty, N.A., Brubaker, C.H.J.: Complexes of copper(II) and some 5-substituted tetrazoles. J. Am. Chem. Soc. **83**, 3779–3782 (1961)
38. Rittenhouse, C.T.: Di-silver aminotetrazole perchlorate. US Patent 3,663,553, 1972
39. Karaghiosoff, K., Klapötke, T.M., Sabaté, C.M.: Energetic silver salt with 5-aminotetrazole ligands. Chemistry **15**, 1164–1176 (2009)
40. Bates, L.R., Jenkins, J.M.: Production of 5-nitrotetrazole salts. US Patent 4,094,879, 1978
41. Bates, L.R., Jenkins, J.M.: Improvements in or relating to production of 5-nitrotetrazole salts. GB Patent 1,519,796, 1978
42. Bates, L.R.: The potential of tetrazoles in initiating explosives systems. In: Proceedings of 13th Symposium on Explosives and Pyrotechnics, pp. III1–III10, 1986
43. Koldobskii, G.I., Soldatenko, D.S., Gerasimova, E.S., Khokhryakova, N.R., Shcherbinin, M.B., Lebedev, V.P., Ostrovskii, V.A.: Tetrazoles: XXXVI. Synthesis, structure, and properties of 5-nitrotetrazole. Russ. J. Org. Chem. **33**, 1771–1783 (1997)
44. Khmelnitskii, L.I.: Spravochnik po vzryvchatym veshchestvam. Voennaya ordena Lenina i ordena Suvorova Artilleriiskaya inzhenernaya akademiya imeni F. E, Dzerzhinskogo, Moskva (1962)
45. Herz, E.: C-nitrotetrazole compounds. US Patent 2,066,954, 1937
46. Gilligan, W.H., Kamlet, M.J.: Synthesis of Mercuric 5-nitrotetrazole. Report N3WC/WOL TR 76-146. Navel Surface Weapons Center, Silver Spring, MD (1976)
47. Hirlinger, J., Fronabarger, J., Williams, M., Armstrong, K., Cramer, R.J.: Lead azide replacement program. In: Proceedings of NDIA, Fuze Conference, Seattle, Washington, April 5-7, 2005
48. Fronabarger, J., Sanborn, W.B., Bichay, M.: An investigation of some alternatives to lead based primary explosive. In: Proceedings of 37th AIAA 2001-3633; Joint Propulsion Conference and Exhibit, pp. 1–9, Salt Lake City, Utah, 2001

49. Fronabarger, J.W., Williams, M.D., Sanborn, W.B.: Lead-free primary explosive composition and method of preparation. US Patent 2009/0069566 A1, 2009
50. Fronabarger, J.W., Williams, M.D., Sanborn, W.B.: Lead-free primary explosive composition and method of preparation. WO Patent 2008/048351 A2, 2008
51. Klapötke, M., Sabaté, C.M., Welch, J.M.: Alkali metal 5-nitrotetrazolate salts: prospective replacements for service lead(II) azide in explosive initiators. Dalton Trans. 6372–6380 (2008)
52. Millar, R.W.: Lead-free Initiator Materials for Small Electro-explosive Devices for Medium Caliber Munitions. Report QinetiQ/FST/CR032702/1.1, QINETIQ Ltd, Farnborough, 2003
53. Harris, A.D., Herber, R.H., Jonassen, H.B., Wertheim, G.K.: Complexes of iron(II) and some 5-substituted tetrazoles. J. Am. Chem. Soc. **85**, 2927–2930 (1963)
54. Brown, M.E., Swallowe, G.M.: The thermal decomposition of the silver(I) and mercury(II) salts of 5-nitrotetrazole and of mercury(II) fulminate. Thermochim. Acta **49**, 333–349 (1981)
55. Blay, N.J., Rapley, R.J.: The testing of primary explosives for stability and compatibility In: Jenkins, J.M., White, J.R. (eds.) Proceedings of the International Conference on Research in Primary Explosives, vol. 3, pp. 20/1–20/19, Waltham Abbey (1975)
56. Redman, L.D., Spear, R.J.: Mercuric 5-Nitrotetrazole, a Possible Replacement for Lead Azide in Australian Ordnance. Part 1. An Assessment of Preparation Methods. Report MRL-R-901, Department of Defense, Material Research Laboratories, Ascot Vale, 1983
57. Klapötke, M., Sabaté, C.M.: Less sensitive transition metal salts of the 5-nitrotetrazolate anion. Cent. Eur. J. Energetic Mater. **7**, 161–173 (2010)
58. Lee, K., Coburn, M.D.: Binary eutectics formed between ammonium nitrate and amine salts of 5-nitrotetrazole I. Preparation and initial characterization. J. Energetic Mater. **1**, 109–122 (1983)
59. Scott, C.L.: Mercuric 5-nitrotetrazole as a lead azide replacement. In: Proceedings of the International Conference on Research in Primary Explosives, vol. 2, pp 15/1-15/23 Waltham Abbey, (1975)
60. Hirlinger, J.M.: Investigating alternative "green" primary explosives. In: Proceedings of NDIA 39th Annual Gun & Ammunition Missiles & Rockets Conference & Exhibition, Baltimore, MD, 2004
61. Gilligan, W.H., Kamlet, M.J.: Method of preparing the acid copper salt of 5-nitrotetrazole. US Patent 4,093,623, 1978
62. Hirlinger, J.M., Bichay, M.: New Primary Explosives Development for Medium Caliber Stab Detonators. Report SERDP PP-1364, US Army ARDEC, 2004
63. Talawar, M.B., Agrawal, A.P., Asthana, S.N.: Energetic co-ordination compounds: Synthesis, characterization and thermolysis studies on bis-(5-nitro-2H-tetrazolato-N^2)tetraammine cobalt (III) perchlorate (BNCP) and its new transition metal (Ni/Cu/Zn) perchlorate analogues. J. Hazard. Mater. **120**, 25–35 (2005)
64. Spear, R.J., Elischer, P.P.: Studies of stab initiation. Sensitization of lead azide by energetic sensitizers. Aust. J. Chem. **35**, 1–13 (1982)
65. Lee, K., Coburn, M.D.: Ethylenediamine salt of 5-nitrotetrazole and preparation. US Patent 4,552,598, 1985
66. McGuchan, R.: Improvements in primary explosive compositions and their manufacture. In: Proceedings of 10th Symposium on Explosives and Pyrotechnics, San Francisco, 1979
67. Glover, D.J.: Analysis of Mercuric 5-nitrotetrazole. Report NSWC/WOL/TR 77-71. Naval Surface Weapon Center, Silver Spring, MD (1977)
68. Fronabarger, J.W., Williams, M.D.: Preparation of lead-free primary explosive. US Patent 2010/0280254 A1, 2010
69. Scott, C., Leopold, H.S.: Single chemical electric detonator. US Patent 3,965,951, 1976
70. Tompa, A.S.: Thermal stability of an explosive detonator. Thermochim. Acta **80**, 367–377 (1984)
71. Scott, C.L., Leopold, H.S.: Stab-initiated explosive device containing a single explosive charge. US Patent 4,024,818, 1977

72. Duguet, J.: Priming composition which is sensitive to percussion and a method for preparing it. US Patent 4,566,921, 1986
73. Hagel, R., Bley, U.: Anzündmittel. DE Patent 10,221,044, 2002
74. Piechowicz, T.: Les explosifs d'amorcage nouveaux et leurs applications spatiales. In: Proceedings of Utilisation des éléments pyrotechniques et explosifs dans les systèmes spatiaux; Colloque International, pp. 99–104, Tarbes, 1968
75. Henry, R.A., Finnegan, W.G.: An improved procedure for the deamination of 5-aminotetrazole. J. Am. Chem. Soc. **76**, 290–291 (1954)
76. Taylor, G.W.C., Jenkins, J.M.: Progress toward primary explosives on improved stability. In: Proceedings of 3rd Symposium on Chemical Problems Connected with the Stability of Explosives, pp. 43–46, Ystad, 1973
77. Stierstorfer, J., Klapötke, T.M., Hammerl, A., Chapman, R.D.: 5-Azido-1*H*-tetrazole—Improved synthesis, crystal structure and sensitivity data. Zeitschrift für anorganische und allgemeine Chemie **634**, 1051–1057 (2008)
78. Lieber, E., Levering, D.R.: The reaction of nitrous acid with diaminoguanidine in acetic acid media. Isolation and structure proof of reaction products. J. Am. Chem. Soc. **73**, 1313–1317 (1951)
79. Friedrich, W.: Spreng- und Zündstoffe. DE Patent 695,254, 1940
80. Rathsburg, H.: Explosive compound for primers and detonators. US Patent 1,511,771, 1924
81. Friedrich, W., Flick, K.: Verfahren zur Herstellung von Tetrazylazid bzw. seinen Salzen. DE Patent 719,135, 1942
82. Marsh, F.D.: Cyanogen azide. J. Org. Chem. **37**, 2966–2969 (1972)
83. Klapötke, T.M., Stierstorfer, J.: The CN$_7^-$ anion. J. Am. Chem. Soc. **131**, 1122–1134 (2009)
84. Hammerl, A., Klapötke, T.M., Noth, H., Warchhold, M.: Synthesis, structure, molecular orbital and valence bond calculations for tetrazole azide, CHN$_7$. Propellants Explosives Pyrotechnics **28**, 165–173 (2003)
85. Rathsburg, H.: Initial primers and a process for their manufacture. GB Patent 185,555, 1921
86. Thiele, J., Ingle, H.: Ueber einige Derivate des Tetrazols. Justus Liebigs Annalen der Chemie **287**, 233–265 (1895)
87. Thiele, J.: Ueber Isocyantetrabromid. Berichte der deutschen chemischen Gesellschaft **26**, 2645–2647 (1893)
88. Thiele, J., Marais, J.T.: Tetrazolderivate aus Diazotetrazotsäure. Justus Liebigs Annalen der Chemie **273**, 144–160 (1893)
89. Hammerl, A., Holl, G., Klapötke, T.M., Mayer, P., Noth, H., Piotrowski, H., Warchhold, M.: Salts of 5,5'-azotetrazolate. Eur. J. Inorg. Chem. 834–845 (2002)
90. Reddy, G.O., Chatterjee, A.K.: A thermal study of the salts of azotetrazole. Thermochim. Acta **66**, 231–244 (1983)
91. Thiele, J.: Ueber Azo- und Hydrazoverbindungen des Tetrazols. Justus Liebigs Annalen der Chemie **303**, 57–75 (1898)
92. Singh, R.P., Verma, R.D., Meshri, D.T., Shreeve, J.N.M.: Energetic nitrogen-rich salts and ionic liquids. Angew. Chem. **45**, 3584–3601 (2006)
93. Pierce-Butler, M.A.: Structure of bis[hydroxolead(III)] 5,5'-azotetrazolediide. Acta Crystallogr. B **38**, 2681–2683 (1982)
94. Mayants, A.G., Vladimirov, V.N., Razumov, N.M., Shlyanochnikov, V.A.: Razlozhenie solei azotetrazola v kislykh sredakh. Zhurnal organicheskoi khimii **27**, 2450–2455 (1991)
95. Hiskey, M.A., Goldman, N., Stine, J.R.: High-nitrogen energetic materials derived from azotetrazolate. J. Energetic Mater. **16**, 119–127 (1998)
96. Taylor, G.W.C., Thomas, A.T.: Improvements in the manufacture of lead azotetrazole. GB Patent 986,631, 1965
97. Rathsburg, H.: Manufacture of detonating compositions. US Patent 1,580,572, 1926
98. Stollé, R., Roser, O.: Abkömmlinge des Amino-5-tetrazols. Journal für praktische Chemie **139**, 63–64 (1934)
99. Friederich, W.: Disruptive or brisant explosive composition. US Patent 2,170,943, 1939

Chapter 9
Tetrazole Ring-Containing Complexes

Heterocycles with high nitrogen content (triazoles and tetrazoles) have been studied by researchers of energetic materials for a very long time. The triazole derivatives were found to be useful in the area of secondary explosives and propellants, while tetrazole compounds were found to have some potential in priming. Heavy metal salts of a variety of tetrazole derivatives were examined in early studies and some even found practical applications.

One of the reasons stimulating the search for new primary explosives today is the need to replace toxic lead azide with some environmentally benign alternative (other reasons are described in Chap. 1). A suitable replacement needs to have many of lead azide's properties within relatively narrow ranges. It would therefore be desirable to have the ability to influence the properties of the resulting substance by slightly modifying its structure. This goal is not achievable with metal salts alone but seems to be realistic with metal complexes or coordination compounds or, as recommended by IUPAC, coordination entities [1].

The structure of these complexes consists of a central metal atom bonded to a surrounding array of molecules or anions (ligands or complexing agents). The general formula may be described as $[M_x(L)_y](An)_z$ or $(Cat)_z[M_x(L)_y]$ where M is a metal cation, L is a ligand, Cat is a cation and An is an anion of an acid, often with the function of oxidizer. The structure of such compounds enables changes of the physical, chemical, and explosive properties in a relatively wide range by altering the number and type of the ligands, the type of cation, the anion, or the coordinating metal. Research activities on these substances have been reported from Europe, USA, Russia, India, and China.

Transition metals are known to influence the burning behavior of energetic materials. The metal atom in the coordination compound therefore plays two roles (1) it serves as a bonding element for ligands and ions, (2) it influences (catalyzes) burning rate. Burning rate affects the ability of a substance to undergo deflagration to detonation transition (DDT) and therefore directly influences initiating efficiency. Hence a variation in the central metal atom of a complex structure affects its ability to serve as an initiating substance. Both of the above-mentioned properties also depend on the nature of the oxidizing group. Changing the nitrate anion for a perchlorate one usually leads to an increase in burning rate. The type of ligand often determines the sensitivity parameters of the resulting complex. Some methods of selection of particular building blocks of the resulting complex are nicely summarized by Sinditskii and Fogelzang [2]. Careful variation of central atom, ligand, and ions enables design of a structure with properties tailored to specific applications.

9.1 Cobalt Perchlorate Complexes

An interesting insight into primary explosives and their perspectives including some complex salts with substituted tetrazoles in the molecule have been published by Ilyushin [3]. It was stated there that, for tetrazole-containing complexes, initiating efficiency depends on composition and that the length of the predetonation zone is affected by metal (cation), ligand, and anion (oxidizer). Further, the DDT process depends on the enthalpy of formation and it is also faster for complexes with those substituents which exhibit exothermal reactions in the first stages of decomposition.

Almost endless possibilities exist for combining metal atoms, ligands, and ions. For the purpose of this chapter, we have selected only some of the more common complexes with at least some potential in initiating devices.

9.1.1 Pentaamine(5-cyano-2H-tetrazolato-N^2)cobalt(III) perchlorate (CP)

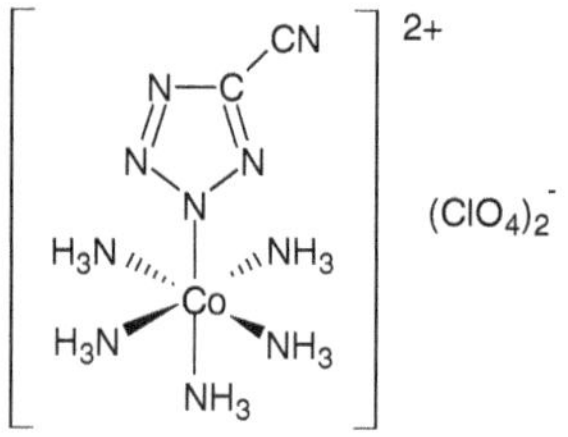

Pentaamine(5-cyano-2H-tetrazolato-N^2)cobalt(III) perchlorate (CP) was developed in the late 1960s as an alternative primary explosive material that has a rapid DDT while maintaining a high level of manipulation safety [4–6].

The structure of pentaamine (5-cyano-2H-tetrazolato-N^2)cobalt(III) perchlorate is shown above. Bonding via the N2 ring nitrogen in the final product has been confirmed by single-crystal X-ray structure determination [7] and ^{15}N NMR analysis [8].

The strength of the coordinate bond between cobalt and the various nitrogen atoms differs. The bond between cobalt and nitrogen in the tetrazole ring is the strongest due to the negative electronic charge of the 5-cyanotetrazole ligand. The four equatorial bonds have equal strength which is weaker than the tetrazole cobalt bond but stronger than the reverse side Co−N bond: bond connecting tetrazole ring > bonds in equatorial position > bond in axial position [9, 10].

9.1.1.1 Physical and Chemical Properties

CP forms yellow monoclinic crystals with crystal density 1.97 g cm^{-3} [11]. It is compatible with most metallic and ceramic materials used in typical detonator design and also epoxies cured by amines or anhydrides. This is interesting since many organic explosives and amine cured epoxy materials have compatibility problems. It is also compatible with PETN [12].

However, it has been found to be incompatible with copper and should not be considered for designs containing this metal [5, 11, 13]. The good compatibility of this material may be jeopardized by impurities from the CP synthesis process (nitrate and perchlorate ions) [11].

The solid decomposition products and impurities from CP preparation (CP co-product 5-carboxamidotetrazolatopentaamminecobalt(III) perchlorate, see Sect. 9.1.1.3) in amounts from 1 to 10 % have no apparent autocatalytic or inhibitive effect on CP decomposition. But impurities from the CP synthesis process (nitrate, perchlorate ions) can lead to compatibility problems. Gaseous decomposition products such as ammonia inhibit the decomposition by pushing the reactions to higher temperatures [11].

9.1.1.2 Explosive Properties

The sensitivity of CP to shock, electrostatic discharge, and hot wire noticeably depends on its form. It is relatively insensitive to these stimuli in bulk form (practically comparable to secondary explosives). However, in the pressed state, CP is as sensitive as primary explosives. This behavior is the main advantage of this compound [4].

The sensitivity of CP to electrostatic discharge (ESD) is highly dependent on density. The sensitivity to ESD further depends on its particle size, with smaller particles showing higher sensitivity (Fig. 9.1).

The overall behavior is therefore controlled by a combination of particle size and density that both contribute to the resulting U-shaped curve dependency of ESD sensitivity (Fig. 9.2). At lower densities, crushing of crystals during pressing and therefore decrease of particle size may contribute to the higher sensitivity (lower initiation energy) overtaken at higher densities by increase of dielectric strength.

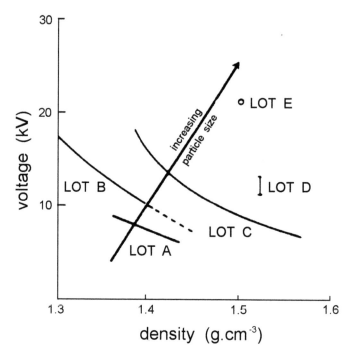

Fig. 9.1 Sensitivity of CP to electrostatic discharge as a function of particle size (increasing particle size from lot A to lot E) [13]. Reprinted by permission of IPSUSA Seminars, Inc.

Loose CP powder and unconfined pellets are not ignitable by the human body equivalent electrostatic discharge (20 kV, 600 pF, 500 Ω) [11].

The comparison of sensitivity to the mechanical stimuli between CP and secondary explosives (PETN, RDX) is shown in Table 9.1 [6]. CP unlike LA or LS does not show sensitivity typical for primary explosives. In bulk form it exhibits sensitivity on the level of secondary explosives. After the thermal initiation it undergoes DDT only when confined [4].

The detonation velocity of CP in a confined state can be obtained from the following formula:

$$D = 0.868 + 3.608\rho$$

where D is detonation velocity (in km s^{-1}) and ρ is the initial density of CP (in g cm^{-3}). The relation is valid for a sample with a 6.35 mm diameter [13].

CP is thermally stable up to 80 °C (Co(II) concentration from CP decomposition was below 800 ppm following 3 years of aging). At 120 °C average 2.2 % of CP decomposes after 1,078 days of aging [11].

9.1 Cobalt Perchlorate Complexes

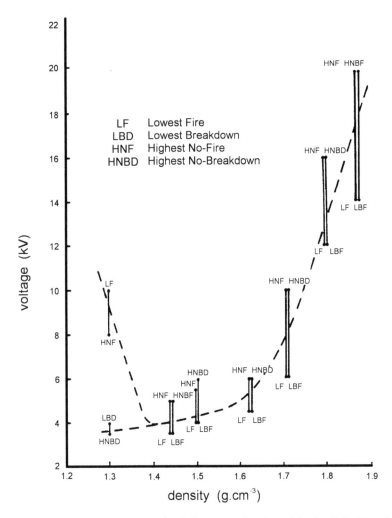

Fig. 9.2 Sensitivity of CP to electrostatic discharge as a function of density [13]. Reprinted by permission of IPSUSA Seminars, Inc.

Table 9.1 Impact (drop-hammer machine, 2.5 kg hammer, 50 % probability) and friction sensitivity (BAM apparatus, 1 event in 10 trials) of powdered CP [6]

Impact sensitivity h_{50} (cm)			Friction sensitivity (kg)	
CP	PETN	RDX	CP	RDX
60.6	15.5	34.5	1	1

9.1.1.3 Preparation

Preparation of 5-Cyanotetrazole

5-Cyanotetrazole is a main intermediate for preparation of CP. The most convenient route for its preparation (and also other substituted tetrazoles) is the 1,3-dipolar cycloaddition reaction of azides with nitriles [14]. Pure 5-cyanotetrazole is prepared by treating a hydrazonic acid with cyanogen via the following reaction [15, 16]:

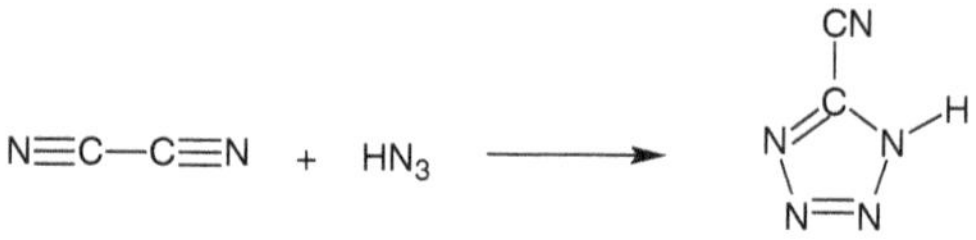

$$N\equiv C-C\equiv N + HN_3 \longrightarrow \text{5-cyanotetrazole}$$

The reaction conditions influence the composition of the final product. The formation of side products 5,5'-bitetrazole or 5-carboxyamidetetrazole is possible with this reaction [17]. The metallic salts of 5-cyanotetrazole can be prepared when metallic azides are used instead of hydrazonic acid in presence of sulfur dioxide [14].

The other method of preparing 5-cyanotetrazole is based on treatment of 1-cyanoformimidic acid hydrazide with nitrous acid. The intermediate 1-cyanoformimidic acid hydrazide could be prepared by the reaction of cyanogen and hydrazine [18, 19].

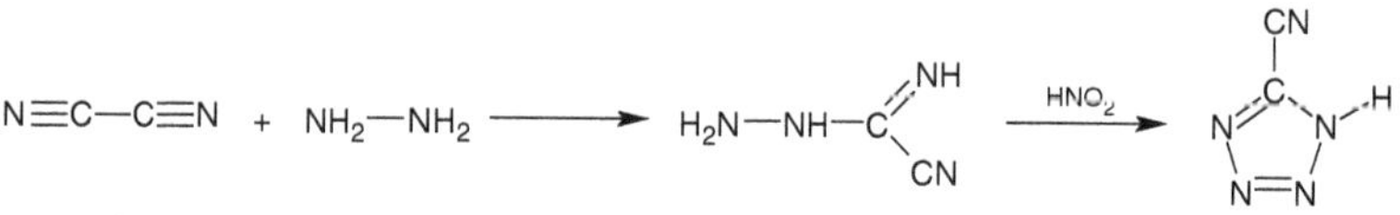

$$N\equiv C-C\equiv N + NH_2-NH_2 \longrightarrow H_2N-NH-C(=NH)(CN) \xrightarrow{HNO_2} \text{5-cyanotetrazole}$$

1-cyanoformimidic acid hydrazide

Preparation of CP

CP is prepared via a batch process described by Fleming et al. (cited by Lieberman [4]). The four steps are summarized below:

$$2\ Co(NO_3)_2 + 8\ NH_3(aq) + 2\ (NH_4)_2CO_3 + 1/2\ O_2 \xrightarrow{H_2O} 2\ [Co(NH_3)_5CO_3]NO_3 + 2\ NH_4NO_3 + H_2O$$
CPCN

$$[Co(NH_3)_5CO_3]NO_3 + 3\ HClO_4 \xrightarrow{H_2O} [Co(NH_3)_5H_2O](ClO_4)_3 + CO_2$$
APCP

$$C_2N_2 + NaN_3 + HCl \xrightarrow{H_2O} \text{5-cyanotetrazole} + NaCl$$

$$\text{5-cyanotetrazole} + [Co(NH_3)_5H_2O](ClO_4)_3 \xrightarrow{H_2O} [\text{Co complex}]^{2+} (ClO_4)_2^-$$

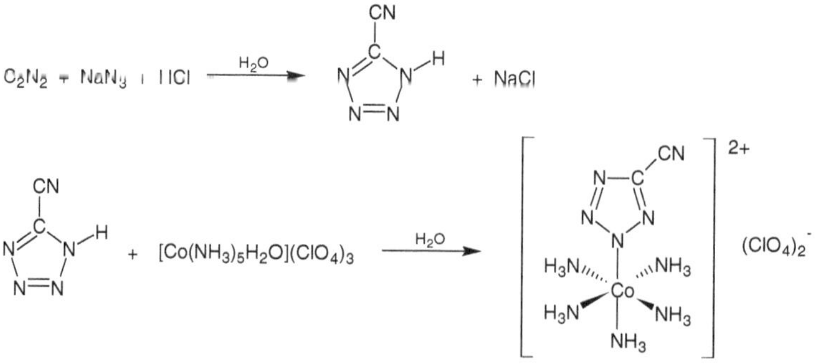

CP

9.1 Cobalt Perchlorate Complexes

The intermediate APCP is purified by recrystallization from an excess of an aqueous perchloric acid. CP is purified by recrystallization from a slightly acidified (by perchloric acid) aqueous solution containing ammonium perchlorate [13].

Two main side products form during the preparation of CP. The first one is 5-carboxamidotetrazolatopentaaminecobalt(III) perchlorate referred to as "amide complex" and the second one referred to as "amidine chelate" of CP (see scheme below). The formation of the amide is assumed to be a simple hydrolysis reaction. The following reaction sequence has been proposed for the formation of these two side products [4, 13]:

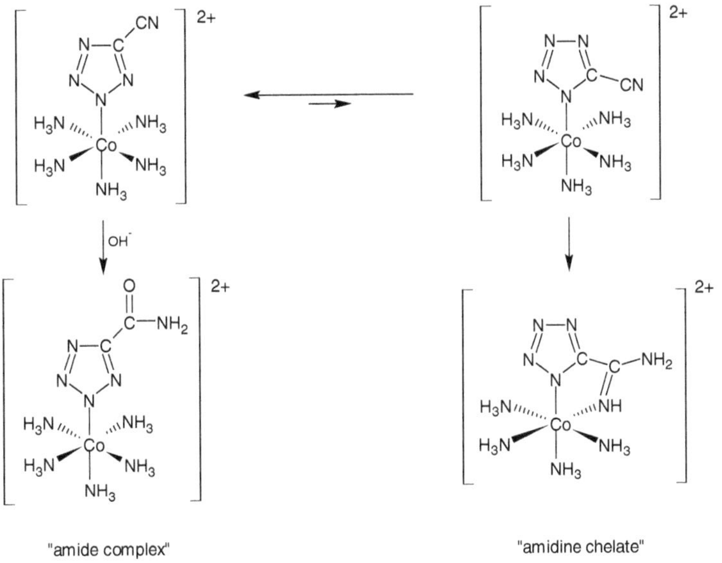

"amide complex"   "amidine chelate"

The "amide complex" is about as soluble as CP [11]. It is however possible to purify CP by recrystallization from an aqueous solution of ammonium perchlorate slightly acidified by perchloric acid. The amount of amide complex decreases from the original content of around 10 % to 2–3 %. In order to obtain a particle size suitable for application in detonators it is necessary to reprecipitate the purified CP from a slightly acidified aqueous solution by controlled addition to chilled stirred isopropyl alcohol [13].

9.1.1.4 Uses

The production of CP began in 1977 and by 1979 the first production of a CP detonator for Department of Energy (USA) had been successfully accomplished [6]. It has been used alone as a primary explosive in "Spark-Safe Low-Voltage Detonators." A generalized CP detonator is shown in Fig. 9.3.

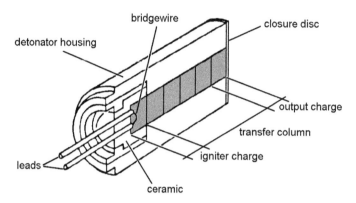

Fig. 9.3 Generalized CP detonator design [4, 20]. Reprinted by permission of IPSUSA Seminars, Inc.

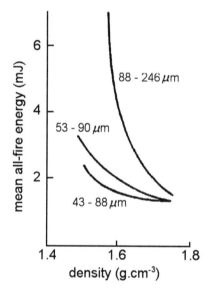

Fig. 9.4 Hot wire ignition sensitivity of CP as a function of density [13]. Reprinted by permission of IPSUSA Seminars, Inc.

CP is pressed in several layers by various loading pressures. The resulting density determines the function of each of the layers as demonstrated in Fig. 9.4 for the all-mean fire energy [20] and in Fig. 9.5 for the run to detonation length [21].

First layer—igniter charge—is initiated by the hot resistive bridgewire. CP in this region is pressed by a higher compacting pressure (at 172–276 MPa) for two reasons: First, to ensure powder-to-bridgewire contact, and secondly to lower the energy necessary for reliable ignition. It can be seen from Fig. 9.4 that the all-fire energy is reduced by increasing CP powder density and decreasing its particle size.

9.1 Cobalt Perchlorate Complexes

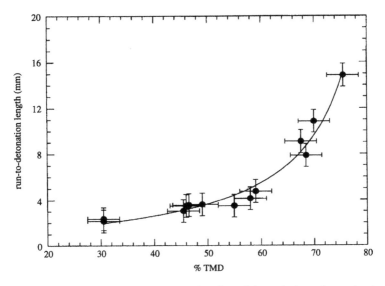

Fig. 9.5 Run to detonation length of CP as a function of theoretical maximum density [21]. Reprinted with permission from Luebcke P. E., Dickson P. M., and Field J. E.: "An experimental study of the deflagration-to-detonation transition in granular secondary explosives", Proceedings of the Royal Society of London, Series A: Mathematical, Physical and Engineering Sciences, 1995, 448, 439–448. Copyright (1995) The Royal Society

For these reasons, a fine CP material (nominally 15 μm average crystal size) is utilized and the ignition regions are loaded to a density of about 1.7 g cm^{-3} [12, 13, 20, 22].

The second layer—transfer column—is loaded by lower pressure (69 MPa). The lower loading pressure of CP permits desired gas flow through the column and thereby promotes the desired deflagration-to-detonation transition (DDT). Reduced density of the powder in this region is required to yield reliable DDT (see Fig. 9.5). Consequently, the DDT region utilizes powder density 1.5 g cm^{-3} (whereas values approaching 1.7 g cm^{-3} fail to yield transition) [12, 13, 20, 22].

The third—an output charge of the detonator—may be the end, or an extension, of the transfer column. If an enhanced output is desired, it can be loaded to higher density (about 1.7 g cm^{-3}) [12, 13, 20, 22].

A high-voltage detonator (HDV) based on CP was also developed for the Titan IV launch vehicle [23]. The production of CP had been discontinued due to its toxicity [21].

9.1.2 CP Analogs

A large number of CP analogs have been synthesized. They varied in either having the cyano group replaced by other groups or the central metal cation replaced with another metal.

The complexes with other 5-substituted (–X is substituent) tetrazole have the general formula $[Co(NH_3)_5N_4CX](ClO_4)_2$ with the same structure as CP. They can be synthesized by the reaction of 5-substituted tetrazole with APCP in an aqueous medium in the same way as CP [24].

Fleming et al. prepared cobalt complexes with 5-substituted tetrazole with the following substituents –X: –CN, –Cl, –H, –NO$_2$, –CH$_3$, and –CF$_3$. The analyses showed the formation of only one isomer of each compound. The substituted tetrazoles are generally N-2 bonded to the cobalt atom. Density values of such substances are in the range from 1.84 to 2.03 g cm^{-3}. All compounds exhibited exothermic decomposition above 250 °C. The increasing electrostatic sensitivity and the increasing ease of transition to detonation correlate with increasing electron-withdrawing ability of the moiety [24].

Johnson et al. studied analogs of CP in which the cobalt atom was replaced by chromium, iridium, rhodium, and ruthenium. The rhodium and chromium complexes can be prepared by reaction of the APCP analog $[Me(NH_3)_5(H_2O)]$ $(ClO_4)_3$ with 5-cyanotetrazole in an aqueous solution, whereas the others have been shown to consist of product mixtures. The rhodium complex is the most thermally stable (comparable to CP) while the iridium, chromium, and ruthenium complexes are significantly less stable (determined by DSC) [4, 25].

Very broad studies into cobalt pentaaminate analogs of CP have been reported from the Saint Petersburg State Technological Institute group of Ilyushin and Tselinskii. Numerous articles have been summarized into two monographs [26, 27].

All synthesized complexes were found to have an octahedral structure with the tetrazole molecule coordinated via the N2 atom. Some of these monosubstituted tetrazoles are summarized in Table 9.2.

Studies of thermal stability of pentaamine cobalt(III) perchlorate complexes during slow heating revealed that only two analogs of CP(–X = –NO$_2$ and –CH=NON=CH–N$_3$) decompose in one macroscopic step. The decomposition of the rest of the complexes is a more complicated multistep process. Due to this it is quite difficult to determine exact decomposition mechanisms. It was, however, found that the process is diffusion controlled. The complexes listed in Table 9.2 undergo pyrolysis if heated in vacuum. Gaseous products evolved in this process do not depend on the type of the substituent in position 5 of the cycle and contain ammonia, nitrogen, water, hydrogen chloride, carbon dioxide, carbon monoxide, oxides of nitrogen, and a small amount of hydrogen cyanide. The perchlorate anion takes part in the oxidation of the ligands only at the end of the decomposition process [27].

9.1 Cobalt Perchlorate Complexes

Table 9.2 Densities, detonation velocities, and temperatures of decomposition of analogs of CP, reprinted from [27]

Substituent (–X)	$\rho_{calculated}$ (g cm^{-3})	$\rho_{experimental}$ (g cm^{-3})	$D_{calculated}$ (km s^{-1}) at ρ	$D_{experimental}$ (km s^{-1}) at ρ	$T_{decomposition}$ (°C)
–H	1.97	1.97	7.14 (1.97)	–	280
–NO$_2$	2.01	2.03	6.30 (1.61)	6.65 (1.61)	265
–CH$_3$	1.90	1.88	6.94 (1.90)	–	282
–N=NO$_2^-$ NH$_4^+$	1.87	1.87	–	6.33 (1.52)	250
–NH$_2$	1.96	1.95	6.14 (1.62)	6.50 (1.62)	270
–C(NO$_2$)$_2^-$ NH$_4^+$	1.88	1.88	–	6.32 (1.48)	201
–CH$_2$N$_3$	1.92	1.94	7.44 (1.94)	–	302 (explosion)
–C(NO$_2$)$_3$	2.01	2.05	8.03 (2.05)	–	132
furazan-NO$_2$	1.99	1.97	7.76 (1.97)	–	280
furazan-N$_3$	1.94	1.95	7.71 (1.95)	–	198
furazan-NH$_2$	–	1.93	7.34 (1.93)	–	280
furazan-N$^-$–NO$_2$ NH$_4^+$	1.87	1.83	7.42 (1.83)	–	255

Sensitivity to impact of 5-substituted CP analogs has been determined by Ilyushin et al. [26] and is reprinted in Table 9.3.

The sensitivity of various analogs of CP to laser monoimpulse was summarized by Ilyushin et al. [28] and more generally in [29].

Best properties from all analogs of CP are reported for pentaamine (5 nitro 2H tetrazolato-N^2)cobalt(III) perchlorate (abbreviated by authors to NKT). It has low sensitivity to electrostatic discharge; predetonation distance is 4.5 mm with charge diameter 6.5 mm and density 1.60–1.63 g cm^{-3}. NKT is thermally stable—heating it at 200 °C in a hermetically sealed container for 6 h does not change its properties. It is not hygroscopic and, unlike CP, it is not too toxic. Careful optimization of reaction conditions gives up to 90 % yields [26].

Table 9.3 Sensitivity to impact, hammer weight 2 kg, sample weight 15 mg [26]

Substituent (–X)	Impact height for 10 % probability of initiation (cm)
–CN	65
–H	95
–NO$_2$	65
–CH$_3$	60
–Cl	55
–CF$_3$	55
RDX	75
PETN	30

9.1.2.1 Uses

NKT has been tested in oil well perforators. It was found that NKT works safely in PVP-1 at temperatures up to 150 °C under 80 MPa in a borehole environment even after being there for 6 h [26].

9.1.3 1,5-Cyclopentamethylenetetrazole Complexes

Experimental work on the preparation of low toxicity primary explosives from the group of cobalt perchlorate complexes has been reported from Russia [26, 30]. The ligand chosen for the complex was "corazol"—1,5-cyclopentamethylenetetrazole:

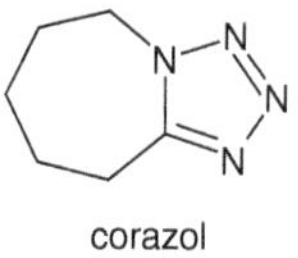

corazol

The first coordination compound published was pentaamine(1,5-cyclopentamethylene-tetrazolato-N^3) cobalt(III) perchlorate (PAC):

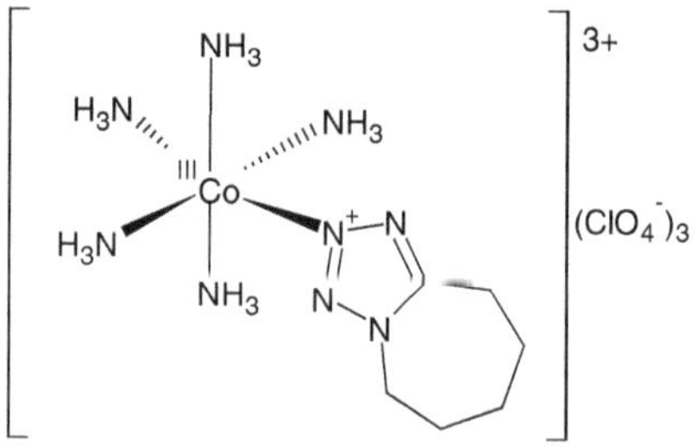

PAC

X-ray crystallography of the complex is not available and the bonding atom of corazol has therefore not been determined. It is however expected to have the above-mentioned general formula. In a recent paper, Ilyushin et al. [31] reported

9.1 Cobalt Perchlorate Complexes

compatibility of organic polynitro compounds with CP analogs having 1,5-cyclopentamethylenetetrazole (PMT) in its molecule. The substances studied included the above-mentioned PAC and its analog with the (1-methyl-5-aminotetrazole-N^2) ligand, and complexes with copper and nickel as central atoms and PMT as ligand:

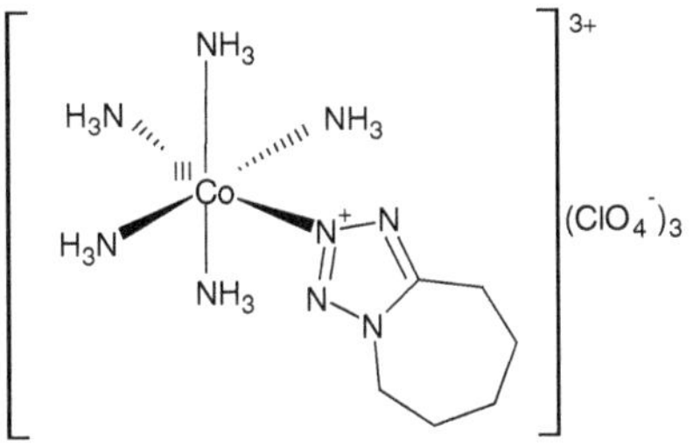

[structures of PAC, Co1M5AT, CuL$_2$, CuL$_4$, NiL$_6$, NiL$_4$]

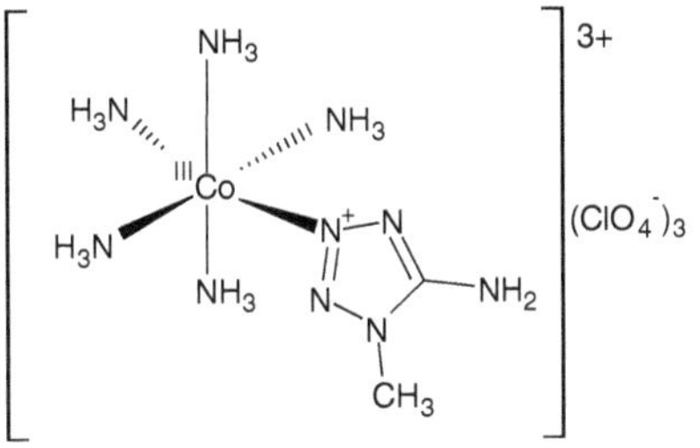

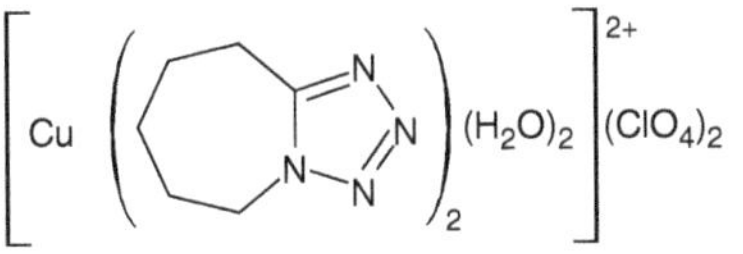

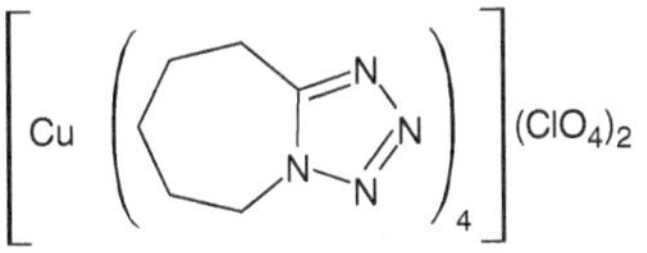

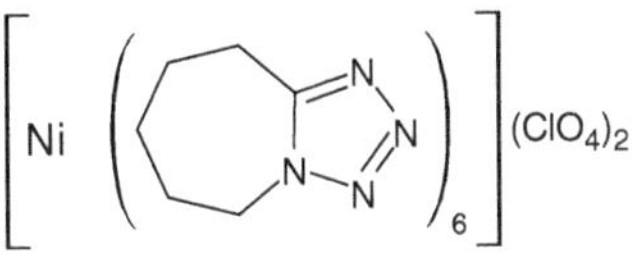

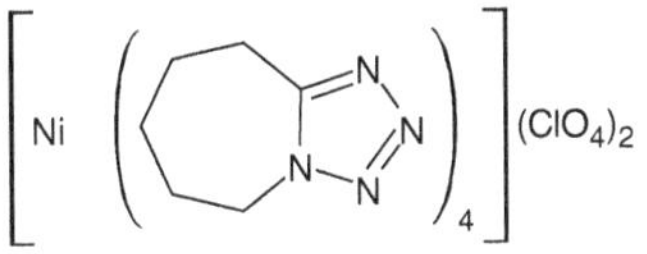

9.1.3.1 Physical and Chemical Properties

The thermal decomposition of the PAC complex determined by nonisothermal heating at 5 °C min^{-1} rate showed relatively good thermal stability. The first mild endothermal process in the form of a broad peak from 95 to 245 °C is accompanied by 14.7 % weight loss. The inner molecules of ammonia break away in this process. At 245 °C melting begins followed by significant exothermal decomposition (250–350 °C) with 71 % mass loss. At even higher temperatures cobalt is oxidized to Co_2O_3 [26]. Thermal stability of CuL$_4$ is similar—onset of fast decomposition is at 230 °C under nonisothermal conditions at a heating rate 5 °C min^{-1}. Some slow exothermal decomposition accompanied by 30 % weight loss is observed in the temperature range from 185 to 235 °C [32].

The thermal stability studies of PAC showed that the decomposition of the complex is a multistep process where the first step is related to a breaking away of internal molecules of ammonia leading to the destruction of the complex. The second step is the decomposition of the corazol ligand supported in later stages by oxidative action of the perchlorate anion [26]. The kinetics of decomposition of the PAC complex as determined by TG, DTG, and DTA provided the following parameters: beginning of the fast decomposition 245 °C; logarithm of preexponential factor in Arrhenius equation 8.6; activation energy 116 kJ mol^{-1} and first reaction order (without diffusion processes) [31]. From these values, it seems that the half-life at 80 °C is around 8 years.

The compatibility (STANAG 4147) of the above-mentioned complexes with tested explosives varies. HMX is compatible with NiL_6 and CuL_4, has some interaction with NiL_4, CuL_2 and PAC and is incompatible with Co1M5AT. CL-20 is compatible with NiL_6, CuL_4, Co1M5AT, PAC, some interaction exists for NiL_4 and minor incompatibility occurs with CuL_2. FOX-7 is compatible with NiL_6, NiL_4, CuL_2, partially compatible with CuL_4 and incompatible with PAC and Co1M5AT. Details of thermal analysis and burning rate studies of mixtures of these complexes with HMX, CL-20, and FOX-7 are reported in the same reference [31–33]. The toxicity of PAC is lower than that of TNT [34].

Oral toxicity was tested on mice. This showed $LD_{50} = 70$ mg kg^{-1} for corazol and $LD_{50} = 300$ mg kg^{-1} for the complex (PAC) placing it in toxicity class 3 (low risk substances). The toxicity of Co(III) ammoniacal complexes is comparable to the toxicity of the Co(III) nitrate (calculated on the metal content). For comparison, it is less toxic than TNT which belongs to class 2 [26].

9.1.3.2 Explosive Properties

Co1M5AT and PAC are known to have short DDT distances. The impact sensitivity of PAC determined by GOST-4545-88 (2 kg hammer, impact height 25 cm) is 16 %. The initiation efficiency for RDX is 0.4 g; the calculated detonation velocity is 6.98 km s^{-1} for density 1.82 g cm^{-3}. This clearly indicates that the complex is a primary explosive (with relatively low initiation efficiency) but with sensitivity to mechanical stimuli comparable to secondary explosives [26].

9.1.3.3 Preparation

Pentaamine(1,5-cyclopentamethylene-tetrazolato-)cobalt(III) perchlorate (PAC) is prepared by the substitution of water by corazol in a slightly acidic aqueous solution according to the following scheme:

$$[Co(NH_3)_5(H_2O)](ClO_4)_3 + (CH_2)_5CN_4 \longrightarrow [Co(NH_3)_5(CH_2)_5CN_4](ClO_4)_3 + H_2O$$

| APCP | corazol | | PAC | |

The reaction is carried out at 90–100 °C for 2–3 h [26].

9.1 Cobalt Perchlorate Complexes

Previous preparation of copper(II) tetrakis-[1,5-cyclopentamethylenetetrazol] perchlorate (CuL$_4$) in a solution of 2,2-dimethoxypropane (yield 96 %) [35] was replaced by a less efficient (yield 80 %) but economically more suitable route using an aqueous solution or isopropyl alcohol. Schematic representation of the preparation of CuL$_4$ from cupric perchlorate hexahydrate and 1,5-cyclopentamethylenetetrazol (PMT) is shown according to the following scheme:

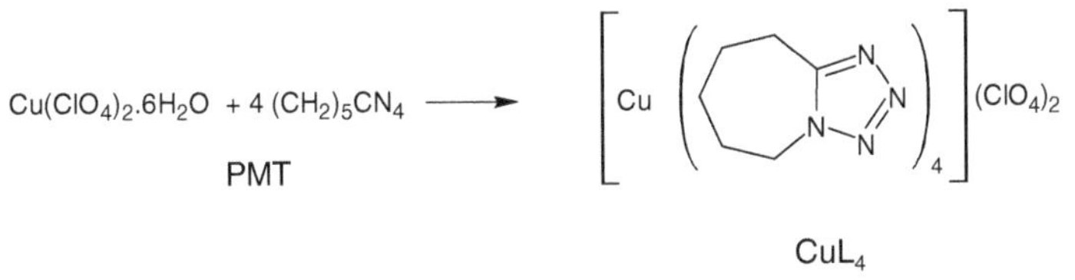

$$Cu(ClO_4)_2 \cdot 6H_2O + 4\ (CH_2)_5CN_4 \longrightarrow [Cu(PMT)_4](ClO_4)_2$$
$$\text{PMT} \qquad\qquad\qquad\qquad\qquad\qquad \text{CuL}_4$$

9.1.3.4 Uses

The complexes are in an experimental stage of development. PAC and Co1M5AT have been tested in model detonators while complexes of nickel and copper have been used as burn rate modifiers in model propellants.

9.1.4 Tetraammine-cis-bis(5-Nitro-2H-tetrazolato-N^2)cobalt(III) perchlorate (BNCP)

In 1986, Bates reported three new similar coordination structures all very much like CP. Most attention was given to tetraammine-*cis*-bis(5-nitro-2*H*-tetrazolato-*N*^{2}) cobalt(III) perchlorate (BNCP) [36].

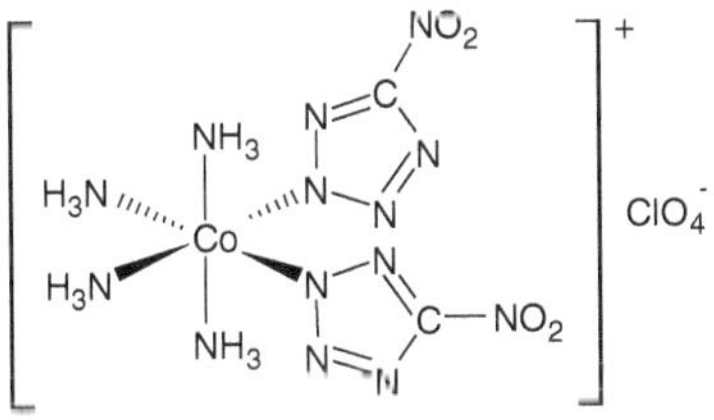

The structure of BNCP is an octahedrally coordinated cobalt atom with four Co−NH$_3$ bonds and two slightly shorter Co−N bonds for the *cis*-tetrazolato ligands [37].

BNCP forms a yellow-gold crystalline solid which is typical for octahedral cobalt(III) complexes having six coordinating nitrogens (as in the case with CP). BNCP crystallizes in a monoclinic system [38]. The density of the BNCP monocrystal is 2.05 g cm^{-3} [27].

9.1.4.1 Explosive Properties

BNCP is slightly more sensitive to impact and friction than CP and exhibits similar sensitivity to electrostatic discharge. Different values of sensitivity of BNCP are published in the literature some of which are summarized in Table 9.4. Ilyushin et al. reported BNCP to be less sensitive to impact than PETN (see Table 9.5). Decreasing particle size decreases sensitivity [27].

BNCP undergoes the DDT process more rapidly than CP under steel confinement and also performs moderately well under aluminum and to a lesser extent also under plastic (Lexan) confinement (Lexan DDT confinement provides velocities of about 70 % of the steel value). CP does not undergo DDT when using either of these two materials for confinement (Al, Lexan) [38, 42].

BNCP and its analogs are sensitive to light and may be detonated by laser pulses. Initial studies [43] report use of a variety of laser sources at 800 and 1,060 nm, with later activities, studying the sensitivity of pressed powders to a single pulse, using solid state neodymium laser (1,060 nm, 1.5 J, 2 ms, beam diameter 1 mm) [44]. BNCP analogs are evaluated as alternatives to substances such as hydrazino tetrazole mercury(II) perchlorate which is probably the most sensitive substance with respect to laser pulses (initiation time around 30 ns at 1.10^{-4} J) [39].

BNCP is also a more powerful explosive than CP (0.66 mm at 80 % TMD versus 0.46 mm at 76 % TMD for CP using depth of dent in a steel witness plate) and further has a very good initiation efficiency—13 mg to initiate PBXN-301 (PETN based PBX). Detonation velocity of CP is 3.2 km s^{-1} at 80 % TMD (1.58 g cm^{-3}) and of BNCP 2.7 km s^{-1} at 76 % TMD (1.54 g cm^{-3}) [38, 45].

Calculated detonation velocity of BNCP is reported to be 8.1 km s^{-1} at density 1.97 g cm^{-3}, beginning of the intensive decomposition is at 269 °C, initiating efficiency is 50 mg for RDX in a number 8 detonator tube, time of the DDT is around 10 µs, and heat of decomposition is 3.32 kJ g^{-1} [27].

9.1.4.2 Preparation

BNCP is prepared by a two-step reaction. Tetraaminecarbonatocobalt(III) nitrate (CTCN) is prepared in the first step followed by its reaction with the sodium salt of 5-nitrotetrazole.

$$4\,Co(NO_3)_2 + 12\,NH_{3(aq)} + 4\,(NH_4)_2CO_3 + O_2 \longrightarrow 4\,[Co(NH_3)_4CO_3]NO_3 + 4\,NH_4NO_3 + 2\,H_2O$$
$$\text{CTCN}$$

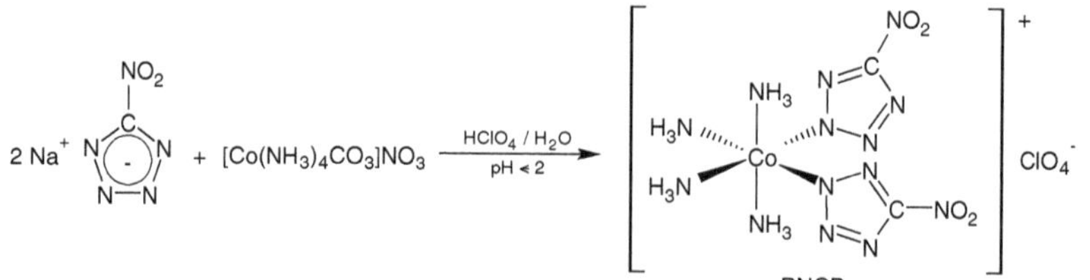

9.1 Cobalt Perchlorate Complexes

Table 9.4 Comparison of sensitivity of BNCP and CP, h_{10} (10 %) and h_{50} (50 %) fire level, confined

	BNCP	CP	Reference
Density (g cm^{-3})	2.03	1.98	[38]
	2.05	–	[39]
Impact sensitivity (cm), 2 kg hammer	50 (h_{10})	55–70^a (h_{10})	[38]
	32 (h_{50})	–	[40]
	Same as PETN	–	[41]
Friction sensitivity^b (g)	600	>1,000	[38]
Electrostatic discharge sensitivity^c (kV)	>25 (loose powder)	>25 (loose powder)	[38]
	4.8^d (79 % TMD)	5.2^d (82 % TMD)	
	8.5^d (89 % TMD)	8.0^d (86 % TMD)	

^{a}Lot-dependent
^{b}Julius Peters BAM tester, threshold
^{c}600 pF, 500 Ω
^{d}Conf

Table 9.5 Comparison of sensitivity of BNCP and some other energetic materials (50 % probability of fire) [27]

	BNCP	CP	LA	PETN	RDX
h_{50} (cm)	17.0	20.9	4.0	12.0	24.0

2,5 kg Hammer, 35 mg sample size

The preparation of the intermediate CTCN is described in inorganic synthesis [46]. The second step in the synthesis of BNCP is carried out using the dihydrate of the sodium salt of 5-nitrotetrazole. The raw BNCP is recrystallized to remove impurities that could compromise its performance and stability [38, 42, 47].

The yield of the reaction is reported to be in the range 50–70 %. Careful analysis of the influence of the reaction conditions revealed that the most important factor is the reaction temperature and the best yield (70 %) is achieved at 90 °C. Alternatively, microwave heating may be employed, which reduces reaction time to one half and increases the yield to 78 % of theoretical [27].

The main drawback of the standard preparation route as described above is the necessity of isolating the highly sensitive intermediate copper 5-nitrotetrazole salt in preparation of sodium 5-nitrotetrazole salt (see salts of 5-nitrotetrazole in Chap. 8.3). A safer way of preparing sodium 5-nitrotetrazole is by a noncatalyzed Sandmayer reaction. It is extremely important to carefully purify the resulting sodium 5-nitrotetrazole from sodium nitrite which otherwise competes with the nitrotetrazole anion in the formation of BNCP leading to complexes with a coordinated nitro group at lower temperatures or to sensitive Co(II) 5-nitrotetrazole monohydrate at higher temperatures. The presence of the latter in BNCP is not acceptable. Details of the reaction conditions and its optimization with respect to suppression of formation of sensitive side products are described by Ilyushin et al. [27].

9.1.4.3 Uses

BNCP was originally proposed as a possible replacement of lead azide and other primary explosives in semiconducting bridge and hot wire detonators [36, 38]. BNCP has also been investigated for application in optical ordinance particularly with regard to laser diode-initiated detonators. Carbon-doped BNCP (1%) is somewhat more sensitive to the diode laser output (800 nm) than comparably doped CP [38, 43].

9.1.5 BNCP Analogs

A number of molecules that could be described as BNCP analogs were studied. Preparation of energetic complexes with $-NO_2$ group bonded to the tetrazole ring replaced by $-NH_2$, $-H$, and $-CH_3$ is possible according to the general equation [44]:

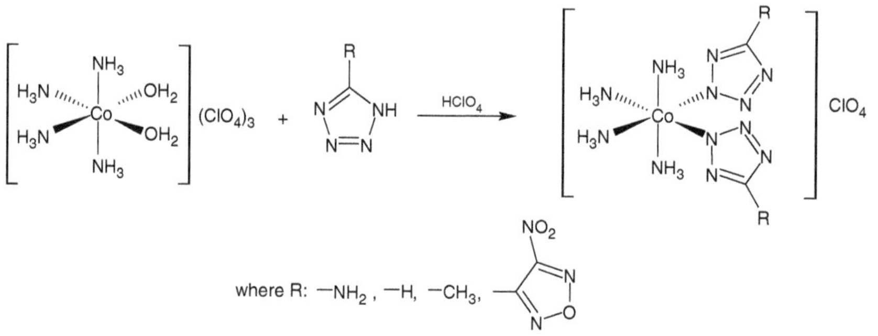

The resulting substances (except for the one resulting from 5(3'-nitro-1',2',5'-oxadiazolyl-4'-)tetrazole) however showed a metal to ligand ratio of 2:3. This would indicate that the general formula of the cation is $[Co_2(NH_3)_8(NT-R)_3]^{3+}$. According to Ilyushin et al., it is believed that one of the three tetrazole rings acts as a bridge between two cobalt atoms giving a structure with the following general formula [27]:

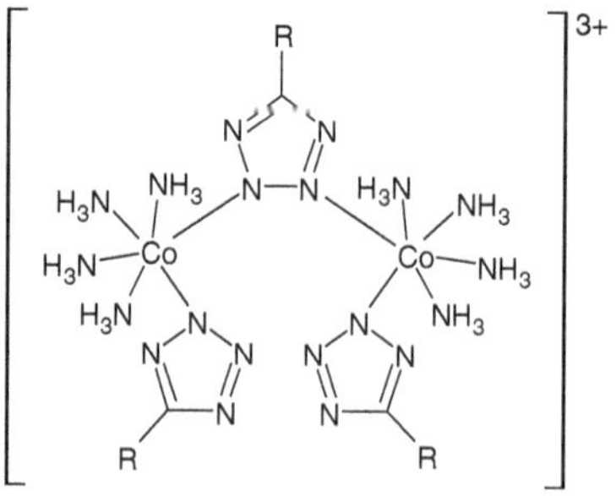

Fig. 9.6 Crystals of 1,5-diaminotetrazole ferrous perchlorate complex [50]. Reprinted by permission of the Department of the Navy employees, M. Bichay, K. Armstrong, R. J. Cramer and the Pacific Scientific employess, J. Fronabarger and M. D. Williams

9.1.6 Perchlorate Complexes of 1,5-Diaminotetrazole

1,5-Diaminotetrazole forms complex compounds with metals. The ferrous (DFeP) and cupric complexes as perchlorate salts have been reported as the most promising candidates for "green" primary explosives [48–50]. The probable chemical formulas of both compounds are shown below.

9.1.6.1 Physical and Chemical Properties

Very little is published about the physical and chemical properties of these 1,5-diaminotetrazole complexes. Crystals of the 1,5-diaminotetrazole ferrous perchlorate complex are shown in Fig. 9.6.

9.1.6.2 Explosive Properties

Sensitivity values of cupric and ferrous perchlorate complexes of 1,5-diaminotetrazole are compared to LS in Table 9.6 and the performance data measured by plate dent test and compared to LA are summarized in Table 9.7.

The effect of the density of the primary explosive on the efficiency of detonators is shown in Table 9.8. Performance of both complexes decreases with increasing loading pressure and density contrary to LA. This phenomenon is greater for the cupric complex. Performance of the ferrous complex is significantly higher at all

Table 9.6 Sensitivity and thermal stability of cupric and ferrous perchlorate complex of 1,5-diaminotetrazole [48, 51]

Compound	Impact sensitivity ball drop (mJ)	Friction sensitivity small BAM (g)		DSC (20 °C min^{-1}) (°C)	
		No-fire level	Low fire level	Onset	Peak
Cu(II) complex	11	100	110	235	240
Fe(II) complex	13	40	50	216	223
LS	25	40	50	290	305

Table 9.7 Plate dent test results of Cu(II) and Fe(II) perchlorate complex of 1,5-diaminotetrazole [48, 50, 51]

Compound	Loading pressure (MPa)	Density (g cm^{-3})	Average dent in Al block (mm)
Cu(II) complex	69	1.412	16.8
Fe(II) complex	69	1.465	27.0
LA (RD 1333)	69	3.119	37.3

Table 9.8 Results of the plate dent test of Cu(II) and Fe(II) perchlorate complex of 1,5-diaminotetrazole compared with LA (aluminum plate) [48, 51]

Loading pressure (MPa)	Density (g cm^{-3})			Average dent in Al block (mm)		
	Cu(II) complex	Fe(II) complex	LA (RD 1333)	Cu(II) complex	Fe(II) complex	LA (RD 1333)
34.5	1.18	1.28	2.62	25.0	35.3	27.7
69	1.33	1.40	2.99	23.7	40.3	33.7
138	1.52	1.52	3.18	14.7	37.7	39.0
276	1.62	1.63	3.55	7.7	23.3	46.3

tested densities than the cupric complex. It surpasses LA at lower densities but brisance of LA is superior to that for the Fe(II) complex at higher densities.

9.1.6.3 Preparation

The metallic complexes are prepared in two steps (a) preparation of the heterocyclic ligand 1,5-diaminotetrazole and (b) preparation of the final complex. 1,5-Diaminotetrazole is prepared by the reaction of thiosemicarbazide, sodium azide, and lead monoxide in a dimethylformamide solvent [48].

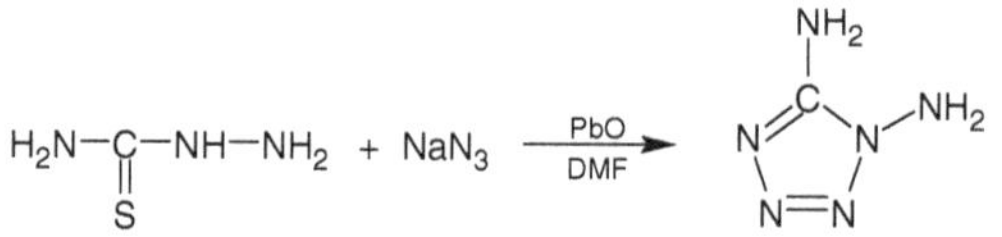

$$H_2N-\underset{\underset{S}{\|}}{C}-NH-NH_2 + NaN_3 \xrightarrow{PbO}{DMF}$$ 1,5-diaminotetrazole

The 1,5-diaminotetrazole then reacts with the appropriate perchlorate to yield the final product [48].

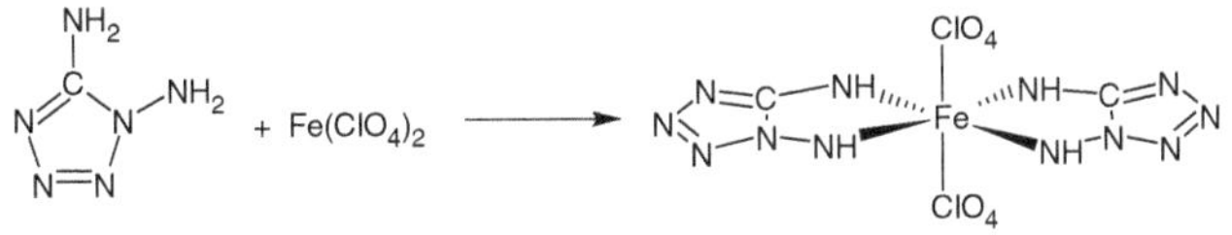

9.1.6.4 Uses

Bichay and Hirlinger have recently reported the ferrous and cupric perchlorate complexes as the most promising candidates for LA replacement [48, 50]. The ferrous complex has been tested as a primary explosive in stab detonators (M59) [50].

9.1.7 Other Perchlorate-Based Complexes

An interesting study on the burning rate of various coordination complexes has been published by Fogelzang et al. [52]. They studied the effect of oxygen-containing oxidant anion (X), organic fuel (L–ligand), and the type of central metal atom (M) of substances of general formula $[M(L)_n]X_2$ on burning behavior of the same. The highest burning rate was observed for complexes with BrO_3^-, ClO_3^-, and ClO_4^- as anion oxidizers. Faster burning rates were also observed for coordination compounds with an easily oxidizable carbohydrazide ligand compared to the ethylenediamine ligand. Variation of the type of oxidant was found to change the burning rate by two orders of magnitude whereas a change in the type of ligand resulted only in a 5- to 10-fold increase. The catalyzing effect was studied for complexes with the following metals: Ni, Co, Cu, Cd, Pb, Zn, Mg, Ca, Ba, and Sr as central atoms. The oxidizer anion was either perchlorate or nitrate and the carbohydrazide was used as the ligand [52].

9.2 Perchlorate-Free Complexes

9.2.1 Iron- and Copper-Based 5-Nitrotetrazolato-N^2 Complexes

A relatively new group of initiating explosives has recently been described by Americans. It is reported as a group of: "previously undescribed green primary explosives based on complex metal dianions and environmentally benign cations,

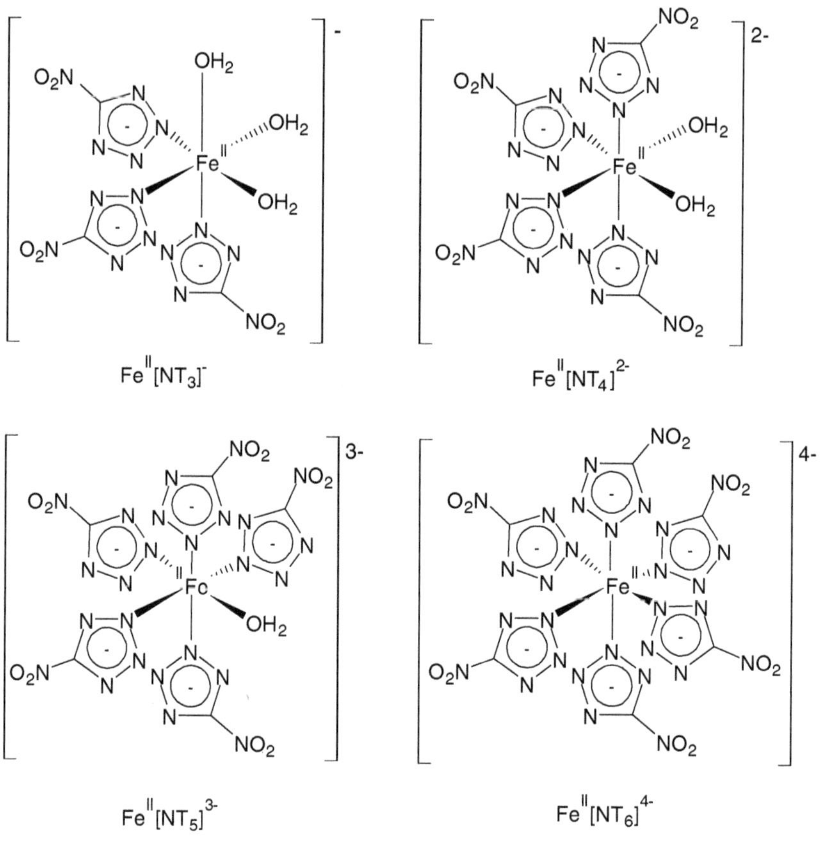

Fig. 9.7 Anions of 5-nitrotetrazolato-N^2-ferrate coordination complexes [53]

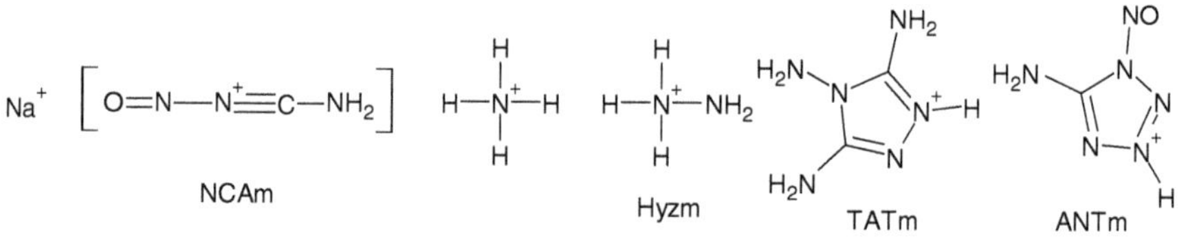

Fig. 9.8 Cations of 5-nitrotetrazolato-N^2-ferrate coordination complexes [53]

$(cat)_2[M^{II}(NT)_4(H_2O)_2]$ (where cat is NH_4^+ or Na^+, M is Fe^{2+} or Cu^{2+}, and NT^- is 5-nitrotetrazolato-N^2)" [53, 54].

The explosive performance is controlled by changing the number of the nitrotetzrazole ligands and the sensitivity by selection of the cation. Four anions with iron as the central atom denominated $Fe^{II}[NT_3]^-$, $Fe^{II}[NT_4]^{2-}$, $Fe^{II}[NT_5]^{3-}$, $Fe^{II}[NT_6]^{4-}$ (Fig. 9.7) and six cations—sodium (Na^+), nitrosocyanaminium (NCAm), ammonium (NH_4^+), hydrazonium (Hyzm), 1,2,5-triamino-1,2,3-triazolium (TATm) and 5-amino-1-nitroso-1,2,3,4-tetrazolium (ANTm) were reported (Fig. 9.8) [53].

9.2 Perchlorate-Free Complexes

Table 9.9 Selected properties and sensitivities of ammonium 5-nitrotetrazolato-N^2-ferrate hierarchies (reprinted from [53, 54])

	$NH_4[Fe^{II}(NT_3)]$	$(NH_4)_2[Fe^{II}(NT_4)]$	$(NH_4)_3[Fe^{II}(NT_5)]$	$(NH_4)_4[Fe^{II}(NT_6)]$
Density (g cm^{-3})	2.10 ± 0.02	2.20 ± 0.03	2.34 ± 0.02	2.45 ± 0.02
Exo. DSC (°C)^a	261	255	253	252
ESD (J)	>0.36	>0.36	>0.36	>0.36
Friction (kg)	4.2	2.8	1.3	0.8
Impact (cm)	15	12	10	8

^{a}Heating rate 5 °C min^{-1}

9.2.1.1 Physical and Chemical Properties

The primary explosives prepared from the above-mentioned cations and anions are sparingly soluble in most common organic solvents and water, are moisture and light resistant, and thermally stable up to 250 °C. The densities of some of the candidates are summarized in Table 9.9.

9.2.1.2 Explosive Properties

All of the above-mentioned primary explosives are insensitive to electrostatic spark regardless of cation type even when dry. When wet they are insensitive to friction, impact, and open flame. The complexes with the greater number of nitrotetrazole ligands are more sensitive and have better performance. Regardless of the number of the ligands, the complexes with a sodium cation detonate while those with an ammonium cation undergo DDT when exposed to open flame. The complexes containing the sodium cation are more sensitive to friction and impact than their ammonium analogs (e.g., friction sensitivity of $(NH_4)_2[Fe^{II}(NT_4)]$ increased from 2,800 g to 20 g for $(Na)_2[Fe^{II}(NT_4)]$). The sensitivity of the ammonium complexes is summarized in Table 9.9 [53, 54].

9.2.1.3 Preparation

The salts are prepared from stoichiometric amounts of NT salt and Fe^{II} salt (e.g., [Fe(H$_2$O)$_6$]Cl$_2$) in refluxing water (or ethanol) for 2 h [55]. Stoichiometric equivalents of the reactants lead to high yields of pure compounds thereby avoiding dangerous purification steps. Further, the preparation leads only to relatively innocuous waste (chlorides and nitrates) [53].

9.2.1.4 Uses

The sodium complex $Na_2[M^{II}(NT)_4(H_2O)_2]$ with M being either Fe^{2+} or Cu^{2+} has been tested in MK1 electric and M55 stab detonators. The tests with MK1 detonator proved that both LA and LS may be fully replaced by the complex (the text does not specify if Fe^{2+} or Cu^{2+} was used). The complex further successfully replaced LA in the transfer charge of an M55 detonator [53].

9.2.2 Perchlorate-Free CP Analogs

Replacing the perchlorate anion in pentaamine cobalt (III) perchlorate complexes seems to be a logical way of eliminating the potentially ecologically problematic anion. The effect of such replacement was studied by Ilyushin et al. using dinitroguanidine which was considered suitable due to its ability to form complexes, its enthalpy of formation, and its oxidizing ability [27].

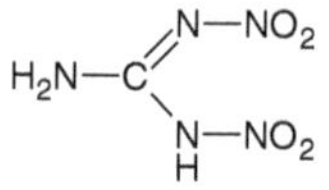

$$H_2N-C\begin{smallmatrix}N-NO_2\\N-NO_2\\H\end{smallmatrix}$$

For comparison, Ilyushin et al. [27] selected complex cobalt(III) pentaaminate perchlorates having temperature of decomposition in the range 265–280 °C and initiating efficiency around 0.2 g for RDX.

The preparation of the complex substance was achieved by dosing of an aqueous solution of the relevant perchlorate complex to a stirred alcoholic solution of dinitroguanidine at room temperature. Excess of dinitroguanidine was about 100 % compared with the theoretical. Over 30–40 min, the dinitroguanidine complex precipitated from the reaction mixture with an almost quantitative yield according to the following reaction scheme [27]:

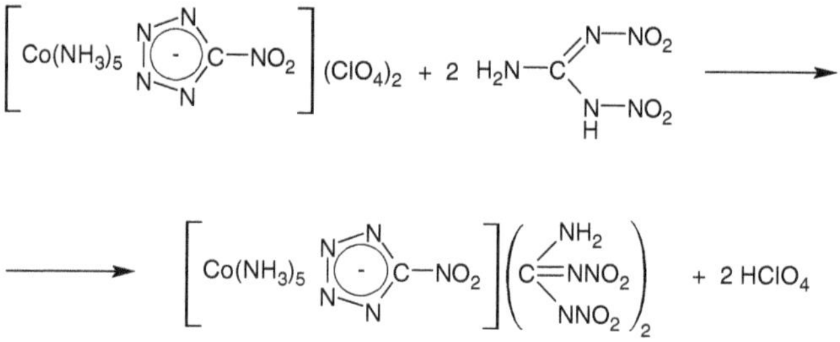

$$[Co(NH_3)_5\text{-tetrazole-}C-NO_2](ClO_4)_2 + 2\ H_2N-C(N-NO_2)(N-NO_2H) \longrightarrow$$

$$\longrightarrow [Co(NH_3)_5\text{-tetrazole-}C-NO_2]\left(C(NH_2)=NNO_2 / NNO_2\right)_2 + 2\ HClO_4$$

I

The following two complexes were prepared from other perchlorate analogs to complement the one mentioned above:

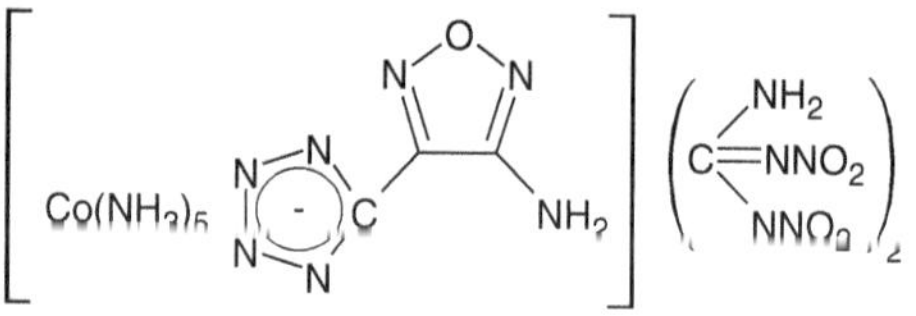

$$[Co(NH_3)_5\text{-tetrazole-}C\text{-furazan-}NH_2]\left(C(NH_2)=NNO_2 / NNO_2\right)_2$$

II

$$[Co(NH_3)_5\text{-tetrazole-}C\text{-furazan-}NH_2](N_3)_2$$

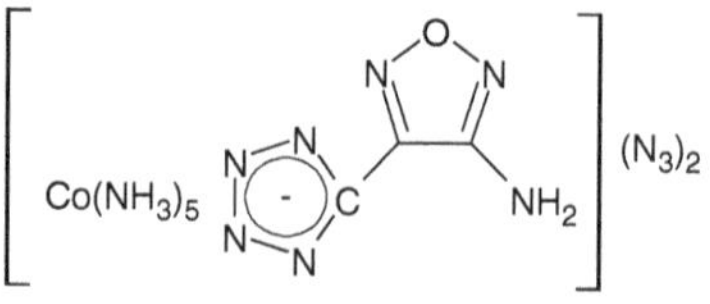

III

Table 9.10 Properties of pentaamine cobalt(III) complex dinitroguanidines [27]

Complex	Calculated TMD (g cm^{-3})	Temperature at start of fast decomposition (°C)	Initiating efficiency (mg)	Experimental detonation velocity (km s^{-1}) at density (g cm^{-3})
I	1.83	185	>500	6.42/1.736
II	1.98	200	480	6.50/1.38
III	1.95	165	>500	–

The temperature of the start of fast decomposition (DTA, 5 °C min^{-1}) decreases by 80–100 °C when the perchlorate anion is exchanged for dinitroguanidine or azide. Unlike pentaamine (5-nitro-2H-tetrazolato-N^2)cobalt(III) perchlorate (NKT), which decomposes in one step, complex I decomposes in four steps when heated by 5 °C min^{-1}. It also loses crystal water around 110 °C.

All three complexes (I–III) listed in Table 9.10 are less sensitive to impact than PETN. Theoretical detonation velocities of all three complexes at their TMD are over 8 km s^{-1}. From the experimental results, it seems that these complexes can be dead pressed. The results in Table 9.10 clearly indicate that replacement of perchlorate by dinitroguanidine or azide leads to a product with insufficient thermal stability and insufficient performance [27].

9.3 Other Transition Metal-Based 5-Nitrotetrazolato-N^2 Complexes

Analogs of the previously mentioned 5-nitrotetrazolato-N^2-ferrate type complexes can be prepared using other transition metals including cobalt (Co^{2+}), copper (Cu^{2+}), manganese (Mn^{2+}), nickel (Ni^{2+}), and zinc (Zn^{2+}). Other types of cations including alkaline metals, alkaline earth metals, aliphatic and catenated high-nitrogen cations, and heterocyclic nitrogen cations have also been patented by Huynh [54]. The same author also patented a number of complexes based on the 1,5-diamino tetrazole ligand [56].

Interesting complexes have been reported to form by reaction of Cu^{2+} ions with bis(tetrazolyl)amine—(H_2bta). Depending on the reaction conditions three substances may form: $Cu(bta)(NH_3)_2$, $Cu(bta)(NH_3)_2 \cdot H_2O$ or $(NH_4)_2Cu(bta)_2 \cdot 2.5 H_2O$. These are all high nitrogen compounds but are surprisingly reported as being insensitive to impact and friction [57].

References

1. Connelly, N.G., Damhus, T., Hartshorn, R.M., Hutton, A.T. (eds.): Red Book: Nomenclature of Inorganic Chemistry: IUPAC Recommendations 2005. Royal Society of Chemistry Publishing, Cambridge (2005)
2. Sinditskii, V.P., Fogelzang, A.E.: Energeticheskie materialy na osnove koordinacionnykh soedinenii. Rossiiskii khimicheskii zhurnal **XLI**, 74–80 (1997)
3. Ilyushin, M.A., Tselinskii, I.V.: Iniciiruyushchie vzryvshchatye veshchestva Sostoyanie i perspektivy. Rossiiskii khimicheskii zhurnal **XLI**, 3–13 (1997)
4. Lieberman, M.L.: Chemistry of (5-cyanotetrazolato-N^2)pentaamminecobalt(III) perchlorate and similar explosive coordination compounds. Ind. Eng. Chem. Prod. Res. Dev. **24**, 436–440 (1985)
5. Massis, T.M., Morenus, P.K., Huskisson, D.H., Merrill, R.M.: Stability and compatibility studies with the inorganic explosive 2-(5-cyanotetrazolato)pentaamminecobalt(III) perchlorate (CP). J. Hazard. Mater. **5**, 309–323 (1982)
6. Weese, R.K., Burnham, A.K., Fontes, A.T.: A study of the properties of CP: Coefficient of thermal expansion, decomposition kinetics and reaction to spark, friction and impact. Int. Annu. Conf. ICT **36**, 37/1–37/12 (2005)
7. Graeber, E.J., Morosin, B.: Structures of pentaammine(5-cyanotetrazolato-N^2)cobalt(III) perchlorate (CP), [Co(C$_2$N$_5$)(NH$_3$)$_5$](ClO$_4$)$_2$, and (5-amidinotetrazolato-N^1, N^5)tetraamminecobalt(III) bromide (ATCB), [Co(C$_2$H$_3$N$_6$)(NH$_3$)$_4$]Br$_2$. Acta Crystallogr. **C39**, 567–570 (1983)
8. Balahura, R.J., Purcell, W.L., Victoriano, M.E., Lieberman, M.L., Loyola, V.M., Fleming, W., Fronabarger, J.W.: Preparation, characterization, and chromium(II) reduction kinetics of tetrazole complexes of pentaamminecobalt (III). Inorg. Chem. **22**, 3602–3608 (1983)
9. Geng, J., Lao, Y.: Electronic structure and thermal decomposition of high energy coordination compound CP. In: Proceedings of 17th International Pyrotechnics Seminar, pp. 410–415, Beijing, China, Oct 28-31, 1991
10. Geng, J., Lao, Y.: The electronic energy-state structure of new explosive 2-(5-cyanotetrazolato) pentaamminecobalt(III) perchlorate. In: Proceedings of 16th International Pyrotechnics Seminar, pp. 304–313, Jönköping, Sweden, June 24-28, 1991
11. Massis, T.M., Morenus, P.K., Huskisson, D.H., Merrill, R.M.: In: Barton, L.R. (ed.) Compatibility of Plastics/Materials with Explosives Processing Explosives, pp. 105–119. Materials Process Division, American Defense Preparedness Association, Washington, DC (1980)
12. Lieberman, M.L.: Bonfire-safe low-voltage detonator. US Patent 4,907,509, 1990
13. Lieberman, M.L., Fronabarger, J.W.: Status of the development of 2-(5-cyano-2 H-tetrazolato) penta ammine cobalt (III) perchlorate for DDT devices. In: Proceedings of 7th International Pyrotechnic Seminar, pp. 322–355, Vail, Colorado, 1980
14. Arp, H.P.H., Decken, A., Passmore, J., Wood, D.J.: Preparation, characterization, X-ray crystal structure, and energetics of cesium 5-cyano-1,2,3,4-tetrazolate: Cs[NCCNNNN]. Inorg. Chem. **39**, 1840–1848 (2000)
15. Oliveri-Mandala, E., Passalacqua, T.: Azione dell'acido azotidrico sul cianogeno. Formazione del ciano-tetrazolo. Gazzetta Chimica Italiana **41**, II, 430–435 (1911)
16. Lifschitz, J.: Synthese der Pentazol-Verbindungen I. Berichte der deutschen chemischen Gesellschaft **48**, 410–420 (1915)
17. Benson, F.R.: The chemistry of the tetrazoles. Chem. Rev. **41**, 1–61 (1947)
18. Matsuda, K., Morin, L.T.: Preparation and reactions of 1-cyanoformimidic acid hydrazide. J. Org. Chem. **26**, 3783–3787 (1961)
19. Morin, L.T., Matsuda, K.: Preparation of 5-cyanotetrazoles. US Patent 3,021,337, 1962
20. Lieberman, M.L., Villa, F.J., Marchi, D.L., Lause, A.L., Yates, D., Fronabarger, J.W.: Review of low voltage detonators. In: Proceedings of 11th Symposium on Explosives and Pyrotechnics, Philadelphia, 1981

21. Luebcke, P.E., Dickson, P.M., Field, J.E.: An experimental study of the deflagration-to-detonation transition in granular secondary explosives. Proc. R. Soc. Lond. A Math. Phys. Eng. Sci. **448**, 439–448 (1995)
22. Lieberman, M.L.: Spark-safe low-voltage detonator. US Patent 4,858,529, 1989
23. Munger, A.C., Tibbitts, E.E., Neyer, B.T., Thomes, J.A., Knick, D.R., Demana, T.A.: Development of a full qualified detonator for the Titan IV launch vehicle. In: Proceedings of 22nd International Pyrotechnics Seminar, pp. 591–615, Fort Collins, CO, 1996
24. Fleming, W., Fronabarger, J.W., Lieberman, M.L., Loyola, V.M.: Synthesis and characterization of 5-substituted tetrazolatopentaamminecobalt(III) perchlorates. In: Proceedings of 2nd Chemical Congress of the North American Continent, Las Vegas, NV, 1980
25. Johnson, R., Fronabarger, J., Fleming, W., Lieberman, M., Loyola, V.: Synthesis and characterization of 5-substituted tetrazolatopentaamminecobalt(III) perchlorates. In: Proceedings of International Chemical Congress of Pacific Basin Societies, Honolulu, HI, 1984
26. Ilyushin, M.A., Tselinskii, I.V., Sudarikov, A.M.: Razrabotka komponentov vysokoenergeticheskikh kompozicii. SPB:LGU im. A.S.Pushkina – SPBGTI(TU), Sankt Peterburg (2006)
27. Ilyushin, M.A., Sudarikov, A.M., Tselinskii, I.V.: Metallokompleksy v vysokoenergeticheskikh kompoziciyakh. SPB:LGU im. A.S.Pushkina – SPBGTI(TU), Sankt Peterburg (2010)
28. Ilyushin, M.A., Tselinskii, I.V., Chernai, A.V.: Svetochustvitelnye vzryvshchatye veshchectva i sostavy i ich iniciirovanie lazernym monoimpulsom. Rossiiskii khimicheskii zhurnal **XLI**, 81–88 (1997)
29. Ilyushin, M.A., Tselinskii, I.V.: Ispolzovanie lazernogo iniciirovaniya energoemkikh coedinenii v nauke i tekhnike (obzor). Zhurnal prikladnoi khimii **73**, 1233–1239 (2000)
30. Ilyushin, M.A., Tselinskii, I.V., Bachurina, I.V., Novoselova, L.O., Konyushenko, E.N., Kozlov, A.S., Gruzdev, Y.A.: Application of energy saturated complex perchlorates. In: Avrorin, E.N., Simonenko, V.A. (eds.) Proceedings of Zababakin Scientific Talks—International Conference on High Energy Density Physics, vol. 849, Snezhinsk, Russia (2005)
31. Ilyushin, M.A., Bachurina, I.V., Smirnov, A.V., Tselinskii, I.V., Shugalei, I.V.: Study of the interaction of polynitro compounds with transition metal complexes with 1,5-pentamethylenetetrazole as a ligand. Cent. Eur. J. Energetic Mater. **7**, 33–46 (2010)
32. Bachurina, I.V., Ilyushin, M.A., Tselinskii, I.V., Gruzdev, Y.A.: High-energy complex copper (II) perchlorate with 1,5-pentamethylenetetrasole as ligand. Russ. J. Appl. Chem. **80**, 1643–1646 (2007)
33. Ilyushin, M.A., Gruzdev, Y.A., Bachurina, I.V., Smirnov, A.V., Tselinskii, I.V., Andreeva, Y.N.: The compatibility of high explosives with energetic coordination complexes of cobalt(II). In: Otis, J., Krupka, M. (eds.) Proceedings of New Trends in Research of Energetic Materials, vol. 2, pp. 640–644. Univerzita Pardubice, Pardubice (2007)
34. Zhilin, A.Y., Ilyushin, M.A., Tselinskii, I.V., Nikitina, Y.A., Kozlov, A.S., Shugalei, I.V.: Complex energetic perchlorates of cobalt(III) amminates, with cyclopentamethylenetetrazole as ligand. Russ. J. Appl. Chem. **78**, 188–192 (2005)
35. Kuska, H.A., Ditri, F.M., Popov, A.I.: Electron spin resonance of pentamethylenetetrazole manganese(II) and copper(II) complexes. Inorg. Chem. **5**, 1272–1277 (1966)
36. Bates, L.R.: The potential of tetrazoles in initiating explosives systems. In: Proceedings of 13th Symposium on Explosives and Pyrotechnics, pp. III1–III10, Hyatt, Palmetto Dunes, Hilton Head Island. South Carolina. December 2-4, 1986
37. Morosin, B., Dunn, G.R., Assink, R., Massis, T.M., Fronabarger, J., Duesler, E.N.: The secondary explosive tetraamine-cis-bis(5 H-nitro-2 H-tetrazolato-N^2)cobalt(III) perchlorate at 293 and 213 K. Acta Crystallogr. **C53**, 1609–1611 (1997)
38. Fronabarger, J., Schuman, A., Chapman, R.D., Fleming, W., Sanborn, W.B., Massis, T.: Chemistry and development of BNCP, a novel DDT explosive. In: Proceedings of 31st AIAA 95-2858; Joint propulsion conference and exhibit, San Diego, CA, 1995

39. Ilyushin, M.A., Tselinskii, I.V., Zhilin, A.Y., Ugryumov, I.A., Smirnov, A.V., Kozlov, A.S.: Coordination complexes as inorganic explosives for initiation systems. Hanneng Cailiao (Energetic Materials) **12**, 15–19 (2004)
40. Talawar, M.B., Agrawal, A.P., Asthana, S.N.: Energetic co-ordination compounds: Synthesis, characterization and thermolysis studies on bis-(5-nitro-2H-tetrazolato-N^2)tetraammine cobalt (III) perchlorate (BNCP) and its new transition metal (Ni/Cu/Zn) perchlorate analogues. J. Hazard. Mater. **120**, 25–35 (2005)
41. Zeman, S.: Technologie základních výbušin. Univerzita Pardubice, Pardubice (2005)
42. Fronabarger, J.W., Sanborn, W.B., Massis, T.: Recent activities in the development of the explosive-BNCP. In: Proceedings of 22nd International Pyrotechnics Seminar, pp. 645–652, Fort Collins, Colorado USA, July 15-19, 1996
43. Merson, J.A., Salas, F.J., Harlan, J.G.: The development of laser ignited deflagration-to-detonation transition (DDT) detonators and pyrotechnic actuators. In: Proceedings of 19th International Pyrotechnics Seminar, pp. 191–206, Christchurch, NEW ZEALAND, Feb 21-25, 1994
44. Zhilin, A.Y., Ilyushin, M.A., Tselinskii, I.V., Kozlov, A.S., Lisker, I.S.: High energy capacity cobalt (II) tetrazolates. Zhurnal prikladnoi khimii **76**, 592–596 (2003)
45. Fyfe, D.W., Fronabarger, J., Brickes, R.W.: BNCP prototype detonator studies using a semiconductor bridge initiator. In: Tulis, A.J. (ed.) International Pyrotechnics Seminar, vol. 20, pp. 341–343. IIT Research Institute, Colorado Springs, CO (1994)
46. Schlessinger, G.: Inorganic Synthesis. McGraw-Hill, New York (1960)
47. Fronabarger, J., Schuman, A., Chapman, R.D., Fleming, W., Sanborn, W.B., Massis, T.: Chemistry and development of BNCP, a novel DDT explosive. In: Proceedings of International Symposium on Energetic Materials Technology, ADPA Meeting #450, pp. 254–258, Orlando, FL, 1994
48. Hirlinger, J.M., Bichay, M.: New Primary Explosives Development for Medium Caliber Stab Detonators. Report SERDP PP-1364. US Army ARDEC, Washington, DC (2004)
49. Hirlinger, J.M.: Ivestigating Alternative 'GREEN' Primary Explosives In: Proceedings of NDIA 39th Annual Gun & Ammunition Missiles & Rockets Conference & Exhibition; Presentation, Baltimore, USA, 2004
50. Hirlinger, J., Fronabarger, J., Williams, M., Armstrong, K., Cramer, R.J.: Lead azide replacement program. In: Proceedings of 49th Annual Fuze Conference. National Defense Industrial Association (NDIA), Seattle, WA (2005)
51. Hirlinger, J.: Investigating alternative "green" primary explosives. In: Proceedings of 39th Annual Gun & Ammunition Conference, Baltimore, MD, 2004
52. Fogelzang, A.E., Sinditskii, V.P., Egorshev, V.Y., Serushkin, V.V.: Effect of structure of energetic materials on burning rate. In: Brill, T.B., Russel, T.P., Tao, W.C., Wardle, R.B. (eds.) Proceedings of Material Research Society Symposia, vol. 418, pp. 151–161. Materials Research Society, Boston, MA (1995)
53. Huynh, M.H.V., Coburn, M.D., Meyer, T.J., Wetzer, M.: Green primaries: Environmentally friendly energetic complexes. Proc. Natl. Acad. Sci. **103**, 5409–5412 (2006)
54. Huynh, M.H.V.: Lead-free primary explosives. Patent WO2008054538, 2008
55. Huynh, M.H.V., Coburn, M.D., Meyer, T.J., Wetzer, M.: Green primary explosives: 5-nitrotetrazolato-N^2-ferrate hierarchies. Proc. Natl. Acad. Sci. **103**, 10322–10327 (2006)
56. Huynh, M.H.V.: Explosive complexes. US Patent 20080200688A1, 2008
57. Friedrich, M., Gavez-Ruiz, J.C., Klapötke, T.M., Mayer, P.: BTA copper complexes. Inorg. Chem. **44**, 8044–8052 (2005)

Chapter 10
Organic Peroxides

10.1 Peroxides of Acetone

The reaction of acetone with hydrogen peroxide gives various types of organic peroxides of acetone. The simple linear hydroperoxides and hydroxyperoxides form at first and then condense to the linear dimer and trimer analogs. The linear dihydroperoxide 2,2-dihydroperoxypropane (I) and its dimer bis(2-hydroperoxypropane)peroxide (II) form as a main product in an acetone and hydrogen peroxide mixture, in the absence of an acidic catalyst [1].

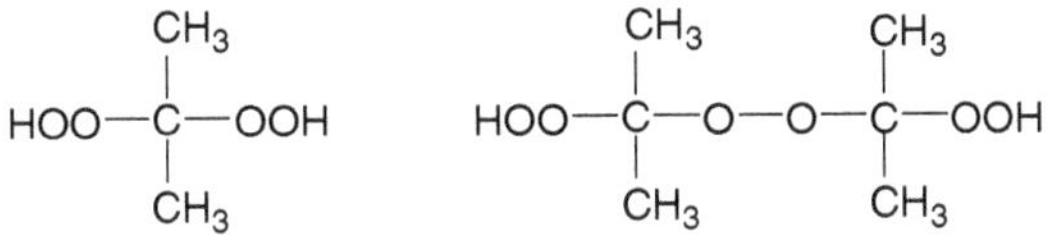

$$\underset{I}{HOO-\underset{\underset{CH_3}{|}}{\overset{\overset{CH_3}{|}}{C}}-OOH} \qquad \underset{II}{HOO-\underset{\underset{CH_3}{|}}{\overset{\overset{CH_3}{|}}{C}}-O-\underset{\underset{CH_3}{|}}{\overset{\overset{CH_3}{|}}{C}}-OOH}$$

Depending on the reaction conditions—primarily pH, concentration of reagents—the cyclic dimer, trimer form from their linear analogs. Cyclic dimer and trimer crystallize from the reaction mixture when catalyzed by an acid. The reaction mechanisms of cyclic peroxide formation have been investigated by several authors and are schematically summarized below [1–7].

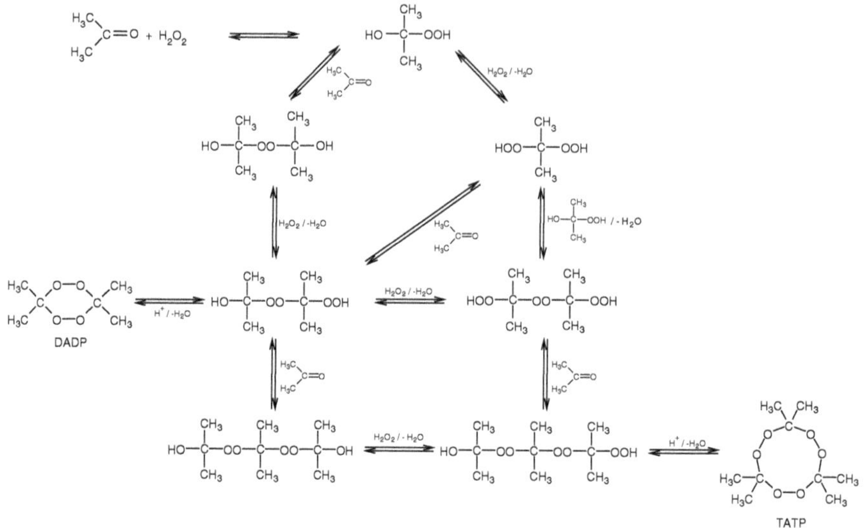

[Scheme showing equilibria between acetone + H_2O_2 and various peroxide products including DADP and TATP]

The reaction conditions for the preparation of dimer and trimer peroxides are quite similar. This may seem a little confusing, as the same reactants and catalysts are used for preparation of TATP as well as for DADP. The issue of particular reaction conditions leading to dimer and trimer remains without a clear explanation. We found that the key factor governing the composition of the final product (which is either DADP, TATP, or a mixture of both) is the acidity of the reaction mixture. The trimer forms in a slightly or moderately acidic solution while the dimer (or mixture of both) forms in a highly acidic environment [8].

The formation of higher linear peroxides (oligoperoxides) of acetone as by-products in the preparation of the trimer has recently been published. Sigman et al. proved the presence of peroxides terminated by n hydroperoxygroups where $n = 1$–8. These peroxides have low thermal stability and convert into the trimer upon heating [9, 10].

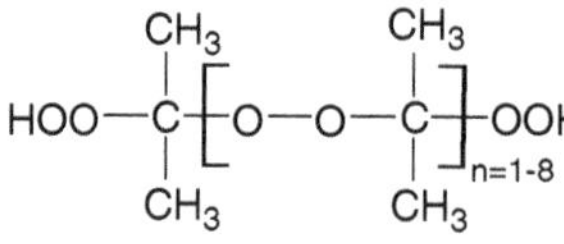

$$HOO-C(CH_3)_2-[O-O-C(CH_3)_2]_{n=1-8}-OOH$$

Linear peroxides of acetone have never been used as primary explosives and therefore we have decided to focus exclusively on cyclic organic peroxides in the following sections.

10.1.1 Diacetone Diperoxide

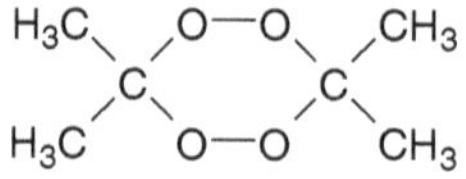

[Structure of diacetone diperoxide: six-membered ring with two $C(CH_3)_2$ groups connected by two $O-O$ bridges]

Diacetone diperoxide (3,3,6,6-tetramethyl-1,2,4,5-tetroxane; DADP) was first synthesized by Baeyer and Villiger in 1899 by the reaction of acetone with "Caro's" reagent (H_2SO_5, prepared by rubbing potassium persulfate with concentrated sulfuric acid and then adding potassium sulfate) [11]. This substance showed similar properties (identical elemental analysis, similar solubility and explosiveness) as already 4 years known "Wollfenstein's superoxide" (TATP) but melted at a significantly higher temperature (132–133 °C vs. 90–94 °C for TATP). Baeyer and Villiger have therefore continued their research and, based on the results of cryoscopic measurements, concluded that this substance is a dimer [12].

10.1.1.1 Physical and Chemical Properties

The molecule of diacetone diperoxide exhibits chain conformation in the solid state (see Fig. 10.1) [13, 14]. These conformers are bonded by intermolecular hydrogen bonds which result in the formation of layers of molecules as shown in Fig. 10.2. This picture further clearly shows that all of the oxygen atoms of DADP are involved in hydrogen bond bridges [13].

Diacetone diperoxide forms colorless crystals (see Fig. 10.3) with crystal density 1.33 g cm^{-3} (X-ray) [13, 14] and melting point published mostly in the range 130–132 °C [11, 15–18].

DADP is a very volatile substance; vapor pressure of DADP is 17.7 Pa at 25 °C (determined by gas chromatography). The Clapeyron equation, which describes dependence of vapor pressure (P in Pa) on temperature (T in K), is:

$$\ln P = 35.9 - 9,845.1/T$$

DADP is a more volatile substance than TATP [19]. Its vapor pressure is, according to Oxley, about 2.6 times higher than that of TATP at the same temperature. The reason is most likely due to a lower molecular weight. The heat of sublimation of DADP is 81.9 kJ mol^{-1} (calculated from the Clapeyron equation) [20].

Egorshev et al. [19] described the dependence of the vapor pressure (P in atm) on temperature (T in K) by the following equations:

$$\ln P = 20.82 - 8,569.1/T \quad \text{(for temperature range } 67 - 120\,°C)$$
$$\ln P = 13.5 - 5,600/T \quad \text{(DADP in liquid state – above } 120\,°C)$$

Heat of sublimation, evaporation, and melting for DADP are 71.3 kJ mol^{-1}, 46.5 kJ mol^{-1}, and 24.7 kJ mol^{-1} respectively. These values were determined from the above-mentioned equations by Egorshev et al. [19].

DADP is soluble in most organic solvents and practically insoluble in water and dilute acids and bases. DADP does not react with water even when boiling. It shows very good compatibility with metals [17, 18] and only slowly decomposes by the action of a solution of zinc sulfate or cupric chloride in combination with metallic

Fig. 10.1 3D structure of the DADP molecule [13]. Reprinted with permission from Dubnikova et al., Decomposition of TATP Is an Entropic Explosion, J. Am. Chem. Soc. 2005, 127, 1146–1159. Copyright (2005) American Chemical Society

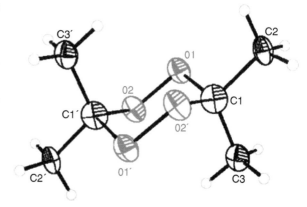

Fig. 10.2 Hydrogen-bonded network via $C(sp^3)$–H···O bridges in the plane [13]. Reprinted with permission from Dubnikova et al., Decomposition of TATP Is an Entropic Explosion, J. Am. Chem. Soc. 2005, 127, 1146–1159. Copyright (2005) American Chemical Society

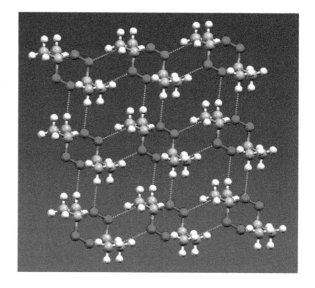

zinc or copper at room temperature within several hours [21]. In solutions, DADP slowly decomposes on heating. Cafferata et al. studied thermal decomposition of DADP in benzene in the temperature range 135.5–165 °C. Acetone, oxygen, and toluene are major decomposition products; methyl isopropyl ether, methyl acetate, bibenzyl, methanol, methane, ethane, and carbon dioxide are minor products. The authors also proposed a chemical mechanism for thermolysis [22].

10.1.1.2 Explosive Parameters

The impact sensitivity of DADP reported in Fedoroff, Shefield, and Kaye's encyclopedia is 7 cm for 2 kg hammer [17]; the Zhukov encyclopedia reported lower sensitivity than lead azide [18]. Egorshev et al. measured the impact

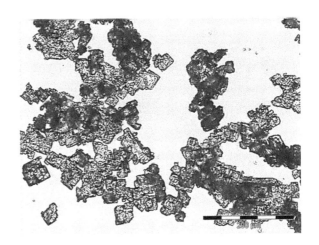

Fig. 10.3 DADP crystals

sensitivity of DADP and TATP and compared it with tetrazene (GNGT). Results are summarized in Table 10.1. DADP is almost two times less sensitive than TATP, and the sensitivity of GNGT is in-between the sensitivity of TATP and DADP [19].

The friction sensitivity of DADP is according to own results about the same as the sensitivity of TATP and slightly lower than LS (see Fig. 2.19) [23]. Sensitivity of DADP to electrostatic discharge was measured and found to be lower than that for its trimer analog and other common primary explosives (see Fig. 2.21) [24].

Ignition of small amounts of DADP by flame produces high flames similar to those produced by combustion of nitrocellulose. In comparison to the trimer, combustion is slower and less violent without any sound effect and producing a bit more soot. The dependence of burning rate of DADP and TATP on pressure has recently been studied by Egorshev et al. (see Fig. 10.4). The burning rate increases proportionally with pressure for both peroxides to reach about 1 cm s^{-1} at atmospheric pressure for TATP and 0.2 cm per second for DADP. Diperoxide burns approximately five times slower than TATP. The flame temperature of TATP exceeds that of DADP [19].

Brisance of DADP determined by the sand test is 30.1 g (0.4 g of peroxide initiated by 0.2 g mercury fulminate) which corresponds to 63 % TNT [17]. The initiation efficiency of DADP reported by Zhukov encyclopedia is between MF and LA [18]. This however disproved by Egorshev et al. [19] who tested DADP samples from 0.1 to 0.5 g pressed at 30 MPa with RDX as secondary charge in a blasting cap tube. Even 0.5 g DADP was not able to induce detonation of RDX, unlike TATP which caused full detonation of RDX in an amount of 0.1 g in the same conditions [19].

Table 10.1 Sensitivity of DADP, TATP and GNGT to impact (measured on Russian K 44 1 impact machine for primary explosives; h_{50}; 200 g weight; 12 mg pressed sample) [19]

	h_{50} (cm)	E_{50} (J)[a]
DADP	21.2 ± 1.5	0.42
GNGT	16 ± 2	0.31
TATP	11.7 ± 0.8	0.23

[a]Drop height converted to drop energy.

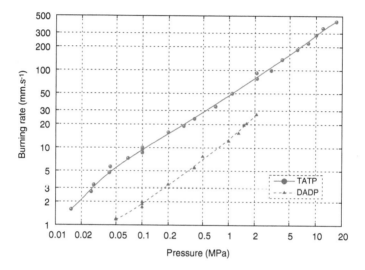

Fig. 10.4 Dependence of burning rate of TATP and DADP on various ambient pressures [19]

10.1.1.3 Preparation

DADP may be prepared in several different ways. The most common one is the reaction of acetone with hydrogen peroxide using acid catalysis. This may be carried out either in an aqueous or in a nonaqueous environment. A method suitable for relatively easy preparation of smaller scale laboratory samples is based on the transformation of TATP to DADP. It is further possible to prepare DADP by ozonolysis which is, however, not very practical.

The most common way of preparing DADP, by the reaction of acetone with hydrogen peroxide using acid catalysis, may be summed up by the following simplified equation:

$$(CH_3)_2C=O + H_2O_2 \xrightarrow{\text{catalyst}} \text{DADP}$$

TATP forms by a similar reaction. As mentioned previously, our investigations have led us believe that the most important factor affecting the composition of the product is the acidity of the reaction mixture. DADP forms in a highly acidic environment while TATP forms in less acidic conditions. The high concentrations and amounts of the acid catalyst necessary for the formation of DADP may however lead to decomposition of the reaction mixture. The reaction mixture was observed to change color locally to light yellow just before runaway occurs [8].

The preparation of DADP in an aqueous environment in the presence of large amounts of sulfuric acid as catalyst has been published by Bayer and Villager, Phillips and Pastureau [17, 25, 26] and in presence of other mineral acids by us [8]. Preparation in a nonaqueous environment has been reported from acetonitrile [2, 22, 27, 28], ether [11], dichloromethane [13], or acetic anhydride [29]. Other catalysts including Caro's acid [11] and methanesulfonic acid [13] have also been reported.

It was observed that TATP spontaneously transforms to DADP in an acidic environment both in solution and solid state, while acid-free TATP (purified by recrystallization) does not transform. Pure DADP can be prepared by the reaction of TATP with *p*-toluenesulfonic acid or trifluoroacetic acid in a dichloromethane solution for several hours or days. In these conditions, TATP transforms completely into DADP [13, 30].

In our investigations, we have found that raw TATP also spontaneously transforms to DADP in solid state if certain types of catalyst (such as sulfuric, perchloric, or methanesulfonic acids) are used. Acid probably remains closed in the crystal lattice of raw TATP and causes the transformation. The rate of transformation depends on temperature and on the reaction conditions used during the preparation of TATP. The rate of transformation increases with increasing temperature and is higher for TATP samples where the catalyst content during preparation was higher [31, 32].

Cyclic organic peroxides (including DADP) further form by ozonolysis of alkenes. Ozonide forms at first and cyclic organic peroxide is the final product of these reactions which typically yields many by-products [15, 16, 33–36].

10.1.1.4 Uses

DADP has been proposed for only a small number of applications: in the explosives industry as a substitute for mercury fulminate in detonators and igniters [17], in the petrochemical industry as an additive for Diesel fuels [37] and in the polymer industry as a polymerization initiator for methyl methacrylate [27]. It has, however, never been used practically [38] due to its low physical stability.

10.1.2 Triacetone Triperoxide

[Structure of TATP: nine-membered ring with three C(CH$_3$)$_2$ groups connected by O-O peroxide linkages]

Triacetone triperoxide (3,3,6,6,9,9-hexamethyl-1,2,4,5,7,8-hexaoxonane; TATP) was first synthesized by Richard Wolffenstein in 1895 [39]. He discovered the compound accidentally while working on his thesis called "Effects of hydrogen peroxide treatment on alkaloids" [40]. During his investigation of the effects of hydrogen peroxide treatment on coniine alkaloid, Wolffenstein used acetone as a solvent (as the author puts it, he used acetone because of its indifference toward hydrogen peroxide). However, in this particular case, instead of the 1 % solution of hydrogen peroxide that had been previously used, he used a higher concentration (10 %). After letting the solution stand for several days, crystals of a new, so far undiscovered, compound were formed, and Wolffenstein carried out an analysis of this new compound. He successfully identified the compound on the basis of elemental analysis and cryoscopic determination of its molecular weight. He described its physical properties and recognized it as an explosive compound. During his subsequent experiments, he increased the yields of TATP by adding phosphoric acid to the hydrogen peroxide and acetone reaction mixture. His method, i.e., reaction of acetone with hydrogen peroxide in an acidic medium, is still a common method for obtaining TATP [39].

The nine-membered ring of the TATP molecule forms a "twist boat chair" conformation (Fig. 10.5). The crystal packing (Fig. 10.6) consists of stacks around the molecular threefold axis with no apparent C–H⋯O interactions [13]. Theoretical studies of other possible conformations have been published by Yavari et al. [41].

10.1.2.1 Physical and Chemical Properties

Triacetone triperoxide forms colorless crystals which appear white when finely ground (Fig. 10.7). Its crystal density is 1.272 g cm^{-3} (X-ray) [13, 42]; melting point is reported in the range 95–98.8 °C and decomposition above 150–160 °C [17, 39, 43–46]. The heat of formation of TATP is 90.8 kJ mol^{-1} [47]. Triacetone triperoxide is nonhygroscopic and very volatile [17].

It spontaneously recrystallizes at room or even at subambient temperatures (0 °C) into nicely formed cubes as can be seen in Fig. 10.8. The detail of the surface of the crystal during recrystallization is presented in Fig. 10.9. The storage temperature and the type of acid used for TATP preparation both have significant impacts on the rate of TATP recrystallization. The process is much faster for TATP

10.1 Peroxides of Acetone 263

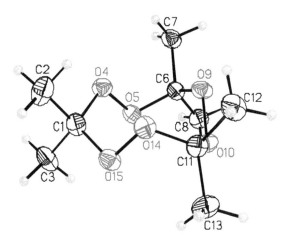

Fig. 10.5 3D structure of TATP molecule [13]. Reprinted with permission from Dubnikova et al., Decomposition of TATP Is an Entropic Explosion, J. Am. Chem. Soc. 2005, 127, 1146–1159. Copyright (2005) American Chemical Society

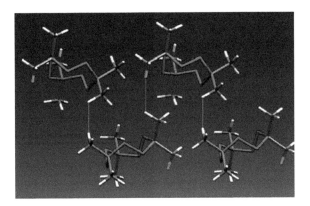

Fig. 10.6 Packing of TATP as capped-stick model along the crystallographic screw axis in a perpendicular view [13]. Reprinted with permission from Dubnikova et al., Decomposition of TATP Is an Entropic Explosion, J. Am. Chem. Soc. 2005, 127, 1146–1159. Copyright (2005) American Chemical Society

prepared using sulfuric or perchloric acid than it is if hydrochloric or nitric acid is used. Increased temperature speeds up the recrystallization in general.

The type of acid used during TATP preparation has a significant impact on its long-term stability. TATP prepared using sulfuric or perchloric acid decomposes and lose weight more quickly than pure TATP (by recrystallization). The loss of weight of TATP prepared using hydrochloric or nitric acid is much slower and about the same as observed for pure TATP and is related to sublimation rather than decomposition [48].

Reany et al. recently published that TATP forms at least six different polymorphic crystals. The type of acid used as a catalyst for TATP preparation and the solvent used for TATP crystallization have an impact on the crystal morphology of the product.

Fig. 10.7 Spontaneous recrystallization of powdery TATP at room temperature (after 1 month)

Fig. 10.8 Same sample of TATP crystals *left*—as prepared, *center*—2 weeks and *right*—2 months at room temperature

Fig. 10.9 Detail of surface of TATP crystals (2 months old sample)

10.1 Peroxides of Acetone

Three different crystal forms are present in raw TATP: needles, prisms, and plates. X-ray crystallography revealed that these three forms represent three different polymorphs: needles form monoclinic, prisms form orthorhombic, and plates form triclinic crystal systems. Strong acids (sulfuric, nitric, and hydrochloric) produce a mixture of polymorphs where the major one is needle crystals while weak acids (acetic, citric) produce a mixture of needles and plates. Another three polymorphs (all are monoclinic) form by crystallization from various organic solvents [30].

The vapor pressure of TATP has recently been determined by Oxley et al. by gas chromatography. The vapor pressure is 6.95 Pa [49] or 7.87 Pa [20] at 25 °C which is about 10^4 times higher than TNT. The Clapeyron equation which describes dependence of vapor pressure (P in Pa) on temperature (T in K) is:

$$\ln P = 31.4 - 8{,}719.9/T$$

The heat of sublimation of TATP is 72.5 kJ mol^{-1} (calculated from the Clapeyron equation) [20].

Egorshev et al. [19] described dependence of vapor pressure (P in atm) on temperature by the following equations:

$$\ln P = 22.73 - 9{,}695.5/T \quad \text{(for temperature range } 12 - 97\,°\text{C)}$$
$$\ln P = 15.29 - 6{,}978.8/T \quad \text{(temperature above}$$
$$\text{melting point, TATP in liquid state)}$$

Heat of sublimation, evaporation, and melting of TATP are 80.6 kJ mol^{-1}, 58.1 kJ mol^{-1}, and 22.6 kJ mol^{-1} respectively. These values were determined from the above-mentioned equations by Egorshev et al. [19].

TATP does not react with water (even when boiling) nor with common metals (except a slight reaction with lead) [50]. It is practically insoluble in water and aqueous ammonia but soluble in most organic solvents (see Table 10.2) [17, 43, 50, 51].

TATP does not decompose by action of diluted bases and its decomposition by diluted, but strong, inorganic acids proceeds to completion only at higher temperatures. Concentrated hydroiodic acid causes it to decompose violently without flames [18, 52] however in contact with concentrated sulfuric acid it ignites. TATP also slowly decomposes by the action of concentrated hydrochloric acid and 35 % sulfuric acid. The chemistry of decomposition by the action of these acids has been studied by Armitt et al. [53]. TATP decomposes both on exposure to acid vapors and suspended in liquid acid. Acetone, DADP, chloroacetone, and 1,1-dichloroacetone are the major products of decomposition by hydrochloric acid in the vapor phase. As the decomposition progresses, DADP decomposes and other poly-chlorinated acetones form (as far as hexachloroacetone). When the decomposition took place in liquid hydrochloric acid the main product was 1,1-dichloroacetone and poly-chlorinated acetones were only minor products. Sulfuric acid causes TATP to decompose to DADP and acetone, and DADP

Table 10.2 Solubility of TATP in organic solvents

Solvent	Solubility (g/100 ml solvent)	
	At 17 °C[a] [50]	At room temperature [43]
Acetone	7.96	16.5
Benzene	19.3	–
Carbon disulfide	11.2	–
Carbon tetrachloride	52.5	–
Chloroform	110	111
Diethylether	4.2	–
Ethanol	0.12	3.5
Hexane	–	11.1
Methanol	–	3.8
Pyridine	17.9	–
Toluene	–	34.7

[a]Units unspecified; assumed to be wt.% in reference; converted to g per 100 ml of solvent in this table.

gradually decomposes to acetone in its turn. Degradation of TATP occurs at a faster rate in liquid acid than by the action of acid vapors [53].

TATP also slowly decomposes by boiling in methanol solutions in presence of stannous chloride. Combustion of a toluene solution of TATP was proposed by Bellamy as a suitable method for disposal [43]. Extensive study of TATP decomposition by the action of chemical agents has recently been published by Oxley et al. [21]. They observed that TATP best decomposes by application of a solution of zinc sulfate or cupric chloride. The process takes only several hours at room temperature if these salts are used in combination with metallic zinc or copper.

The chemical destruction of larger amounts of dry TATP is however hazardous and not recommended, as it is exothermic. The heat liberated during decomposition could possibly ignite any remaining TATP [21, 52].

Stability of TATP in solutions in organic solvents is affected by residual acidity in TATP crystals (comes from synthesis). Pure (recrystallized) TATP without residual acids is relatively stable similar to TATP prepared from hydrochloric acid. In contrast TATP from sulfuric and perchloric acid relatively quickly decomposes. The extensive study focused on problems with stability of TATP solutions (aimed primarily on forensic labs) has been published recently [54].

The formation of TATP complexes with metal ions has been experimentally studied in the gas phase. Two groups of structures were proposed based on the theoretical calculations. The first, cyclic, is similar to crown ethers and the second, is formed by three fragments of TATP bonded to a cation [55, 56]. In the complexes of the first group, the cation (Cu^+, Li^+, Na^+, K^+, Cd^{2+}, Zn^{2+}, and In^{3+}) is centered in and slightly above the binding cavity and is equidistant from the oxygen atoms of TATP (see Fig. 10.10), probably due to steric effects related to the size of cations and the cavity. The bonding energies decrease with increasing ionic radii and their values are similar to the values for 18-crown-6 ether complexes with corresponding cations [56]. This approach of complex formation is often used in the mass spectrometry of TATP. Addition of alkali metals (or small ions, such as NH_4^+) to

10.1 Peroxides of Acetone

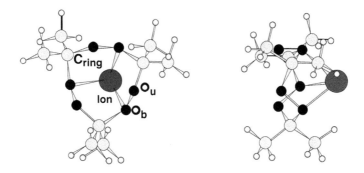

Fig. 10.10 The lowest-energy geometry for the ion–TATP complex [56]. Reprinted with permission from F. Dubnikova et al., Novel Approach to the Detection of Triacetone Triperoxide (TATP): Its Structure and Its Complexes with Ions, The Journal of Physical, *J. Phys. Chem. A* **2002**, *106*, 4951–4956. Copyright (2002) American Chemical Society

the spray solution or mobile phase and subsequent molecular adduct formation in the gas phase can be used to improve the selectivity and sensitivity of TATP detection during the mass spectrometric analysis [55].

Structure of the complexes of the second group is quite different. They are formed by three fragments of TATP (breaking of the three C–O bonds in the TATP molecule) and metal cation with bond –O–O–metal (complex of TATP with Sb^{3+}, Sc^{3+} and Ti^{4}), see Fig. 10.11 [56].

10.1.2.2 Thermal Stability

The thermal decomposition of recrystallized TATP identified as the onset of the DSC curve begins above 160 °C (10 °C min^{-1}, sealed pan) [43].

Thermal stability of raw TATP (but well washed to neutral pH) depends markedly on the type (Fig. 10.12) and amount (Fig. 10.13) of acid used as the catalyst used in its preparation. The decomposition of TATP begins around 145 °C when hydrochloric or nitric acid is used. The amount of catalyst in this case does not have a measurable influence on the thermal stability of prepared TATP (acid to acetone molar ratio n_c/n_a from 2.5×10^{-4} to 5×10^{-1}). A significant influence was however found when using sulfuric or perchloric acid. A low concentration of these two acids ($n_c/n_a \leq 1 \times 10^{-2}$) yields product that decomposes above 145 °C just as in the case of pure TATP. Higher concentrations however yield TATP that decomposes during melting, or even before that, in the solid phase (Fig. 10.13, Table 10.3). It is presumed that the lower thermal stability is a result of a combination of two factors—overall residual acidity within the TATP crystals and acid strength [57].

10.1.2.3 Explosive Properties

Impact sensitivity of TATP is reported by various authors with very wide ranges. Some authors have reported sensitivity of TATP as very high [58] or even as one of the most sensitive explosives known [59]. This is however a bit exaggerated.

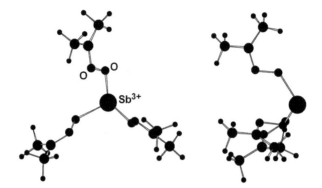

Fig. 10.11 Top view (*left*) and side view (*right*) of the lowest-energy geometry for the Sb^{3+}–TATP complex [56]. Reprinted with permission from F. Dubnikova et al., Novel Approach to the Detection of Triacetone Triperoxide (TATP): Its Structure and Its Complexes with Ions, The Journal of Physical, *J. Phys. Chem. A* **2002**, *106*, 4951–4956. Copyright (2002) American Chemical Society

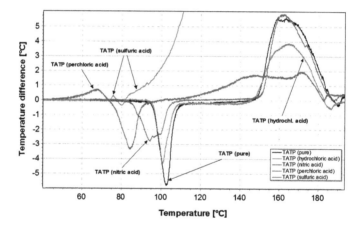

Fig. 10.12 DTA thermograms of TATP prepared using inorganic acids ($n_c/n_a = 2.5 \times 10^{-1}$ for all acids) as catalyst and pure TATP (heating rate 5 °C min^{-1}, 30 mg samples and static air atmosphere) [57]

Our own experimental data clearly indicate that TATP is much less sensitive. The reasons why such a large scatter is reported in the literature is not clear, but may relate to one or more of the following (a) tendency of TATP to recrystallize forming large crystals, (b) spontaneous transformation of TATP to other peroxides, (c) the type of acid used in preparation, (d) crystal size, or (e) experimental setup. Most of the authors who measured the sensitivity of TATP generally did not fully describe the measured samples in terms of shape, size, age, storage temperature, and preparation route. Published values of impact sensitivity are summarized in Fig. 2.15 irrespective of the previously mentioned incomplete description.

The impact sensitivity of pure TATP and its mixtures with other substances has been studied by Mavrodi [60]. His results are summarized in Table 10.4.

10.1 Peroxides of Acetone

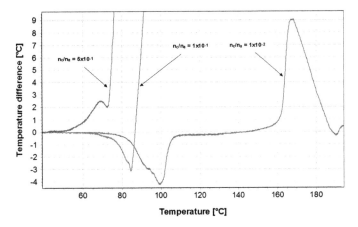

Fig. 10.13 DTA thermograms of TATP demonstrating the influence of the concentration of sulfuric acid in reaction mixture—n_c/n_a on the thermal stability of resulting product (heating rate 5 °C min^{-1}, 30 mg samples and static air atmosphere) [57]

Table 10.3 The dependence of the onset of decomposition of TATP (prepared using various inorganic acids as catalyst) on acetone to acid molar ratio n_c/n_a (heating rate 5 °C min^{-1}) [57]

Molar ratio n_c/n_a	Sulfuric acid	Perchloric acid	Hydrochloric acid	Nitric acid
5×10^{-1} 2.5×10^{-1}	Decomposition starts before melting	Decomposition starts before melting	Decomposition starts above 145 °C	Decomposition starts above 145 °C
1×10^{-1} 5×10^{-2} 2.5×10^{-2}	Decomposition during melting	Decomposition during melting		
1×10^{-2} 5×10^{-3} 2.5×10^{-3} 2.5×10^{-4}	Decomposition starts above 145 °C	Decomposition starts above 145 °C		

Extremely high friction sensitivity reported in Meyer's encyclopedia (TATP 10× more sensitive than LA and 50× more sensitive than MF [58]) seems quite exaggerated. Our measurements are shown along with Meyer's data in Fig. 2.19. Friction sensitivity of TATP is based on our experience comparable to other common primary explosives; it is between LA and MF (see Fig. 2.19) [23, 48, 61]. Neither type of acid used for preparation of TATP nor aging have significant influence on its sensitivity [48].

Comparison of sensitivity to electric discharge with other primary explosives is presented in Fig. 2.21. Yeager tested pure TATP and found that E_{50} is 160 mJ (325 mJ for PETN). When calculating the probability of initiation at lower energies he found that it is statistically possible to initiate TATP at an energy level that can be generated by the human body in approximately 1–2 % of cases. Further, he found that unpurified samples are more sensitive than pure recrystallized ones [62].

Table 10.4 Sensitivity of TATP, TATP mixtures, and reference explosives [60]

	Impact energy (J)	Drop height 2 kg weight (cm)
TATP (fine powder)	3.1	16
TATP (small crystals)	2.4	12
TATP + 20 % glass rubble	1.0	5
TATP + 20 % glass rubble and KClO$_3$	0.49	2.5
TATP + 8–10 % oil	4.3–5.5	22–28
TNT + 7 % TATP	18–19	90–95
Mercury fulminate	0.6	3
Black powder	13	65
Dynamite (75 %)	1.3	6.5
TNT	35	180

Table 10.5 Detonation velocity for TATP

Density (g cm^{-3})	Detonation velocity (m s^{-1})	References
0.47	1,430	[64, 65]
0.68	3,065	[66]
0.92	3,750	[47]
0.95	3,950	[47]
1.18	5,300	[47]

Analysis of detonation products has been carried out by Muraour who found the following gases: carbon dioxide, carbon monoxide, methane, ethane, some higher alkines, hydrogen, and water vapor [47]. Our own experiments led us to believe that at least ethylene is also present.

In small quantities TATP burns quickly after being lit by an open flame (in similar way as dry nitrocellulose); in larger quantities or when sealed it explodes. The dependence of combustion rate on ambient pressure has been measured by Fogelzang et al. [63] and recently by Egorshev et al. [19].

Some values of detonation velocities for TATP for different densities are summarized in Table 10.5.

TATP belongs to the group of highly brisant primary explosives, being superior to both mercury fulminate and lead azide [51]:

- Brisance by send test: 71 % TNT [17]; 46.2–50.5 % TNT (LA 29 %, MF 33 %) [51]
- Plate dent test: depth 3.65 mm—TATP ($\rho = 1.26$ g cm^{-3}), 5.9 mm RDX ($\rho = 1.4$ g cm^{-3}) and 5.43 mm TNT ($\rho = 1.3$ g cm^{-3}) [67]
- Lead block test: TATP 250 cm^3 (88 % TNT) [17, 58, 66]; TATP 320 cm^3 (101 % TNT), TNT with 5 % TATP 370 cm^3 (116 % TNT) [60]

The initiation efficiency of TATP is high (but lower than LA [38]): 0.05 g of TATP initiates PETN (both pressed under 25 MPa) [66], 0.09 g initiates tetryl [68] and 0.16 g [47] or 0.18 g [51, 68] initiates TNT. The comparison with other primary explosives is given in Fig. 2.2. TATP is susceptible to dead pressing. According to Rohrlich and Sauermilch, TATP can be dead pressed at 49 MPa [66].

10.1 Peroxides of Acetone

The explosive properties (detonation velocity, brisance, power) of mixtures of TATP with ammonium nitrate are at the level of common industrial explosives [69, 70]. These mixtures are however highly sensitive to mechanical stimuli; according to our study some of them are more sensitive to impact than pure dry TATP.

10.1.2.4 Preparation

The most commonly employed way of TATP preparation is based on the reaction of acetone with hydrogen peroxide in the presence of an acid catalyst [17, 24, 39, 46, 51, 71]:

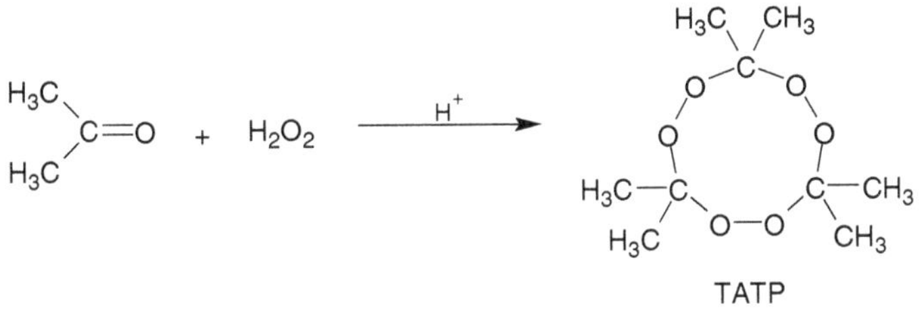

The issue of reaction conditions was addressed above in a previous chapter (DADP) but to summarize it, TATP forms at lower molar ratios of catalyst to acetone and with lower concentrations of catalyst [8]. The formation of chlorinated TATP (3-(chloromethyl)-3,6,6,9,9-pentamethyl-1,2,4,5,7,8-hexoxonane and 3,6-bis(chloromethyl)-3,6,9,9-tetramethyl-1,2,4,5,7,8-hexoxonane) was recently published by Matyáš et al. when hydrochloric acid is used as a catalyst in very large excess. Other hydrohalic acids do not provide similar halogenated cyclic peroxides by this method; TATP forms as a product of the following reaction [72].

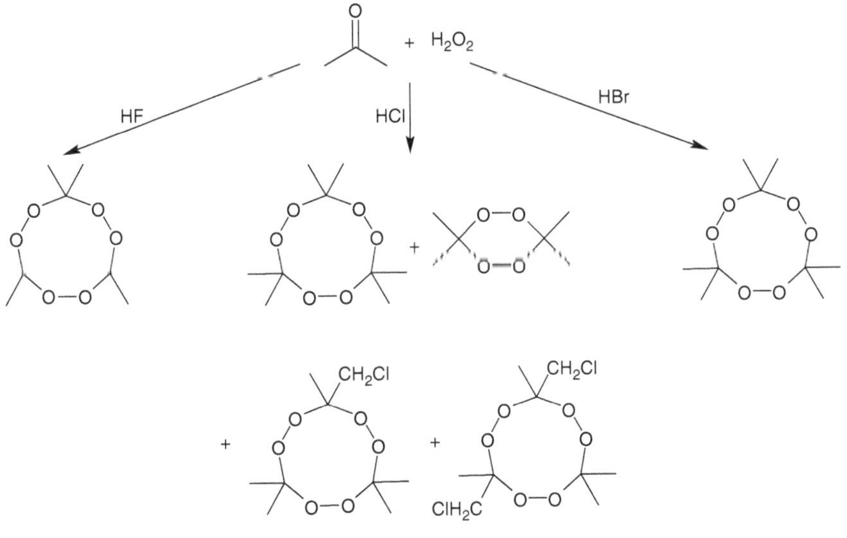

TATP always contains small amount of side products which is reported to be either a tetramer—TeATeP (3,3,6,6,9,9,12,12-octamethyl-1,2,4,5,7,8,10,11-octaoxacyclododecane) [73, 74] or a structural conformer of TATP [75] showing up as a small peak accompanying the main TATP peak in chromatographic studies. This substance may not be eliminated by recrystallization or resublimation (Matyáš and Pachman Unpublished).

Some other less common catalysts are reported for preparation of TATP such as oxone ($2KHSO_5 \cdot KHSO_4 \cdot K_2SO_4$) [76].

The synthesis of TATP in a mixture of acetone and hydrogen peroxide can also be catalyzed by some metals. For example, titania-incorporated mesoporous material leads to TATP formation in acid-free conditions [77].

Just as with DADP, it is possible to obtain TATP by ozonolysis of alkenes. The final product is however not pure as it contains a variety of side products including DADP [16, 33, 35, 44].

TATP also forms by spontaneous transformation of bis(2-hydroperoxypropan-2-yl)ether over several days. Peroxides containing an ether group rearrange spontaneously, whereas peroxides containing peroxy groups do so only in the presence of strong acids [78].

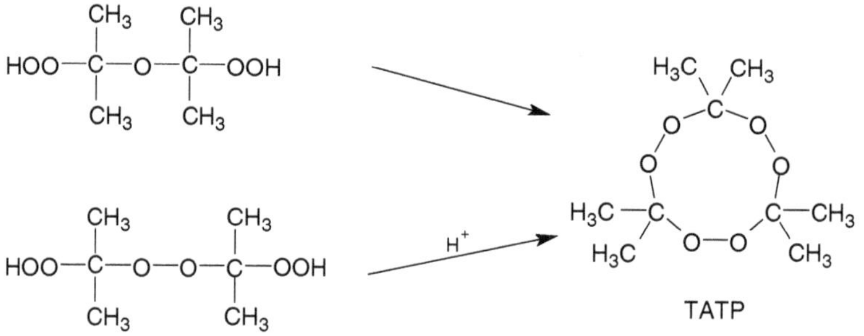

TATP easily sublimates and may therefore be purified by this method. Unlike the spontaneously recrystallized product (see Figs. 10.7 and 10.8), the resublimated one forms beautiful snow white structures resembling tree branches (see Fig. 10.14) (Matyáš and Pachman Unpublished).

Unfortunately, TATP is formed by the action of atmospheric oxygen on diisopropylether. Many explosions have been known to occur during handling or processing (especially distillation) of old diisopropylether due to the formation of TATP [45, 79, 80].

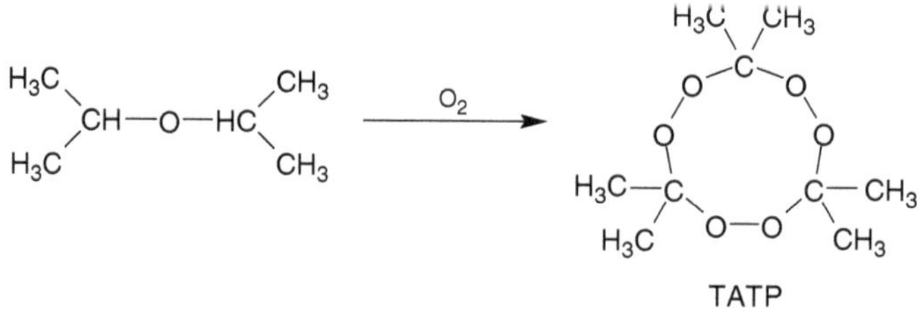

10.1 Peroxides of Acetone

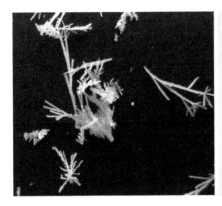

Fig. 10.14 TATP crystals obtained by sublimation (*left*—camera photo, *right*—optical microscopy)

Several other accidental explosions of TATP, unknowingly formed during chemical reactions, have been reported [81–84]. The unwanted formation of acetone peroxides is therefore a problem which needs to be addressed in any industrial scale production. One example of measures for suppressing acetone reaction mixtures resulting from the synthesis of phenol, diphenols, or phenol derivates has been published by Costantini who suggests (a) adjusting pH to a value between 4 and 8; (b) adding a copper compound to the reaction mixture; (c) maintaining reaction temperature between 50 and 150 °C [85]. To avoid unwanted formation of TATP, it is generally recommended never to use acetone as a solvent in reactions of hydrogen peroxide [86].

10.1.2.5 Uses

TATP has been largely studied in the first half of the twentieth century. Due to its simple preparation and sufficient explosive performance, it had been considered for use as a primary charge for detonators [87]. It is astonishing that patent application was registered for the same application of TATP again in 2009 [88]. The reason why it has not been implemented industrially is often incorrectly related to its "high sensitivity." This is however not true, as can be seen from the comparison of sensitivity of peroxides with other primary explosives (see Chap. 2). The real reasons are its high volatility, which is a cause of spontaneous recrystallization (see Figs. 10.7 and 10.8), and unsuitable thermal behavior (melting at low tempera ture) (Matyáš and Pachman Unpublished).

Apart from its application as a primary explosive, TATP has been suggested for use in mixtures or as a part of secondary explosives. A mixture with ammonium nitrate has been suggested by Mavrodi for use in coal mines. This explosive is said to have low detonation temperature and force equivalent to TNT [60].

A number of publications focusing on demonstrating interesting or otherwise visually effective chemical experiments [89, 90] have proposed a relatively effective way of TATP deflagration accompanied with formation of a powerful fireball. However, it is strongly recommended to avoid using TATP for such purposes due to the not insignificant risk of detonation.

Although TATP has found neither industrial nor military applications, it has become very popular among the general public. On top of being very popular among young "scientists," this substance has also attracted attention of criminals, extremists, and terrorists who use it for preparing detonators as well as main charges [91–94]. The main reason for this popularity is the availability of precursors (acetone, hydrogen peroxide, and acids in chemists' stores) [95], wide knowledge of how to make it (broadcasted via internet), and an undemanding procedure for synthesis with high yield.

TATP has been used by terrorists in many bomb attacks around the world (Europe, USA, Middle East, Africa, etc.) [62, 93, 96–99]. Probably the most well-known abuse of TATP was by Richard Reid which used TATP as a primary explosive in his shoe bomb in the attempted bombing of a US airliner on 22 December 2001 [99].

Some organic peroxides, including TATP, have been suggested as additives for diesel fuels [37].

10.1.3 Tetraacetone Tetraperoxide

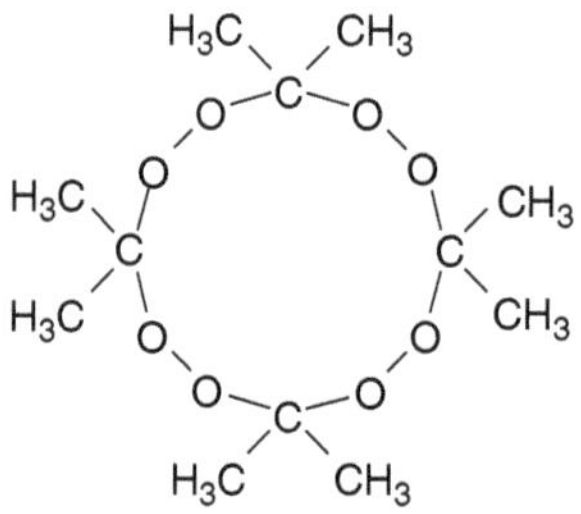

The formation of tetraacetone tetraperoxide (3,3,6,6,9,9,12,12-octamethyl-1,2,4,5,7,8,10,11-octaoxacyclododecane; TeATeP) as a co-product of the reaction creating mainly TATP is reported in several articles [55, 73, 74]. TeATeP is reported to form under conditions very similar to those for DADP or TATP. Peña et al. reported that this peroxide forms in the reaction mixture when the temperature rose above 10 °C [74].

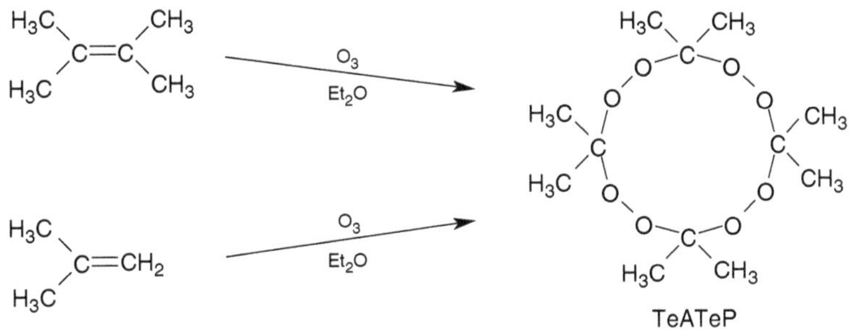

10.2 Hexamethylene Triperoxide Diamine

$$\text{H}_3\text{C}\diagdown\text{C}=\text{C}\diagup\text{CH}_3 \xrightarrow[\text{Et}_2\text{O}]{\text{O}_3}$$

$$\text{H}_3\text{C}\diagdown\text{C}=\text{CH}_2 \xrightarrow[\text{Et}_2\text{O}]{\text{O}_3}$$

TeATeP

Jiang has reported a method of preparation of TeATeP using the reaction of acetone with hydrogen peroxide in presence of tin (II and IV) chloride or hydrochloric acid as the catalyst [100]. This method can however be disputed and according to our experience, normal TATP is formed [8].

The formation of tetraacetone tetraperoxide by ozonolysis (see reaction above) of tetramethylethene or isobutene has been described by Obinokov et al. [101]. These authors report a melting point of 61–65 °C.

10.2 Hexamethylene Triperoxide Diamine

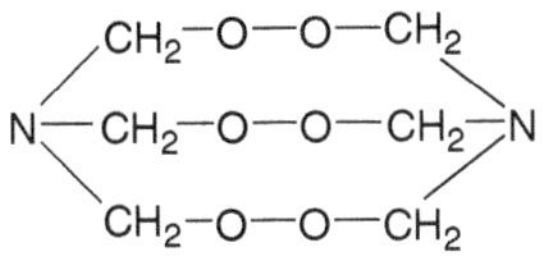

Hexamethylene triperoxide diamine (1,6-diaza-3,4,8,9,12,13-hexaoxabicyclo [4,4,4]tetradecane; HMTD) was first prepared by Legler in 1885. He discovered it while working on the slow combustion of ether. He obtained a solid which, on treatment with ammonia, gave an explosive compound—HMTD [51, 102, 103]. The early history of this substance is described by Bagal [51].

The two nitrogen atoms in the HMTD molecule form an interesting and rare, totally planar, hybridization [104–106]. Two major hypotheses exist which try to explain this planarity; an earlier one based on the steric effect [105] which has subsequently been superseded by a later one, based on an electronic effect [107]. The 3D model of HMTD molecule is shown in Fig. 10.15.

10.2.1 Physical and Chemical Properties

HMTD forms colorless crystals (Fig. 10.16). Freshly prepared HMTD does not smell; older samples acquire an unpleasant odor of rotten fish. Density of HMTD is mostly reported as 1.57 g cm^{-3} [17, 58, 68, 103, 108] but X-ray investigation of

Fig. 10.15 HMTD geometry and atom labeling [105]. Reprinted with permission from A. Wierzbicki and E. Cioffi, Density Functional Theory Studies of Hexamethylene Triperoxide Diamine, J. Phys. Chem. A 1999, 103, 8890–8894. Copyright (1999) American Chemical Society

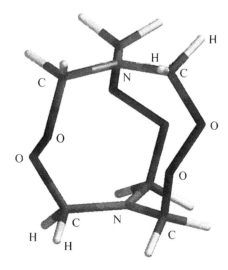

Fig. 10.16 HMTD crystals (*left*—optical microscopy; polarized light; *right*—SEM)

Table 10.6 Solubility of HMTD in water and organic solvents at 22 °C

Solvent	Solubility (g/100 g solvent at 22 °C)	References
Absolute ethanol	<0.01	[103]
Acetone	0.33	[103]
Carbon disulfide	<0.01	[103]
Carbon tetrachloride	0.013	[103]
Chloroform	0.64	[103]
Diethylether	0.017	[103]
Ethane-1,2-diyl diacetate	0.90	[109]
Glacial acetic acid	0.14	[103]
Water	0.01	[103]

HMTD crystal reports 1.597 g cm^{-3} [106]. Heat of formation is -363 kJ mol^{-1} according to Muraour [47] or -335 kJ mol^{-1} according to Danilov et al. [38]. It is slightly hygroscopic and it does not seem to be toxic. HMTD is practically insoluble

10.2 Hexamethylene Triperoxide Diamine

in cold water and most common organic solvents, slightly soluble in acetonitrile, chloroform, dimethyl sulfoxide, dimethyl formamide and glycol acetate (see Table 10.6); HMTD is soluble in hot water [103, 109–111].

HMTD is less chemically stable than TATP; it is commonly reported as unstable in storage and it easily decomposes, particularly at higher temperatures [17, 46]. Low stability of HMTD has been reported for example by Sülzle and Klaeboe. They proved by IR spectroscopy that the compound slowly decomposes on storage even when purified by crystallization from ethanol [112]. Nevertheless, Danilov et al. reported that it can be stored for long periods without decomposing [38].

HMTD slightly hydrolyzes in cold water if left standing for several months, slowly decomposes at 40 °C and more quickly in boiling water according to the following equation [38, 108]:

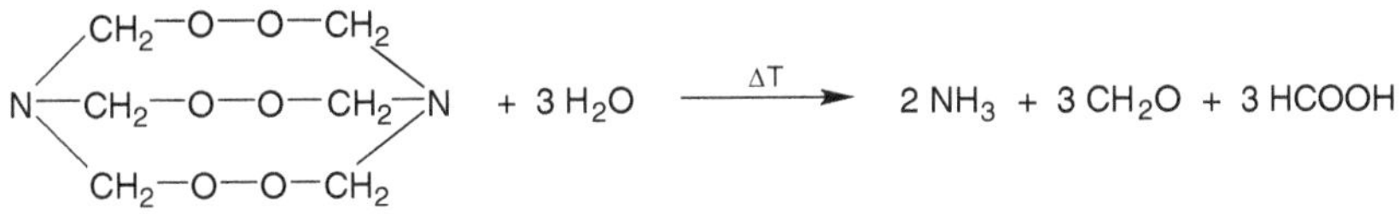

$$\text{HMTD} + 3\,H_2O \xrightarrow{\Delta T} 2\,NH_3 + 3\,CH_2O + 3\,HCOOH$$

According to Bagal and Ilyushin [51, 108], HMTD is stable when stored under water for about 4 weeks at 30 °C.

HMTD decomposes by the action of diluted acids (H_2SO_4, HBr, HCl) at room temperatures; aqueous sodium hydroxide decomposes HMTD at higher temperatures. It violently reacts with concentrated hydroiodic acid and explodes in contact with bromine [52, 103, 113]. Examination of HMTD decomposition by the action of chemical agents has recently been published by Oxley et al. They observed that HMTD decomposes by application of inorganic salts (e.g., $KMnO_4$, $SnCl_2$, $ZnCl_2$, $ZnSO_4$, KI, NaBr) in an acidic environment at room temperature within several hours. The chemical destruction of larger amounts of dry HMTD is however not recommended due to the exothermic nature of the decomposition.

Dry HMTD reacts with most common metals (e.g., zinc, copper and lead) according to Ficheroulle and Kovache [50]. However, Danilov et al. have reported that dry HMTD does not react with metals, but rather rapidly reacts when wet [38, 108].

By the action of nitric acid, in presence of ammonium nitrate, HMTD could be nitrated yielding hexogen (RDX). The total yield of RDX is reported to be 26 % of the theoretical. The reaction can be described by the following equation [114]:

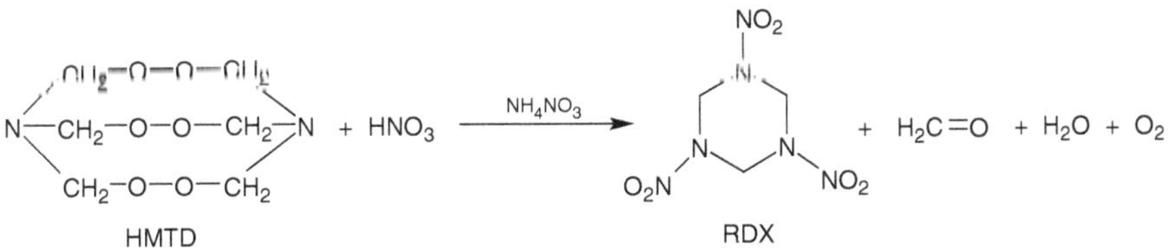

$$\text{HMTD} + HNO_3 \xrightarrow{NH_4NO_3} \text{RDX} + H_2C{=}O + H_2O + O_2$$

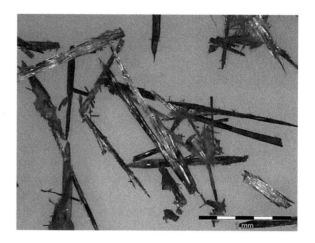

Fig. 10.17 HMTD—recrystallized from chloroform

10.2.2 Explosive Properties

HMTD is sensitive to impact not only in its dry state but also when wet [17]. Its impact sensitivity is mostly referred to as slightly [24, 103, 115, 116] or significantly [51] lower than that of MF. Only Meyer reports it to be significantly more sensitive than MF [58] while Yeager reported surprisingly low sensitivity; between PETN and RDX [62]. The values of impact sensitivity are summarized in Fig. 2.15 from which is clear that it is about the same as for MF.

HMTD is furthermore highly sensitive to friction. Yeager reported HMTD as one of the most friction-sensitive explosives [62]; extremely high sensitivity of HMTD is reported in Meyer's encyclopedia ($10\times$ more sensitive than LA and $50\times$ more sensitive than MF [58]). According to our own results is friction sensitivity of HMTD not extreme; it is slightly lower than LA but significantly higher than MF (see Fig. 2.19).

Sensitivity of HMTD to electric discharge is about the same as that of lead azide, according to our measurements (see Fig. 2.21).

HMTD crystals obtained by recrystallization from chloroform have been reported to undergo spontaneous explosions at room temperature without any other external stimuli [117]. This type of behavior has not been reported for the raw material. The shape of the crystals obtained by recrystallization from chloroform can be seen in Fig. 10.17. The current authors have not encountered any spontaneous explosions while preparing the material for this picture.

The dependence of detonation velocity of HMTD on density is summarized in Table 10.7.

HMTD belongs to the group of highly brisant and powerful primary explosives. The values of brisance, power, and ignition efficiency are summarized below:

- Brisance by send test: 71 % TNT [17]; superior 2.5–3 times MF [103]
- Lead block test: 242 cm^3 [109, 118], 330 cm^3 (110 % TNT) [58]
- Initiation efficiency: 0.05 g with reinforcing cap and 0.06 without it for tetryl; 0.08 g with reinforcing cap and 0.10 without it for TNT [103]; comparison with

Table 10.7 Detonation velocity of HMTD

Density (g cm^{-3})	Detonation velocity (m s^{-1})	References
0.38	2,820	[64, 65]
0.88	4,511	[103]
1.05	4,500	[47]
1.15	5,100	[47]

other primary explosives see Fig. 2.2. The initiating efficiency is three to four times better than that of mercury fulminate.

The equation of HMTD explosion (ignited by impact) is, according to Muraour, the following [47]:

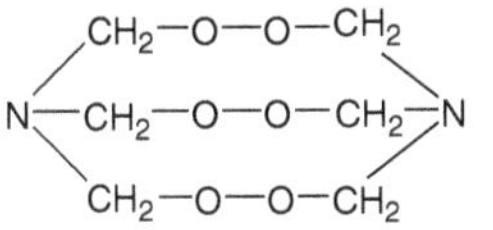

$$N\begin{pmatrix}CH_2-O-O-CH_2\\CH_2-O-O-CH_2\\CH_2-O-O-CH_2\end{pmatrix}N \longrightarrow 0.35\,CO_2 + 3.88\,CO + 0.03\,C_2H_2 + 0.55\,CH_4 + 0.58\,C$$
$$+ 2.44\,H_2 + 1.42\,H_2O + 0.48\,NH_3 + 0.47\,N_2 + 0.58\,HCN$$

Volume of gaseous products is 1,097 dm^3 kg^{-1} according to above-mentioned equation [47]. Ilyushin et al. published different result referring volume of gaseous products of HMTD explosion to be 813 dm^3 kg^{-1} and the heat of explosion 1,058 kJ mol^{-1} [108].

The dependence of burning rate on ambient pressure is linear and is about the same as for TATP [63]. Ignition of small amounts of HMTD by flame results in formation of an orange flame with muffled sound (Matyáš and Pachman Unpublished) [115]. Immediate ignition (explosion) occurs at 200 °C, 0.05 g sample ignited at 149 °C in three seconds [103]. Metz reported that HMTD ignites in the temperature range 125–140 °C [119]. The ignition temperature of HMTD by heating is 123–136 °C (when inorganic acids are used as a catalyst for HMTD preparation) and 146–148 °C (when citric acid is used as a catalyst) at heating rate 5 °C min^{-1} (Matyáš and Pachman Unpublished); Girsewald reported 139 °C at heating rate 20 °C min^{-1} [118].

Pressed or confined HMTD detonates after ignition by flame with a strong acoustic effect. HMTD does not become dead-pressed under 76 MPa [103] or even at 294 MPa [50, 51]. Explosions however occur when pressing coarse crystalline HMTD by a pressure 20 MPa or even more at 49 MPa [51].

10.2.3 Preparation

The easiest way for preparing HMTD is by the reaction of hexamethylenetetramine with hydrogen peroxide in presence of an acidic catalyst [68, 116, 118, 120]. Citric acid is often used as the catalyst but other organic and inorganic acids could be used as well [24, 46, 121].

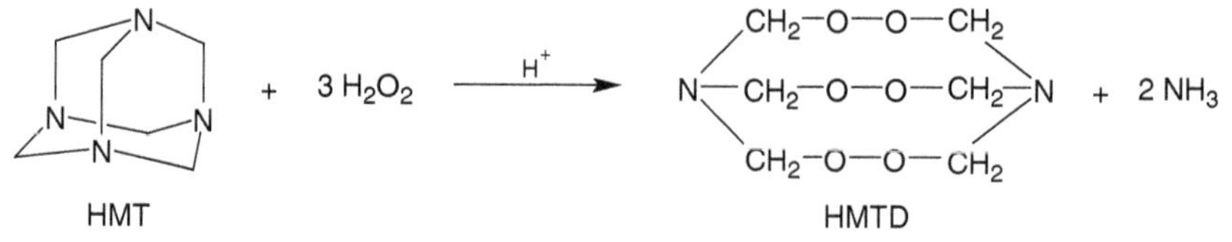

$$\text{HMT} + 3\,H_2O_2 \xrightarrow{H^+} \text{HMTD} + 2\,NH_3$$

Other starting materials that produce hexamethylene tetramine in situ such as formaldehyde with aqueous ammonia or formaldehyde with an ammonium salt (e.g., sulfate) can be used instead of hexamethylenetetramine [68, 122]. The yield of HMTD is, however, low for the latter mentioned example (Matyáš and Pachman Unpublished). Ethanol (in large excess; about 1 l per 1 g HMTD) is recommended for purification of HMTD by crystallization [112].

10.2.4 Uses

HMTD has been patented as a primary explosive for detonators [123] but, due to its low physical and chemical stability and incompatibility with metals, has never been used [17, 50]. However, Danilov et al. have reported that it can be used in war situations when short turnaround time is expected [38]. Recently, investigations of HMTD as a "green" primary explosive for laser-initiated applications have been carried out by Ilyushin et al. [108, 124].

HMTD is, just like TATP, a very popular improvised explosive and as such is often prepared by young "scientists," criminals, extremists, and terrorists [91, 125, 126]. This compound is for example reported to be used by the terrorists in the London bombings of 2005 [99, 127]. The main reason for this popularity is the availability of precursors (hexamethylene tetramine as a firelighter, hydrogen peroxide as hair bleach or disinfectant and citric acids as a seasoning) [95], generally good knowledge of its preparation (broadcasted via internet) and an undemanding procedure for its synthesis with high yield.

HMTD was also investigated as a potential antimalarial drug [128].

10.3 Tetramethylene Diperoxide Dicarbamide

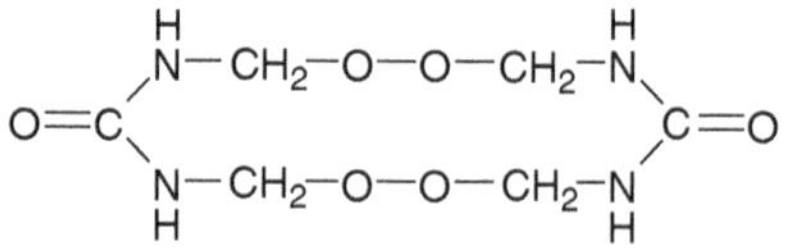

Tetramethylene diperoxide dicarbamide (1,2,8,9-tetraoxa-4,6,11,13-tetraazacyclotetradecane-5,12-dione; TMDD) was first synthesized by Girsewald in 1914 by the reaction of urea and formaldehyde with hydrogen peroxide in presence of nitric

10.3.1 Physical and Chemical Properties

TMDD forms white crystals with a melting point of approximately 180 °C [130]. It is insoluble in water, methanol, ethanol, chloroform, pyridine, and other organic solvents [129–131]. It is, however, soluble in concentrated sulfuric and nitric acids; TMDD can be precipitated back from these solutions by addition of water. TMDD decomposes to hydrogen peroxide, formaldehyde, urea, ammonia, and carbon dioxide by boiling in diluted acids. Similarly, it decomposes by boiling in diluted bases forming hydrogen peroxide and formaldehyde (respectively formic acid due to oxidation by hydrogen peroxide) [129].

10.3.2 Preparation

TMDD is prepared by dissolving urea and formaldehyde in hydrogen peroxide in presence of concentrated nitric acid [129].

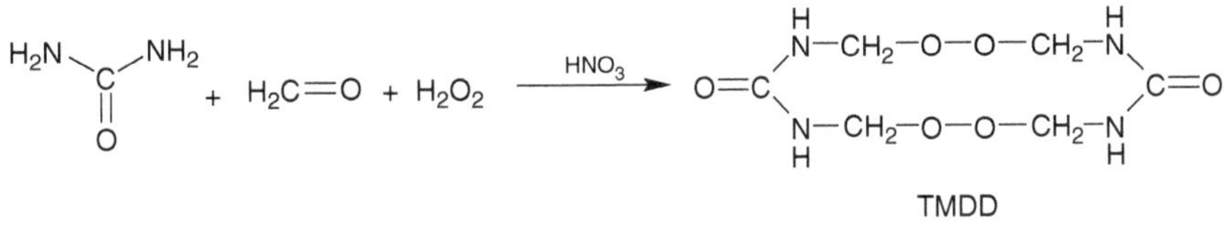

$$H_2N-\underset{\underset{O}{\|}}{C}-NH_2 + H_2C=O + H_2O_2 \xrightarrow{HNO_3} O=C\underset{\underset{H}{N}-CH_2-O-O-CH_2-N\underset{H}{}}{\overset{\overset{H}{N}-CH_2-O-O-CH_2-N\overset{H}{}}{}}C=O$$

TMDD

The reaction mixture with dissolved reactants is then placed in a refrigerator for several days during which time TMDD precipitates [110, 129, 131]. The highest yield is, according to Peña-Quevedo, obtained after 3 days of the reaction [130]. This type of peroxide cannot be purified by crystallization due to its very low solubility in all solvents; organic solvents only remove soluble impurities in TMDD [110, 130, 131].

10.3.3 Use

TMDD was suggested in 1934 as a component of priming mixtures for increasing the combustibility of the mixture [132] and as a primary explosive for blasting caps [133].

References

1. Milas, N.A., Golubovič, A.: Studies in organic peroxides. XXVI. Organic peroxides derived from acetone and hydrogen peroxide. J. Am. Chem. Soc. **81**, 6461–6462 (1959)
2. McCullough, K.J., Morgan, A.R., Nonhebel, D.C., Pauson, P.L., White, G.J.: Ketone-derivated peroxides. Part I. Synthetic methods. J. Chem. Res. (S), 34 (1980)
3. Pacheco-Londoño, L., Peña, Á.J., Primera, O.M., Hernández-Rivera, S.P., Mina, N., García, R., Chamberlain, R.T., Lareau, R.: An experimental and theoretical study of the synthesis and vibrational spectroscopy of triacetone triperoxide (TATP). Proc. SPIE **5403**, 279–287 (2004)
4. Rieche, A.: Über Peroxyde der Äther, der Carbonyl-Verbindungen und die Ozonide. Angew. Chem. **70**, 251–266 (1958)
5. Sauer, M.C.V., Edwards, J.O.: The reactions of acetone and hydrogen peroxide. II. Higher adducts. J. Phys. Chem. **76**, 1283–1288 (1972)
6. Jensen, L., Mortensen, P.M., Trane, R., Harris, P., Berg, R.W.: Reaction kinetics of acetone peroxide formation and structure investigations using Raman spectroscopy and X-ray diffraction. Appl. Spectrosc. **63**, 92–97 (2009)
7. Schulz, M., Kirschke, K.: Advances in Heterocyclic Chemistry. Cyclic Peroxides. Academic Press, London (1967)
8. Matyáš, R., Pachman, J.: Study of TATP: Influence of reaction conditions on product composition. Propellants Explosives Pyrotechnics **35**, 31–37 (2010)
9. Sigman, M.E., Clark, C.D., Caiano, T., Mullen, R.: Analysis of triacetone triperoxide (TATP) and TATP synthetic intermediates by electrospray ionization mass spectrometry. Rapid Commun. Mass Spectrom. **22**, 84–90 (2008)
10. Sigman, M.E., Clark, C.D., Painter, K., Milton, C., Simatos, E., Frisch, J.L., McCormick, M., Bitter, J.L.: Analysis of oligomeric peroxides in synthetic triacetone triperoxide samples by tandem mass spectrometry. Rapid Commun. Mass Spectrom. **23**, 349–356 (2009)
11. Baeyer, A., Villiger, V.: Einwirkung des Caro'schen Reagens auf Ketone. Berichte der deutschen chemischen Gesellschaft **32**, 3625–3633 (1899)
12. Baeyer, A., Villiger, V.: Ueber die Einwirkung des Caro'schen Reagens auf Ketone. Berichte der deutschen chemischen Gesellschaft **33**, 858–864 (1900)
13. Dubnikova, F., Kosloff, R., Almog, J., Zeiri, Y., Boese, R., Itzhaky, H., Alt, A., Keinan, E.: Decomposition of triacetone triperoxide is an entropic explosion. J. Am. Chem. Soc. **127**, 1146–1159 (2005)
14. Gelalcha, F.G., Schulze, B., Lönnecke, P.: 3,3,6,6-Tetramethyl-1,2,4,5-tetroxane: A twinned crystal structure. Acta Crystallogr. C Cryst. Struct. Commun. **C60**, o180–o182 (2004)
15. Milas, N.A., Davis, P., Nolan, J.N.: Organic peroxides. XX. Peroxides from the ozonization of olefins in the presence of carbonium ions. J. Am. Chem. Soc. **77**, 2536–2541 (1955)
16. Murray, R.W., Agarwal, S.K.: Ozonolysis of some tetrasubstitued ethylenes. J. Org. Chem. **50**, 4698–4702 (1985)
17. Fedoroff, B.T., Sheffield, O.E., Kaye, S.M.: Encyclopedia of Explosives and Related Items. Picatinny Arsenal, New Jersey (1960–1983)
18. Zhukov, B.P.: Energeticheskie kondesirovannye sistemy. Izdat. Yanus-K, Moskva (2000)
19. Egorshev, V., Sinditskii, V., Smirnov, S., Glinkovsky, E., Kuzmin, V.: A comparative study on cyclic acetone peroxides. In: 12nd Seminar on New Trends in Research of Energetic Materials, pp. 113–123, Pardubice, Czech Republic, 2009
20. Oxley, J.C., Smith, J.L., Luo, W., Brady, J.: Determining the vapor pressure of diacetone diperoxide (DADP) and hexamethylene triperoxide diamine (HMTD). Propellants Explosives Pyrotechnics **34**, 539–543 (2009)
21. Oxley, J.C., Smith, J.L., Jiaorong, H., Wei, L.: Destruction of peroxide explosives. J. Forensic Sci. **54**, 1029–1033 (2009)
22. Cafferata, L.F.R., Eyler, G.N., Mirifico, M.V.: Kinetics and mechanism of acetone cyclic diperoxide (3,3,6,6-tetramethyl-1,2,4,5-tetraoxane) thermal decomposition in benzene solution. J. Org. Chem. **49**, 2107–2111 (1984)

23. Matyáš, R., Šelešovský, J., Musil, T.: Sensitivity to friction for primary explosives. J. Hazard. Mater. **213–214**, 236–241 (2012)
24. Matyáš, R.: Investigation of properties of selected organic peroxides. University of Pardubice, Ph.D. Thesis, Pardubice (2005)
25. Baeyer, A., Villiger, V.: Ueber die Einwirkung des Caro'schen Reagens auf Ketone. Berichte der deutschen chemischen Gesellschaft **33**, 124–126 (1900)
26. Pastureau, M., Haller, M.A.: Sur un mode de formation d´acétol et d´acide pyruvique par oxydation directe de l´acétone. Comptes rendus de l'Académie des sciences **88**, 1591–1593 (1905)
27. Lockley, J.E., Ebdon, J.R., Rimmer, S., Tabner, B.J.: Cyclic diperoxides as sources of radicals for the initiation of the radical polymerization of methyl methacrylate. Macromol. Rapid Commun. **21**, 841–845 (2000)
28. Lockley, J.E., Ebdon, J.R., Rimmer, S., Tabner, B.J.: Polymerization of methyl methacrylate initiated by ozonates of tetramethylethene. Polymer **42**, 1797–1807 (2001)
29. Dilthey, W., Inckel, M., Stephan, H.: Die Oxydation der Ketone mit Perhydrol. Journal für praktische Chemie **154**, 219–237 (1940)
30. Reany, O., Kapon, M., Botoshansky, M., Keinan, E.: Rich polymorphism in triacetone-triperoxide. Cryst. Growth Des. **9**, 3661–3670 (2009)
31. Matyáš, R., Pachman, J., Ang, H.G.: Study of TATP: Spontaneous transformation of TATP to DADP. Propellants Explosives Pyrotechnics **34**, 484–488 (2009)
32. Matyáš, R., Pachman, J., Ang, H.G.: Study of TATP: Spontaneous transformation of TATP to DADP. Propellants Explosives Pyrotechnics **33**, 89–91 (2008)
33. Griesbaum, K., Volpp, W., Greinert, R., Greunig, H.-J., Schmid, J., Henke, H.: Ozonolysis of tetrasubstituted ethylenes, cycloolefins, and conjugated dienes on polyethylene. J. Org. Chem. **54**, 383–389 (1989)
34. Murray, R.W., Kong, W., Rajadhyaksha, S.N.: The ozonolysis of tetramethylethylene. Concentration and temperature effects. J. Org. Chem. **58**, 315–321 (1993)
35. Criegee, R.: Über den Verlauf der Ozonspaltung. Justus Liebigs Annalen der Chemie **583**, 1–36 (1953)
36. Murray, R.W., Story, P.R., Kaplan, M.L.: Nuclear magnetic resonance study of conformational isomerization in acetone diperoxide. J. Am. Chem. Soc. **88**, 526–529 (1966)
37. Thiemann, A.E.: Über Kraftstoffzusätze in Dieselölen. Automobiltechnische Zeitschrift **45**, 454–457 (1942)
38. Danilov, J.N., Ilyushin, M.A., Tselinskii, I.V.: Promyshlennye vzryvchatye veshchestva; chast I. Iniciiruyushchie vzryvshchatye veshchestva. Sankt-Peterburgskii gosudarstvennyi tekhnologicheskii institut, Sankt-Peterburg (2001)
39. Wolffenstein, R.: Ueber die Einwirkung Wasserstoffsuperoxyd auf Aceton und Mesityloxyd. Berichte der deutschen chemischen Gesellschaft **28**, 2265–2269 (1895)
40. Scheibler, H.. Richard Wolffenstein †. Zeitschrift für angewandte Chemie **42**, 1149–1151 (1929)
41. Yavari, I., Hosseini-Tabatabaei, M.R., Nasiri, F.: Semiempirical SCF MO study of ring inversion in 1,1,4,4,7,7-tetramethylcyclononane and trimeric acetone peroxide. J. Mol. Struct. Theochem **538**, 239–244 (2001)
42. Duin, A.C.T., Zeiri, Y., Dubnikova, F., Kosloff, R., Goddard, W.A.: Atomistic-scale simulations of the initial chemical events in the thermal initiation of triacetonetriperoxide. J. Am. Chem. Soc. **127**, 11053–11062 (2005)
43. Bellamy, A.J.: Triacetone triperoxide: Its chemical destruction. J. Forensic Sci. **44**, 603–608 (1999)
44. Keul, H., Griesbaum, K.: Ozonolysis of olefins containing monochloro substituted double bonds. Can. J. Chem. **58**, 2049–2054 (1980)
45. Rieche, A., Koch, K.: Die Oxydation des Diisopropyläthers. Berichte der deutschen chemischen Gesellschaft **75**, 1016–1028 (1942)
46. Khmelnitskii, L.I.: Spravochnik po vzryvchatym veshchestvam. Voennaya ordena Lenina i ordena Suvorova Artilleriiskaya inzhenernaya akademiya imeni F. E. Dzerzhinskogo, Moskva (1962)

47. Muraour, H.: Sur la théorie des réactions explosives. Cas particulier des explosifs d'amorçage. Mémories présentés a la Société chimique **51**, 1152–1166 (1932)
48. Matyáš, R., Šelešovský, J., Musil T.: Study of TATP: Mass loss and friction sensitivity during ageing. Cent. Eur. J. Energ. Mater. **9**, 251–260 (2012)
49. Oxley, J.C., Smith, J.L., Shinde, K., Moran, J.: Determination of the vapor density of triacetone triperoxide (TATP) using a gas chromatography headspace technique. Propellants Explosives Pyrotechnics **30**, 127–130 (2005)
50. Ficheroulle, H., Kovache, A.: Contribution à l'étude des explosifs d'amorçage. Mémorial des poudres **31**, 7–27 (1949)
51. Bagal, L.I.: Khimiya i tekhnologiya iniciiruyushchikh vzryvchatykh veshchestv. Mashinostroenie, Moskva (1975)
52. Matyáš, R.: Chemical destruction of triacetone triperoxide and hexamethylenetriperoxidediamine. In: Proceedings of 6th Seminar on New Trends in Research of Energetic Materials, pp. 164–173, Pardubice, Czech Republic, 2003
53. Armitt, D., Zimmermann, P., Ellis-Steinborner, S.: Gas chromatography/mass spectrometry analysis of triacetone triperoxide (TATP) degradation products. Rapid Commun. Mass Spectrom. **22**, 950–958 (2008)
54. Pachman, J., Matyáš, R.: Study of TATP: Stability of TATP solutions. Forensic Sci. Int. **207**, 212–214 (2011)
55. Cotte-Rodriguez, I., Chen, H., Cooks, R.G.: Rapid trace detection of triacetone triperoxide (TATP) by complexation reactions during desorption electrospray ionization. Chem. Commun. **9**, 953–955 (2006)
56. Dubnikova, F., Kosloff, R., Zeiri, Y., Karpas, Z.: Novel approach to the detection of triacetone triperoxide (TATP): Its structure and its complexes with ions. J. Phys. Chem. A **106**, 4951–1159 (2002)
57. Matyáš, R., Pachman, J.: Thermal stability of triacetone triperoxide. Sci. Technol. Energetic Mater. **68**, 111–116 (2007)
58. Meyer, R., Köhler, J., Homburg, A.: Explosives. Wiley-VCH, Weinheim (2002)
59. Yinon, J.: Forensic and Environmental Detection of Explosives. Wiley, New York (1999)
60. Mavrodi, G.E.G.: Improvements in or relating to explosives of the organic peroxide class. GB Patent 620,498, 1949
61. Šelešovský, J., Matyáš, R., Musil, T.: Using of the probit analysis for sensitivity tests - sensitivity curve and reliability. In: Proceedings of 14th Seminar on New Trends in Research of Energetic Materials, pp. 963–967, Pardubice, Czech Republic, 2011
62. Yeager, K.: Trace Chemical Sensing of Explosives. Wiley, New Jersey (2007)
63. Fogelzang, A.E., Egorshev, V.Y., Pimenov, A.Y., Sinditskii, V.P., Saklantii, A.R., Svetlov, B.S.: Issledovanie statsionarnogo goreniya initsiiruyushchikh vzryvchatykh veshchestv pri vysokykh davleniyakh. Dokl. Akad. Nauk SSSR **282**, 1449–1452 (1985)
64. Kuzmin, V.V., Kozak, G.D., Solovev, M.Y., Tuzkov, Y.B.: Forensic investigation of some peroxides explosives. In: Proceedings of 11th Seminar on New Trends in Research of Energetic Materials, pp. 387–393, Pardubice, Czech Republic, 2008
65. Kuzmin, V.V., Solovev, M.Y., Tuzkov, Y.B., Kozak, G.D.: Forensic investigation of some peroxides explosives. Cent. Eur. J. Energetic Mater. **5**, 77–85 (2008)
66. Rohrlich, M., Sauermilch, W.: Sprengtechnische Eigenschaften von Trizykloazetonperoxyd. Zeitschrift für das gesamte Schiess- und Sprengstoffwesen **38**, 97–99 (1943)
67. Lefebvre, M.H., Falmagne, B., Smeets, B.: Sensitivities and performances of non-regular explosives. In: Proceedings of 7th Seminar on New Trends in Research of Energetic Materials, pp. 164–173, Pardubice, Czech Republic, 2004
68. Bubnov, P.F.: Initsiruyushchie vzryvchatye veshchestva i sredstva initsirovaniya. Gosudarstvennoe izdatelstvo oboronnoi promyshlennosti, Moskva (1940)
69. Matyáš, R., Šelešovský, J.: Power of TATP based explosives. J. Hazard. Mater. **165**, 95–99 (2009)

70. Matyáš, R., Trzciński, W., Cudzilo, S., Zeman, S.: Detonation performance of TATP/AN-based explosives. Propellants Explosives Pyrotechnics **33**, 296–300 (2008)
71. Groth, P.: Crystal structure of 3,3,6,6,9,9-hexamethyl-1,2,4,5,7,8-hexa-oxacyclononane ("trimeric acetone peroxide"). Acta Chem. Scand. **23**, 1311–1329 (1969)
72. Matyáš, R., Jirásko, R., Lyčka, A., Pachman, J.: Study of TATP: Formation of new chloroderivates of triacetone triperoxide. Propellants Explosives Pyrotechnics 219–224, **36** (2011)
73. Schulte-Ladbeck, R., Kolla, P., Karst, U.: Trace analysis of peroxide-based explosives. Anal. Chem. **75**, 731–735 (2003)
74. Peña, Á.J., Pacheco-Londoño, L., Figueroa, J., Rivera-Montalvo, L.A., Román-Velazquez, F. R., Hernández-Rivera, S.P.: Characterization and differentiation of high energy cyclic organic peroxides by GC/FT-IR, GC-MS, FT-IR, and Raman microscopy. Proc. SPIE **5778**, 347–358 (2005)
75. Widmer, L., Watson, S., Schlatter, K., Crowson, A.: Development of an LC/MS method for the trace analysis of triacetone triperoxide (TATP). Analyst **127**, 1627–1632 (2002)
76. Murray, R.W., Jeyaraman, R.: Dioxiranes: Synthesis and reactions of methyldioxiranes. J. Org. Chem. **50**, 2847–2853 (1985)
77. Kim, T.J., Heo, N.H., Kim, J.-H., Seo, G.: Formation of acetone cyclic triperoxide over titania-incorporated mesoporous materials. React. Kinet. Catal. Lett. **79**, 287–293 (2003)
78. Belič, I., Suhadolc, T.: Peroxides of diisopropylether. Experientia **25**, 473 (1969)
79. Acree, F., Haller, H.L.: Trimolecular acetone peroxide in isopropyl ether. J. Am. Chem. Soc. **65**, 1652 (1943)
80. Kharasch, M.S., Gladstone, M.: Ether peroxides. J. Chem. Educ. **16**, 498 (1939)
81. Noponen, A.: Violent explosion. Chem. Eng. News **55**, 5 (1977)
82. Stirling, C.J.M.: Explosion warning. Chem. Brit. **5**, 36 (1969)
83. Brewer, A.D.: Peroxide/acetone mixture hazard. Chem. Brit. **11**, 335 (1975)
84. Micetich, R.G.: 6-Aminopenicillanic acid sulphoxide and ampicillin sulphoxide. Chem. Brit. **13**, 163 (1977)
85. Costantini, M.: Destruction of acetone peroxides. US Patent 5,003,109, 1991
86. Davies, A.G.: Organic Peroxides. Butterworths & Co., London (1961)
87. Pyl, G.: Verfahren zur Herstellung von Initialzündmitteln. DE Patent 423,176, 1925
88. Holtmeier, U., Holtmeier, L.: Sprengzünder. DE Patent DE 10,2009,007,178 A1, 2009
89. El-Awady, A.A., Prell, L.J.: Exothermic reactions and unstable compounds: a demonstration of fires, flames, and smoke In: 2nd Annual Symposium of Chemistry Demonstrations, Western Illinois University, Macomb, USA, May, 1979
90. Shakhashiri, B.Z.: Chemical Demonstrations. A Handbook for Teachers of Chemistry. The University of Wiskonsin Press, Madison (1983)
91. Šarapatka, J.: Analýza událostí za rok 2002; Bulletin ochranné služby Policie České republiky, Report. Ministry of the Interior of the Czech Republic, Praha (2003)
92. IntelCenter. London Tube Bus Attack (LTBA) v 1.5, Report: http://www.intelcenter.com/LTBA-PUB-v1-5.pdf (Online 29 Jan 2007)
93. Michalske, T., Edelstein, N., Sigman, M., Trewhella, J.: Basic Research Needs for Countering Terrorism, Report: http://www.er.doe.gov/bes/reports/files/NCT_rpt_screen.pdf (Online 29 Jan 2007), 2007
94. Philpott, D.: The London Bombing. Homeland Defence Journal: http://www.homelanddefensejournal.com/pdfs/LondonBombing_SpecialReport.pdf (Online 29 Jan 2007), 2006
95. Dudek, K., Matyáš, R., Dorazil, T.: DIEPE – Detection and identification of explosive precursors and explosives. In: Proceedings of 14th Seminar on New Trends in Research of Energetic Materials, pp. 595–600, Pardubice, Czech Republic, 2011
96. Ember, L.R.: Biochemist arrested in London bombing. Chem. Eng. News **83**, 11 (2005)
97. Stambouli, A., El Bouri, A., Bouayoun, T., Bellimam, M.A.: Headspace-GC/MS detection of TATP traces in post-explosion debris. Forensic Sci. Int. 146S, S191–S194 (2004)
98. Matyáš, R.: Improvizované výbušiny. In: 2nd Mezinárodní pyrotechnický seminář, pp. 100–109, Praha, Czech Republic, 2003

99. Marshall, M., Oxley, J.C.: Aspects of Explosives Detection. Elsevier, Oxford (2009)
100. Jiang, H., Chu, G., Gong, H., Qiao, Q.: Tin chloride catalysed oxidation of acetone with hydrogen peroxide to tetrameric acetone peroxide. J. Chem. Res. (S), 288–289 (1999)
101. Obinokov, V.N., Botsman, L.P., Ishmupatov, L.J., Tolstikov, G.A.: Ozonoliz alkenov i izuchenie reakcii polifunkcionalnykh coedinenii. XVIII. Issledovanije novogo ozonoliticheskogo sinteza karbonovykh kislot. Zhurnal organicheskoi khimii **16**, 524–537 (1980)
102. Legler, L.: Ueber Producte der langsamen Verbrennung des Aethyläthers. Berichte der deutschen chemischen Gesellschaft **18**, 3344–3351 (1885)
103. Taylor, C.A., Rinkenbach, W.H.: H. M. T. D. A new detonating explosive. Army Ordnance **5**, 463–466 (1924)
104. Schaefer, W.P., Fourkas, J.T., Tiemann, B.G.: Structure of hexamethylene triperoxide diamine. J. Am. Chem. Soc. **107**, 2461–2463 (1985)
105. Wierzbicki, A., Cioffi, E.: Density functional theory studies of hexamethylene triperoxide diamine. J. Phys. Chem. A **103**, 8890–8894 (1999)
106. Wierzbicki, A., Salter, E.A., Cioffi, E.A., Stevens, E.D.: Density functional theory and X-ray investigations of P- and M-hexamethylene triperoxide diamine and its dialdehyde derivative. J. Phys. Chem. A **105**, 8763–8768 (2001)
107. Guo, C., Persons, J., Harbison, G.S.: Helical chirality in hexamethylene triperoxide diamine. Magn. Reson. Chem. **44**, 832–837 (2006)
108. Ilyushin, M.A., Tselinsky, I.V., Sudarikov, A.M.: Razrabotka komponentov vysokoenergicheskikh kompozitsii. SPB:LGU im. A. S. Pushkina – SPBGTI(TU), Sankt-Peterburg (2006)
109. Konrad: Die Verwendung von Hexamethylentetramin in der Sprengstoffindustrie. Nitrocellulose **5**, 123–124 (1934)
110. Peña-Quevedo, A.J., Mina-Calmide, N., Rodríguez, N., Nieves, D., Cody, R.B., Hernández-Rivera, S.P.: Synthesis, characterization and differentiation of high energy amine peroxides by MS and vibrational microscopy. Proc. SPIE **6201**, 62012E/62011–62012E/62010 (2006)
111. Peña-Quevedo, A.J., Hernández-Rivera, S.P.: Mass spectrometry analysis of hexamethylene triperoxide diamine by its decomposition products. Proc. SPIE **7303**, 730303/730301–730303/730311 (2009)
112. Sülzle, D., Klaeboe, P.: The infrared, Raman and NMR spectra of hexamethylene triperoxide diamine. Acta Chem. Scand. A **42**, 165–170 (1988)
113. Marotta, D., Alessandrini, M.E.: Ricerche sull'esametilen-tetrammina. - I. Esametilentetrammina e perossido di idrogeno. Gazzetta Chimica Italiana **59**, 942–946 (1929)
114. Szyc-Lewańska, K., Urbański, T.: Chemistry of cyclonite. Nitration of hexamethylene triperoxide diamine. Bulletin de l'Académie polonaise des sciences; Serie des sciences chimiques, geologiques et geographiques **6**, 165–167 (1958)
115. Stettbacher, A.: Porokha i vzryvshchatye veshchestva. ONTI Glavnaya redakciya khimicheskoi literatury, Moskva (1936)
116. Urbański, T.: Chemistry and Technology of Explosives. PWN—Polish Scientific Publisher, Warszawa (1967)
117. Fogelzang, A.E., Serushkin, V.V., Sinditskii, V.P.: O spontannom vzryvje geksametilentriperoksiddiamina. Fizika goreniya i vzryva **25**, 129–131 (1989)
118. Girsewald, C.: Beiträge zur Kenntnis des Wasserstoffperoxyds. Über die Einwirkung des Wasserstoffperoxyds auf Hexamethylentetramin. Berichte der deutschen chemischen Gesellschaft **45**, 2571–2576 (1912)
119. Metz, L.: Die Prüfung von Zündhütchen (Initialsprengstoffen) auf Schlagempfindlichkeit und Flammenwirkung. Zeitschrift für das gesamte Schiess- und Sprengstoffwesen **23**, 305–308 (1928)
120. Girsewald, C.: Verfahren zur Darstellung von Hexamethylentriperoxyddiamin. DE Patent 263,459, 1912
121. Pease, J., Newell, F.: Process of producing a new organic compound and the compound when produced thereby. GB Patent 339,024, 1929

122. Baeyer, A., Villiger, V.: Ueber die Nomenclatur der Superoxyde und die Superoxyde der Aldehyde. Berichte der deutschen chemischen Gesellschaft **33**, 2479–2487 (1900)
123. Girsewald, C.: Verwendung von Hexamethylentriperoxidyddiamin zur Herstellung von Initialzündern. DE Patent 274,522, 1912
124. Ilyushin, M.A., Tselinsky, I.V., Shugalei, I.V., Chernay, A.V., Toftunova, V.V.: "Green" polymer-bound explosive (PBX) for laser initiation. In: Proceedings of 9th Seminar on New Trends in Research of Energetic Materials, pp. 602–607, Pardubice, Czech Republic, 2006
125. Byall, E.T.: Explosives Report 1998–2001. In: Proceedings of 13th INTERPOL Forensic Science Symposium, Lyon, France, 2001
126. Matyáš, R.: Grant of the Ministry of Internal Affairs of Czech Republic, Report RN 20012003003. University of Pardubice, Pardubice (2002)
127. Persons, J., Harbison, G.S.: The ^{14}N quadrupole coupling in hexamethylene triperoxide diamine (HMTD). Magn. Reson. Chem. **45**, 905–908 (2007)
128. Vennerstrom, J.L.: Amine peroxides as potential antimalarials. J. Med. Chem. **32**, 64–67 (1989)
129. Girsewald, C., Siegens, H.: Beiträge zur Kenntnis des Wasserstoffperoxyds. II. Tetramethylen-diperoxyd-dicarbamid. Berichte der deutschen chemischen Gesellschaft **47**, 2464–2469 (1914)
130. Peña-Quevedo, A.J.: Cyclic organic peroxides identification and trace analysis by Raman microscopy and open-air chemical ionization mass spectroscopy. University of Puerto Rico, PhD thesis, Puerto Rico (2009)
131. Peña-Quevedo, A.J., Cody, R., Mina-Calmide, N., Ramos, M., Hernández-Rivera, S.P.: Characterization and differentiation of high energy amine peroxides by direct analysis in real time TOF/MS. Proc. SPIE **6538**, 653828-1, 2007
132. Weale, A., Renfrew, A.: Improvements in or relating to explosive priming compositions. GB Patent 415,779, 1934
133. Spaeth, C.P.: Ignition composition. US Patent 1,984,846, 1934

Chapter 11
Nitrogen Halides

The nitrogen halides are substances containing three atoms of one of the halogens bound together with one atom of nitrogen. The most common one is without question nitrogen triiodide which is often used for demonstration purposes due to its extreme sensitivity to mechanical impulse and impressive purple cloud formed as a result of the explosion. Nitrogen trichloride and nitrogen tribromide are both quite sensitive and explode easily. The last nitrogen halide in the family is a fluorine analog which does not possess explosive behavior and is therefore not mentioned in the following sections.

11.1 Nitrogen Trichloride

The earliest report on the preparation of nitrogen trichloride (NCl_3) dates back to the beginning of the nineteenth century and is attributed to Dulong [1, 2]. Sir Humphry Davy, in his letter to the Royal Society of London, mentioned information received from France 12 months earlier (meaning in 1811) concerning a new compound in the form of a heavier-than-water oil which "explodes by gentle heat with all the violence of fulminating metals" [3]. The discoverer, not named in this letter, most likely Dulong, lost an eye and three fingers during the early experiments with the substance. The details of the discovery and an earlier history of this dangerous substance are described by Snelders [4].

11.1.1 Physical and Chemical Properties

Nitrogen trichloride forms a bright yellow oily liquid with a density of 1.653 g cm^{-3}. It is very volatile [5–9] and has a very unpleasant odor sometimes described as nauseous or pungent [3, 6, 7, 10–12]. The vapors irritate the eyes. In its frozen state, it exists as a rhombohedral crystalline solid. The melting point is reported between -40

and −27 °C, boiling point 71–73 °C [11, 12], and its vapors explode if heated above 93–95 °C [6, 9, 10, 13]. In spite of this fact, distillation of nitrogen trichloride in the open air at 71 °C without explosion was reported by Walke [9]. It is insoluble in water. However, if submerged it slowly decomposes by hydrolysis. In cold water, this process takes several hours [8, 10]. Solubility is on the other hand relatively good in solvents such as benzene, chloroform, carbon tetrachloride, ether, carbon disulfide, and phosphorus trichloride [7, 10, 11]. Sunlight causes the decomposition of nitrogen trichloride solutions or its reaction with solvents. For example, benzene solution gives hexachlobenzene when exposed to sunlight; the chloroform solution decomposes when chlorine, hydrogen, ammonium chloride, and traces of hexachlorethane form (but no carbon tetrachloride) [14]. Nitrogen trichloride solutions in diffuse daylight or in darkness can be stored unchanged for weeks [6].

Nitrogen trichloride violently explodes in contact with many substances (especially with substances which have an affinity for chlorine), such as phosphorus (even in solution), arsenic, many organic substances (e.g., oils, fats, turpentine, naphtha), concentrated alkali hydroxides, ozone, solid iodine, and potassium cyanide [3, 6, 7, 10, 12, 13, 15]. An interesting laboratory demonstration of the preparation and later explosion of nitrogen trichloride is described by Rai [16]. In this experiment, NCl_3 is first prepared by reaction of a concentrated solution of bleaching powder and a saturated solution of ammonium chloride at low temperature. The reaction product is partly present on the surface of the aqueous solution and partly, due to its high density, at the bottom of the test tube. Addition of a few drops of oil or turpentine causes instantaneous explosion of the surface layer of nitrogen trichloride followed by a second explosion of the product at the bottom some time later. Explosion of this substance may be caused even by grease from the fingers [6]. Nitrogen trichloride decomposes in contact with sulfur, acids, and reducing agents [3, 6, 15]. It can be slowly decomposed by dilute ammonium chloride solution to give nitrogen and hydrochloric acid [17]. Using a concentrated solution leads to a rapid decomposition giving the same products.

$$NCl_3 + NH_4Cl \longrightarrow N_2 + 4 HCl$$

Nitrogen trichloride is a photosensitive substance which may explode when exposed to sunlight or magnesium torch light [6, 7, 10, 12]. Photolysis in the gaseous state [18] or in tetrachloromethane solution [19] yields nitrogen and chlorine. The use of other solvents usually leads to their chlorination [14]. Nitrogen trichloride is an interesting chemical agent (aminating and halogenating). Its chemistry was summarized extensively by Kovacic et al. [20].

11.1.2 Explosive Properties

Nitrogen trichloride is an extremely sensitive substance and it explodes violently with the slightest impact or friction [7, 12]. The explosion itself is characterized by

11.1 Nitrogen Trichloride

a brilliant flash [3]. It is a very hazardous substance in its pure form and caused serious injuries to early investigators—Mr. Dulong lost three fingers and one eye [6] and Mr. Davy received severe wounds to an eye when an amount "scarcely as large as a grain of mustard seed" shattered the glass tube containing it [3]. Although very sensitive its performance is relatively low [6, 12].

Nitrogen trichloride decomposes to nitrogen and chlorine. Decomposition may be represented by the following equation [9]:

$$2\ NCl_3 \longrightarrow N_2 + 3\ Cl_2$$

The volume of gaseous product is 370.5 l per 1 kg of nitrogen trichloride [9].

11.1.3 Preparation

Warning—all equipment that may come into contact with nitrogen trichloride must be washed by alkali in order to clean it from grease. Not doing so will lead to explosion when contact between the two occurs.

It is believed that nitrogen trichloride was first obtained accidentally by Dulong who introduced chlorine gas into a solution of ammonium chloride at 8 °C [1–3, 6–8, 16, 21].

$$NH_4Cl + 3\ Cl_2 \longrightarrow NCl_3 + 4\ HCl$$

Aqueous ammonia may also be used as a starting material; however in this case concentration and pH are extremely important. The normal reaction between chlorine and ammonia is obtained only when ammonia is used in stoichiometric proportions [22]:

$$12\ NH_3 + 6\ Cl_2 \longrightarrow NCl_3 + N_2 + 9\ NH_4Cl$$

In excess ammonia, nitrogen trichloride reacts giving partly free nitrogen and partly probably ammonium hypochlorite [22]:

$$NCl_3 + NH_3 \longrightarrow N_2 + 3\ NH_4Cl$$

$$NCl_3 + 2\ NH_4OH + H_2O \longrightarrow 3\ NH_4ClO$$

The reaction of ammonium chloride or aqueous ammonia with chlorine provide product in the form of extremely dangerous oily drops at the bottom of the aqueous phase or as a film on the solution surface. Direct separation of nitrogen trichloride is a dangerous operation and it is therefore recommended to collect it by extraction. A suitable solvent for this operation is tetrachloromethane which cannot be further chlorinated and therefore acts as a good inert medium [14]. The extraction from the aqueous solution into an organic solvent greatly reduces the risk of explosion. Other ammonium salts may also be used as starting materials instead of aqueous ammonia. It is however important to note that some ammonium salts are more suitable than others. The preferred choice among ordinary chemicals is ammonium sulfate since the reaction between ammonium chloride and chlorine is reported to be reversible [10, 21, 23]. The effects of various reaction conditions used during preparation of nitrogen trichloride have been summarized by Mellor [6]. Ammonium nitrate was the substance of choice of early experimenters such as Mr. Burtom or Mr. Children who used it in aqueous solution and let it react with chlorine gas [3].

The formation of nitrogen trichloride has also been observed in the gaseous phase. In this case reaction of gaseous ammonia with chlorine first yields nitrogen trichloride and later, if there is an excess of ammonia, reacts to form ammonium chloride [8, 18].

$$NH_3 + 3\,Cl_2 \longrightarrow NCl_3 + 3\,HCl$$

$$3\,HCl + 3\,NH_3 \longrightarrow 3\,NH_4Cl$$

Nitrogen trichloride may also be produced by electrolysis of acidic aqueous liquids in which ammonium and chloride ions are present [6, 8]. Kolbe found that electrolysis of a concentrated solution of ammonium chloride gives hydrogen on the negative pole but neither oxygen nor chlorine on the positive pole. The platinum positive electrode was however covered by small yellowish oily drops of what proved later to be nitrogen trichloride [24, 25]. The most favorable conditions for formation of nitrogen trichloride are high electrolyte concentration, absence of light, high anodic current density, and separation of anodic and cathodic volumes by diaphragm [24]. The acidity of the electrolyte further determines the composition of the product. Nitrogen trichloride is formed in an acidic environment (hydrochloric acid is recommended) while chloramines ($NHCl_2$ and NH_2Cl) are formed in very weakly acidic, neutral, or alkaline solutions [26]. Nitrogen trichloride can be removed from the electrolyte by passing air (or other gases) through the electrolyte with suction applied over the electrolyte [27, 28].

Another method of nitrogen trichloride preparation was published by Balard, who proved that nitrogen trichloride can be obtained when a solution of hypochlorous acid comes into contact with aqueous ammonia or ammonium salts [6, 8, 15, 29]:

$$3\,HClO + NH_3 \longrightarrow NCl_3 + 3\,H_2O$$

$$3\,HClO + NH_4Cl \longrightarrow NCl_3 + 3\,H_2O + HCl$$

Hentschel [30] modified the above-mentioned preparation route of Balard [29] by replacing hypochlorous acid with sodium hypochlorite and extracting the product with benzene:

11.2 Nitrogen Tribromide

$$2\,NH_4Cl + 7\,NaClO \longrightarrow NCl_3 + 6\,NaCl + NaNO_3 + 4\,H_2O$$

Nitrogen trichloride forms only if the molar ratio of ammonium chloride to chlorine is equal to or greater than 2:3. Even a ratio of 16:3 is reported to give the same product. On the other hand, lower amounts of ammonium chloride result in formation of a solution of chlorine in ammonium chloride instead of the NCl_3.

According to Rai [16], nitrogen trichloride can also be prepared by reaction of bleaching powder (calcium hypochlorite) solution on a saturated solution of ammonium chloride. The reaction is very vigorous and must be carried out at low temperatures. Ice bath cooling is recommended and the reaction temperature should not exceed 0 °C. The reaction does not require special equipment or extraction of the product and is therefore suitable for small-scale preparation or class demonstration of the explosive properties of nitrogen trichloride [16].

Nitrogen trichloride is also formed by passing a mixture of nitrosyl chloride and phosphorus pentachloride through a tube heated to a high temperature. The nitrosyl chloride is prepared by heating a mixture of sodium nitrite and phosphorus pentachloride [31]:

$$NaNO_2 + PCl_5 \longrightarrow NOCl + POCl_3 + NaCl$$

$$NOCl + PCl_5 \longrightarrow NCl_3 + POCl_3$$

Nitrogen trichloride can be relatively safely stored in solution. It has, however, been reported that spontaneous explosion of a 12 % solution in di-*n*-butylether occurred shortly after preparation and violent decomposition of 12 % solution of nitrogen trichloride was reported during preparation without cooling (3–5 % solutions can be prepared safety without cooling) [32]. Solutions of nitrogen trichloride slowly decompose during storage into nitrogen and chlorine even when stored in the dark [17].

11.1.4 Use

The extreme sensitivity of nitrogen trichloride, together with its low physical and chemical stability, prevents it from being used in practical applications as an explosive. It has, however, been used in its gaseous form as an agent for bleaching and maturing of flour [5, 33]. It had also been studied as a chemical agent (aminating and halogenating) for synthesis in organic chemistry [20, 34–39].

11.2 Nitrogen Tribromide

There is very little information about nitrogen tribromide (NBr_3) in open literature and to make the situation even more difficult these literature sources tend to contradict each other.

11.2.1 Physical and Chemical Properties

Nitrogen tribromide (NBr_3) is reported as a red, oily, volatile liquid with offensive odor [6, 8]. From later work it appears, however, that nitrogen tribromide is a solid substance of a deep red color with nitrogen to bromine ratio 1:3 [40]. This extremely temperature-sensitive substance "explodes even at $-100\ °C$ in Nujol-pentane suspension (1:2.7) with the slightest mechanical disturbance" [40].

It slowly hydrolyses in water and explodes in contact with phosphorus or arsenic [6]. Nitrogen tribromide is soluble in chloroform [41] and at temperatures below $-80\ °C$ in polar solvents that do not undergo bromination or oxidation [40]. Nitrogen tribromide reacts with ammonia instantly giving solid, dark violet monobromamine [40]:

$$NBr_3 + 2\ NH_3 \xrightarrow[-87°C]{CH_2Cl_2} 3\ NH_2Br$$

With N-bases (pyridine, urotropine), it directly forms bromine adducts. The step which would lead to nitrogen tribromide adducts, similar to those observed in reaction of nitrogen triiodide adducts in the case of the reaction of $NI_3 \cdot NH_3$ with tertiary bases, is absent [40].

Iodine reacts with nitrogen tribromide in a dichloromethane solution at $-87\ °C$ giving red-brown solid nitrogen dibromideiodide (NBr_2I) which is stable up to $-20\ °C$ [40].

$$NBr_3 + I_2 \xrightarrow[-87°C]{CH_2Cl_2} NBr_2I + IBr$$

The infrared and Raman spectroscopic characterization of the nitrogen tribromide is quite problematic due to its high thermal and mechanical sensitivity. It was reported by Jander [40] that orange-red mixtures of solid nitrogen tribromide and ammonium bromide or acetate have been evaluated instead of the pure substance. The gas phase spectra were obtained by Klapötke [42].

11.2.2 Preparation

The preparation of nitrogen tribromide, as cited in the oldest literature, is based on the action of aqueous potassium or sodium bromide on nitrogen trichloride. It is reported that the initial yellow color slowly changes to red, which becomes deeper and deeper in tint finally giving a product a dense black oily appearance having a chlorine-like odor [6, 9, 12, 13, 43]. This product has similar properties to the nitrogen trichloride (e.g., explodes by contact with phosphorus and arsenic [9]) and was considered to be nitrogen tribromide. The composition of such product was however not checked by serious analytical methods. Galal-Gorchev and Morris [41] therefore raised questions disputing the previously reported formula. Roozeboom [44] found that the above-mentioned reaction leads to an ammonium tribromide NH_4Br_3 in the form of orange crystals. It should however be noted that ammonium bromide was used instead of the sodium or potassium bromide used in the original procedure [44].

Galal-Gorchev and Morris reported that they prepared nitrogen tribromide by reaction of aqueous bromine with ammonia and they confirmed the formation of NBr_3 by extraction into chloroform and analyzed the chloroform solution by UV spectroscopy [41]. Jander [40] reports that the first preparation of nitrogen tribromide was achieved by Thiedemann by reaction of an acidic aqueous ammonia salt solution (pH = 4) with bromine. The UV spectroscopy is mentioned as the method of the product analysis. The isolation of pure nitrogen tribromide from these aqueous solutions was done by extraction with ether and subsequent precipitation in Freon 12 at $-110\,°C$.

Some other methods of preparation of NBr_3 by direct bromination of ammonia are further summarized by Jander [40]. Another possible way for preparation of pure solid NBr_3 is by reaction of bromine or hypobromus acid with an aqueous solution of ammonia [41, 45]. The concentration of reaction products is a function of temperature, acidity, and bromine to ammonia ratio. Nitrogen tribromide, NBr_3 forms from neutral solutions when the molar ratio of bromine to ammonia is greater than 1.5 [45].

Pure solid nitrogen tribromide may be prepared by reaction of bistrimethylsilyl-bromamine and bromine chloride at $-87\,°C$ in pentane according to Jander [40]:

$$[(CH_3)_3Si]_2NBr + 2\,BrCl \xrightarrow{-87°C} NBr_3 + 2\,(CH_3)_3SiCl$$

According to some more recent investigations by Klapötke, nitrogen tribromide also forms by reaction of bromine azide and elemental bromine in gaseous form [42]:

$$NaN_3 + Br_2 \xrightarrow{0-20°C} BrN_3 + NaBr$$

$$BrN_3 + Br_2 \longrightarrow NBr_3 + N_2$$

11.3 Nitrogen Triiodide

Nitrogen triiodide adduct with one molecule of ammonia ($NI_3 \cdot NH_3$) was first prepared by Bernard Courtois, discoverer of iodine, in 1813. It was among the first derivatives of iodine ever made [46, 47]. The history of the investigation of nitrogen triiodide during the nineteenth century is summarized in review by Chattaway [47].

11.3.1 Structure

The generally reported cases of "nitrogen triiodide" mostly refer to one of the nitrogen triiodide adducts with ammonia. The composition of such a substance can be summarized by the general formula $NI_3 \cdot xNH_3$. The number of the ammonium groups varies depending on the reaction conditions but at least the following substances are reported to exist: $NI_3 \cdot NH_3$, $NI_3 \cdot 2NH_3$, $NI_3 \cdot 3NH_3$, $NI_3 \cdot 5NH_3$, and $NI_3 \cdot 12NH_3$. The ammonium adducts with more than one ammonium group are stable only at low temperatures and all of them lose ammonia over a period of time to eventually yield $NI_3 \cdot NH_3$ as a relatively stable final product. Attempts to remove the last NH_3 group resulted in a complete break-up of the molecule [48].

The above-mentioned facts resulted in a common belief that it is impossible to prepare pure nitrogen triiodide—NI_3 without an ammonium molecule—due to its low stability. The preparation of pure NI_3 is however mentioned in at least two papers. The first is the work of Cremer and Duncan [46] from 1930. They reported preparation of nitrogen triiodide by reaction of dry ammonia with less stable dibromoiodides:

$$3\ KIBr_2 + 4\ NH_3 \longrightarrow NI_3 + 3\ KBr + 3\ NH_4Br$$

The product they obtained by this reaction contained iodine and nitrogen in molar ratio 1:3.04 (mass ratio 1:27.54; calc. 1:27.18). The resulting substance can be sublimed in a vacuum at room temperature and condensed by liquid air [46]. This paper seems almost forgotten as we found only one brief reference to it by Špičák and Šimeček [12].

In 1990, a second paper reporting the first-ever preparation of "adduct-free nitrogen triiodide" was published by Tornieporth-Oetting and Klapötke [49]. The authors allowed boron nitride to react with iodine monofluoride in $CFCl_3$ at low temperature (-30 °C) which yielded pure nitrogen triiodide according to the following reaction:

$$BN + 3\ IF \xrightarrow{CFCl_3} NI_3 + BF_3$$

The substance formed was reported to decompose rapidly in solution at 0 °C and within several hours at -60 °C; the solid substance was found to be stable at 196 °C, [49]. It is reported that the preparation of adduct-free nitrogen triiodide is possible only in the absence of ammonia which is in disagreement with the previous method by Cremer [46] and will require further confirmation.

Nitrogen triiodide ($NI_3 \cdot NH_3$) has a polymeric structure $[NI_3 \cdot NH_3]_n$ in which tetrahedral NI_4 units are corner-linked into infinite chains of $-N-I-N-I-$ (2.15 and 2.30 Å) which in turn are linked into sheets by I–I interactions (3.36 Å) in the c-direction; in addition, one I of each NI_4 unit is also loosely attached to an NH_3

11.3 Nitrogen Triiodide

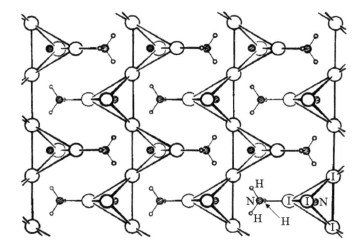

Fig. 11.1 Projection of the monoclinic crystal structure of nitrogen triiodide [40]. Reprinted from Advances in Inorganic Chemistry and Radiochemistry, Volume 19, J. Jander, Recent chemistry and structure investigation of nitrogen triiodide, tribromide, trichloride, and related compounds, p. 1–63, Copyright (1976), with permission from Elsevier

(2.53 Å) that projects into the space between the sheets of tetrahedra [40, 50]. The structure is shown in Fig. 11.1.

The estimate of strength of the bond holding NI_3 to NH_3 is only 16.7 kJ mol^{-1}. The strength of the N–I bond is 201 kJ mol^{-1} [51].

11.3.2 Physical and Chemical Properties

Nitrogen triiodide ($NI_3 \cdot NH_3$) forms black-colored crystals (see Fig. 11.2) with a density of 3.5 g cm^{-3} insoluble in water [11]. Its enthalpy of formation is 146 kJ mol^{-1} [52]. Light causes decomposition of nitrogen triiodide producing nitrogen [53]. It also decomposes in hot water [11]. Nitrogen triiodide decomposes by action of many chemical agents, by action of diluted acids (in the case of sulfuric and hydriodic acids it forms solid iodine and ammonia salt) and it explodes in contact with concentrated acids, bromine, chlorine, ozone, and hydrogen sulfide. It also decomposes in aqueous solutions of alkalis, sulfites, thiosulfates, cyanides, or thiocyanates [11, 47, 54]. Chattaway pointed out that nitrogen triiodide decomposes in aqueous ammonia with the decomposition rate depending on concentrations [47].

The NI_3 molecule forms adducts similar to those with ammonia also with N-bases such as pyridine (forming $NI_3 \cdot py$) [55], quinoline ($NI_3 \cdot 3C_9H_7N$) [56], or urotropine ($NI_3 \cdot I_2 \cdot C_6H_{12}N_4$) [57].

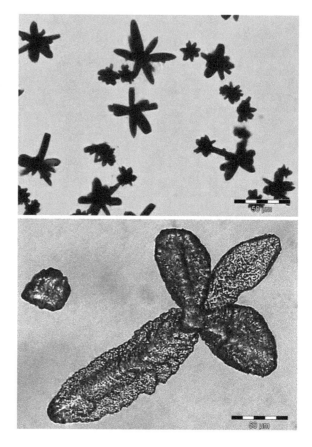

Fig. 11.2 Star crystals of nitrogen triiodide (*Top*—prepared by diffusion of iodine solution into aqueous ammonia, *Bottom*—prepared by diffusion of alkali ICl into aqueous ammonia)

11.3.3 Explosive Properties

Nitrogen triiodide is an extremely sensitive compound when dry. It explodes upon the slightest stimulus as follows:

- Slightest touch (such as a touch by a feather) [11]
- Upon very mild heating (such as a warm air stream) [11]
- With sufficiently strong light (e.g., an electronic flash [58] or light from a spark discharge [11])
- By bombardment with α-particles [59, 60] or by other fission products [61, 62] but not by β-particles [59] and neutrons [61]
- By the sound of a tuning fork [63]
- Even spontaneously [62]

Due to its extreme sensitivity, nitrogen triiodide must be stored only in "wet" conditions, preferably under ether [11]. It retains its sensitivity to a certain extent even under water where it can be initiated by friction [13].

From the studies of the relationship between the light energy required for ignition and time lag between the beginning of the light flash and the beginning

11.3 Nitrogen Triiodide

of explosion, it seems that the initiation of $NI_3 \cdot NH_3$ explosion is caused by a thermal mechanism when light energy changes to thermal energy [11].

The explosion of $NI_3 \cdot NH_3$ produces a cloud of purple–brown iodine vapor. For the mechanism of the $NI_3 \cdot NH_3$ explosion, two equations were proposed. Explosion in air was determined using the following equation [11]:

$$8\, NI_3.NH_3 \longrightarrow 6\, NH_4I + 5\, N_2 + 9\, I_2$$

The equation below was published for explosion of $NI_3 \cdot NH_3$ in a helium environment [51] or for its thermal decomposition at low pressures [48].

$$2\, NI_3.NH_3 \longrightarrow 2\, NH_3 + N_2 + 3\, I_2$$

Tudela [52] reported this equation for the explosion of $NI_3 \cdot NH_3$ without further details. The enthalpy of $NI_3 \cdot NH_3$ explosion, according to the second equation, is -99 kJ mol^{-1} (calculated) [52] and -88.1 ± 4.2 kJ mol^{-1} (experimental) [51]. Activation energy of this decomposition is 75.4–79.6 kJ mol^{-1} [11].

The presence of ammonia stabilizes $NI_3 \cdot NH_3$, because ammonia retards the decomposition of $NI_3 \cdot NH_3$ to pure NI_3. In a concentrated ammonia solution, $NI_3 \cdot NH_3$ cannot be ignited even with a very strong flash. Its stability and sensitivity are also reduced in an ammonia atmosphere. Removal of the NH_3 causes spontaneous explosion of nitrogen triiodide [11] and it therefore explodes immediately in high vacuum when dry [64]. Under pure water local explosions of $NI_3 \cdot NH_3$ do occur, but the explosion does not spread to the surrounding material [11, 58].

11.3.4 Preparation

Nitrogen triiodide mono-ammonium adduct is usually prepared by action of aqueous ammonia on iodine dissolved in a suitable solvent such as aqueous potassium iodide [46–48, 51, 61, 64].

$$3\, I_2 + 5\, NH_3 \longrightarrow NI_3.NH_3 + 3\, NH_4I$$

Nitrogen triiodide is also formed when iodine is used in the solid state [51, 53, 59, 65]. But this method is not very suitable—the iodine, being in particles of some size, not only reacts slowly, but also becomes surface-coated with the iodide which is difficult to remove [47]. An alcohol or chloroform solution of iodine can be used as well [9, 54], however using alcohol solutions of iodine or ammonia is not suitable according to Chattaway. Alcohol rapidly decomposes the nitrogen triiodide-forming triiodomethane which is not easy to remove from the product [47].

Crystalline nitrogen triiodide can be prepared by reaction of aqueous iodine monochloride (ICl) alkalized by potassium hydroxide with aqueous ammonia. On cooling, nitrogen triiodide forms as blackish-brown shiny needles, up to 2 mm long [54, 66]. Nitrogen triiodide is also formed by addition of ammonia to alkaline solutions of potassium hypoiodite (KIO) [54]. Other iodizing agents such as iodine monobromides or dibromoiodides can be used as well [46].

Nitrogen triiodide also forms through an ozonation reaction of KI in an aqueous solution of ammonia (product contains NHI_2 and NH_2I side products) [67]. Formation of gaseous NI_3 was observed in the case of reaction of boron nitride with iodine monofluoride [52].

$$BN + 3\,IF \longrightarrow NI_3 + BF_3$$

11.3.5 Use

Nitrogen triiodide has never been used in any practical industrial application due to its extreme sensitivity. Despite the inherent danger of severe injuries during its preparation and handling, it is particularly popular with young people who use it as a pyrotechnic toy that explodes by a simple touch.

References

1. Dulong, M.: Ueber die verpuffende Verbindung des Salz - und Stick - Gases I. Schweigger's Journal für Chemie und Physik **8**, 302–308 (1811)
2. Dulong, M., Thenard, M.M., Berthollet, C.L.: Sur un Mémoire de M. Dulong, sur une nouvelle substance détonante. Annales de chimie **86**, 37 (1813)
3. Davy, H.: On a new detonating compound. Philos. Trans. R. Soc. Lond. **103**, 1–7 (1813)
4. Snelders, H.A.M.: De ontdekking van het stikstoftrichloride. Chemie Techniek revue **22**, 457–459 (1967)
5. Baker, J.C.: Improvements in process of bleaching and maturing flour. GB Patent 159,166, 1922
6. Mellor, J.W.: A Comprehensive Treatise on Inorganic and Theoretical Chemistry. Longmans Green, London (1958)
7. Krauz, C.: Technologie výbušin. Vědecko-technické nakladatelství, Praha (1950)
8. Whiteley, M.A. (ed.): Halides of Nitrogen. Longman, Green, London, (1949)
9. Walke, W.: Lectures on Explosives. Wiley, New York (1897)
10. Patnaik, P.: Handbook of Inorganic Chemicals. McGraw-Hill, New York (2003)
11. Fedoroff, D.T., Sheffield, O.E., Kaye, S.M.: Encyclopedia of Explosives and Related Items Picatinny Arsenal, New Jersey (1960–1983)
12. Špičák, S., Šimeček, J.: Chemie a technologie třaskavin. Vojenská technická akademie Antonína Zápotockého, Brno (1957)
13. Wisser, J.P.: Explosive Materials—The Phenomena and Theories of Explosion. D. Van Nostrand, New York (1907)
14. Hentschel, W.: Ueber Chlorstickstoff. Berichte der deutschen chemischen Gesellschaft **30**, 1434–1437 (1897)
15. Selivanoff, T.: Chlorides and iodides of nitrogen. J. Chem. Soc. **66**, 312–313 (1894)

References

16. Rai, H.: Note on nitrogen chloride, with a convenient method for its preparation. Chem. News **117**, 253 (1918)
17. Dowell, C.T., Bray, W.C.: Experiments with nitrogen trichloride. J. Am. Chem. Soc. **39**, 896–905 (1917)
18. Griffiths, J.G.A., Norrish, R.O.W.: Photosensitized decomposition of nitrogen trichloride and the induction period of hydrogen-chlorine reaction. Trans. Faraday Soc. **27**, 451–461 (1931)
19. Bowen, E.J.: The photochemistry of unstable substances. J. Chem. Soc. **123**, 1199–1206 (1923)
20. Kovacic, P., Lowery, M.K., Field, K.W.: Chemistry of N-bromamines and N-chloramines. Chem. Rev. **70**, 639–665 (1970)
21. Noyes, W.A.: Inorganic Syntheses. McGraw-Hill Book, New York (1939)
22. Noyes, W.A., Lyon, A.C.: The reaction between chlorine and ammonia. J. Am. Chem. Soc. **23**, 460–463 (1901)
23. Noyes, W.A.: The interaction between nitrogen trichloride and nitric oxide. Reactions of compounds with odd electrons. J. Am. Chem. Soc. **50**, 2902–2910 (1928)
24. Kolbe, H.: Beobachtungen über die oxydirene Wirkung des Sauerstoffs, wenn derselbe mit Hülfe einer elektrischen Säule entwickelt wird. Journal für praktische Chemie **41**, 137–139 (1847)
25. Kolbe, H.: Observations on the oxidizing power of oxygen when disengaged by means of voltaic electricity. Mem. Proc. Chem. Soc. Lond. **3**, 285–287 (1845)
26. Van der Lee, G.: Process for producing nitrogen trichloride. US Patent 2,248,650, 1941
27. Staudt, E.: Process for the manufacture of nitrogen trichloride. US Patent 2,118,903, 1938
28. Staudt, E.: Process for producing nitrogen trichloride. US Patent 2,199,942, 1940
29. Balard, A.J.: Recherches sur la nature des combinaisons décolorantes du chlore. Annales des Chimie et des Physique, 225–304 (1834)
30. Hentschel, W.: Ueber die Zusammensetzung des Chlorstickstoffs. Berichte der deutschen chemischen Gesellschaft **30**, 1792–1795 (1897)
31. Noyes, W.A.: An attempt to prepare nitro-nitrogen trichloride, an electromer of ammono-nitrogen trichloride. J. Am. Chem. Soc. **35**, 767–775 (1913)
32. Schlessinger, G.G.: Nitrogen trichloride-ether mix explodes. Chem. Eng. News **44**, 46 (1966)
33. Baker, J.C.: Process of producing nitrogen trichloride. US Patent 1,510,132, 1924
34. Coleman, G.H., Craig, D.: Nitrogen trichloride and unsaturated ketones. J. Am. Chem. Soc. **49**, 2593–2596 (1927)
35. Coleman, G.H., Craig, D.: Nitrogen trichloride and unsaturated ketones. II. J. Am. Chem. Soc. **50**, 1816–1820 (1928)
36. Coleman, G.H., Howells, H.P.: Addition of nitrogen trichloride to unsaturated hydrocarbons. I. J. Am. Chem. Soc. **45**, 3084–3089 (1923)
37. Coleman, G.H., Mullins, G.M.: Nitrogen trichloride and unsaturated acids. J. Am. Chem. Soc. **51**, 937–910 (1929)
38. Coleman, G.H., Mullins, G.M., Pickering, E.: Nitrogen trichloride and unsaturated hydrocarbons. II. J. Am. Chem. Soc. **50**, 2739–2741 (1928)
39. Coleman, G.H., Buchanan, M.A., Paxson, W.L.: Reaction of nitrogen trichloride with Grignard reagent. J. Am. Chem. Soc. **55**, 3669–3672 (1933)
40. Jander, J.: Recent chemistry and structure investigation of nitrogen triiodide, tribromide, trichloride, and related compounds. Adv. Inorg. Chem. Radiochem. **19**, 1–63 (1976)
41. Galal-Gorchev, H., Morris, J.C.: Formation and stability of bromamide, bromimide, and nitrogen tribromide in aqueous solution. Inorg. Chem. **4**, 899–905 (1965)
42. Klapötke, T.: The reaction of bromine azide with bromine. Polyhedron **16**, 2701–2704 (1997)
43. Performance-oriented packaging standards; changes to classification, hazard communication, packaging and handling requirements based on UN standards and agency initiative. Fed. Reg. **55**, 52402–52729 (1990)
44. Roozeboom, H.W.B.: Ammonium tribromide. J. Chem. Soc. **42**, 140 (1882)
45. Inman, G.W., Lapointe, T.F., Johnson, J.D.: Kinetics of nitrogen tribromide decomposition in aqueous solution. Inorg. Chem. **15**, 3037–3042 (1976)
46. Cremer, H.W., Duncan, D.R.: Nitrogen triiodide. J. Chem. Soc. 2750–2754 (1930)

47. Chattaway, F.D.: The constitution of the so-called "Nitrogen iodide". J. Chem. Soc. **69**, 1572–1583 (1896)
48. Meldrum, F.R.: The thermal decomposition of nitrogen iodide. Proc. R. Soc. Lond. **A 174**, 410–424 (1940)
49. Tornieporth-Oetting, I., Klapöke, T.: Nitrogen triiodide. Angew. Chem. Int. Ed. **29**, 677–679 (1990)
50. Greenwood, N.N.: Chemistry of the Elements. Elsevier, Oxford (1997)
51. Andrews, M.V., Shaffer, J., McCain, D.C.: Nitrogen-iodine bond strength. Thermochemistry of nitrogen triiodide ammine. J. Inorg. Nucl. Chem. **33**, 3945–3947 (1971)
52. Tudela, D.: Nitrogen triioidide. J. Chem. Educ. **79**, 558 (2002)
53. Chattaway, F.D., Orton, K.J.P.: The action of light on nitrogen iodide. Am. Chem. J. **24**, 159–167 (1900)
54. Chattaway, F.D., Orton, K.J.P.: The formation and constitution of nitrogen iodide. Am. Chem. J. **24**, 342–355 (1900)
55. Hartl, H., Ullrich, D.: Die Kristallstruktur von Stickstofftrijodid-1-Pyridin $NI_3 \cdot C_5H_5N$. Zeitschrift für anorganische und allgemeine Chemie **409**, 228–236 (1974)
56. Jander, J., Bayersdorfer, L., Höhne, K.: Nitrogen-iodine compounds. V. Adducts of nitrogen triiodide with *N*-bases and their application for constitution determination of nitrogen iodide [nitrogen to iodine ratio 2:3]. Chem. Abst. CAN **68**, 110940 (1968)
57. Pritzkow, H.: Die Kristallstruktur von Stickstofftrijodid-1-Dijod-1-Hexamethylentetramin $NI_3 \cdot I_2 \cdot C_6H_{12}N_4$. Zeitschrift für anorganische und allgemeine Chemie **409**, 237–247 (1974)
58. Meerkämper, B.: Zum Verhalten des Jodstickstoffs beim Belichten mit Strahlung verschiedener Intensität. Zeitschrift für Elektrochemie und Angewandte Physikalische Chemie **58**, 387–416 (1954)
59. Henderson, G.H.: α-Particles as detonators. Nature **109**, 749 (1922)
60. Poole, H.H.: The detonating action of α-particles. Sci. Proc. R. Dublin Soc. **17**, 93–95 (1922)
61. Bowden, F.P., Singh, K.: Irradiation of explosives with high-speed particles and the influence of crystal size on explosion. Proc. R. Soc. Lond. **A 227**, 22–37 (1954)
62. Feenenberg, E.: The detonation of nitrogen iodide by nuclear fission. Phys. Rev. **55**, 980–981 (1939)
63. Hanus, M.: Méně známé třaskaviny. Synthesia a.s., VUPCH, Pardubice (1996)
64. Garner, W.E., Latchem, W.E.: Note on the decomposition of nitrogen iodide. Trans. Faraday Soc. **32**, 567–569 (1936)
65. Hambly, G.F., Peters, R.: Explosion of nitrogen triiodide: A safer and cleaner demonstration. J. Chem. Educ. **70**, 943 (1993)
66. Bärnighausen, H., Hartl, H., Jander, J.: Kristalldaten von Stickstofftrijodid-1-Ammoniak $NI_3 \cdot NH_3$. Zeitschrift für Naturforschung. Teil B. Anorganische Chemie, organische Chemie, Biochemie, Biophysik, Biologie **21**, 591 (1966)
67. Chudnov, A.F.: Sposob polucheniya iodictogo azota. SU Patent 1,212,935, 1986

Chapter 12
Acetylides

Acetylides are salts of acetylene, which is, under normal conditions, a gas with a slightly acidic character (pK_a is 25, for comparison pK_a of acetic acid is 4.76). Due to their acidic nature, one or both of the hydrogen atoms can be substituted by a metal atom. Furthermore, acetylene forms so-called metallo-addition compounds usually containing the acetylene molecule and an added metal compound ($C_2H_2 \cdot MX$) [1].

The acetylides of alkali metals and salts of alkaline earths are reactive compounds, which violently decompose in contact with water or even in contact with moisture in the air, releasing acetylene. These acetylides do not have the characteristics of primary explosives and are widely used in organic synthesis as a source of the acetylene (ethyn-1,2-diyl) group.

Acetylides of heavy metals (silver, copper, gold, mercury) do not react with moisture and are stable in contact with air. These acetylides are very sensitive to mechanical stimuli and have the characteristics of primary explosives. Their explosive power is considered to be the same as that of azides and fulminates [2].

12.1 Silver Acetylides

Silver acetylide comes in either a simple form (Ag_2C_2), in various double salts or even more complicated complexes depending on the reaction conditions of the preparation (mainly presence of other salts in the reacting mixture and pH). The presence of an oxidizing group (such as nitrate, perchlorate, and also halogens) in complex acetylides appears to increase the explosiveness, whereas the presence of poorly or nonoxidizing anions (e.g., phosphate, sulfate, organic acid residues) appears to decrease the explosiveness [3]. An overview of many complex silver acetylides is copiously described in the literature [4, 5]. Historically, there are only two groups of acetylides which have been studied in detail for possible industrial application (a) silver acetylide itself Ag_2C_2 and (b) its double salts with silver nitrate especially $Ag_2C_2 \cdot AgNO_3$, less frequently $Ag_2C_2 \cdot 6AgNO_3$ [6].

12.1.1 Silver Acetylide

$$Ag_2C_2$$

Simple silver acetylide was first prepared by Quet in 1858 even before acetylene itself had been identified. The procedure was based on the introduction of a gas obtained from decomposition of alcohol by electric discharge into an ammoniacal solution of silver chloride [5, 7, 8].

In the same year, Vogel and Reischaur obtained silver acetylide by introduction of coal–gas into a solution of the silver salt [9]. The determination of the exact structure of the molecule, however, took a relatively long time. The variety of structures considered and their historical development has been summarized by many authors including, for example, Keiser [7] or Stettbacher [9].

12.1.1.1 Physical and Chemical Properties

Silver acetylide forms crystals with most likely a hexagonal structure, crystal density 4.47 g cm^{-3} [10] and heat of formation 357.6 ± 5.0 kJ mol^{-1} [11] (as compared to 36.8 kJ mol^{-1} for sodium acetylide). Silver acetylide, Ag_2C_2, is a white solid substance that is sensitive to light. It darkens on exposure to light and turns black on exposure to direct sunlight for just a few hours [12]. The rate of photodecomposition is much greater than for SA [12]. Silver acetylide is practically insoluble in water and common organic solvents [5, 10, 13]. In solutions of inorganic acids, depending on the type and concentration of the acid, it undergoes hydrolysis to acetylene and the appropriate salt [12, 14]. Acetylene is also produced if it comes into contact with potassium cyanide or sulfite [5, 15]. Action of chlorine, bromine, or some organic nitrocompounds results in very violent or even explosive decomposition [12]; it also explodes in contact with hydrogen sulfide [16]. Silver acetylide easily forms double or triple salts by reaction with concentrated aqueous solutions of silver salts. With a solution of silver acetylide, based on the reaction conditions, it produces various double salts of Ag_2C_2 and $AgNO_3$ [17] (chapter 12.1.2-12.1.4); under the action of a solution of silver perchlorate it forms the double salt silver acetylide-perchlorate with the following formula: $Ag_2C_2 \cdot 2AgClO_4 \cdot 2H_2O$ [18], while the action of silver fluoride solution results in $Ag_2C_2 \cdot 8AgF$ [19]; the triple salt $2Ag_2C_2 \cdot AgF \cdot 9AgNO_3 \cdot H_2O$ is formed by action of silver acetylide on an aqueous solution of silver nitrate and silver tetrafluoroborate (molar ratio 1:1) [20]. An overview of many other double salts is given in the review by Vestin and Ralf [5].

12.1.1.2 Explosive Properties

Simple silver acetylide is generally quite sensitive to mechanical stimuli and definitely more sensitive than its double salt $Ag_2C_2 \cdot AgNO_3$ [2]. Sensitivity of

12.1 Silver Acetylides

Table 12.1 Impact sensitivity of silver acetylides [21]

	Ammonia concentration in reaction mixture (m/m %)	Impact sensitivity	
		Impact height 0.5 kg hammer (cm)	Recalculated energy (J)
Ag_2C_2	0.57	15	0.74
Ag_2C_2	2.0	66	3.2
$Ag_2C_2 \cdot AgNO_3$[a]	–	43	2.1
$Ag_2C_2 \cdot AgNO_3$[b]	–	79	3.9
$Ag_2C_2 \cdot AgNO_3$[b]	–	64	3.1
MF	–	24	1.1
LA	–	43	2.1

[a]Prepared in neutral solution.
[b]Prepared in acidic solution (nitric acid).

simple silver acetylide is about the same as MF according to Stettbacher [9]. Impact sensitivity of silver acetylide has been studied in detail by Taylor and Rinkenbach who found that it is highly affected by reaction conditions, mainly the concentration of ammonia in the reaction mixture. The results are compared with MF and LA in Table 12.1 [21].

Friction sensitivity of silver acetylide is generally high and exceeds sensitivity of most other primary explosives including MF [22] or cuprous acetylide [12] and it is even slightly more sensitive than LA [21]. Köhn [12] reports that significant decomposition is observed even at friction force of 0.2 N (lowest possible force of the machine) when using ceramic plates of BAM type. Our measurements confirm the results of above-mentioned authors. Sensitivity of silver acetylide to friction is one of the highest compared to other common industrial and improvised primary explosives (see Fig. 2.19) [23]. Aging further increases the friction sensitivity as explosion of half year old silver acetylide occurred even during sweeping of sample from ceramic plate by brush (Matyas Unpublished). Friction sensitivity is, just like the impact sensitivity, influenced by the preparation route [21].

Ignition temperatures of silver acetylide reported in various literature sources cover a very broad range spanning from 140 °C to 200 °C. The reason is most likely reflecting different reaction conditions and hence different products. It is therefore very important to always check what was really prepared [12]. The summary of published values of ignition temperatures of silver acetylide in comparison with silver acetylide–silver nitrate is summarized in Table 12.2.

It was observed by Köhn that the ignition temperature changes as the material ages. Some values determined at heating rate 1 °C s^{-1} are presented in Table 12.3 for acetylides of silver and copper [12]. The results are the lowest values obtained in a series of measurements. The sensitivity of silver acetylide to impact, friction, and hot wire does not change with time [12].

The heat of explosion of silver acetylide is 294 kJ mol^{-1} according to Stadler [29]. For a long time it was assumed that only solid products form from the explosion of simple silver acetylide [1]. Later it was, however, shown that some gaseous products do indeed form (2.4 ml from 1 g of sample according to Stadler [29]). Eggert found that the gaseous products of explosion contain water, carbon

Table 12.2 The ignition temperature of simple Ag_2C_2 compared with the double salt $Ag_2C_2 \cdot AgNO_3$

Ignition temperature (°C)		Conditions of determination	References
Ag_2C_2	$Ag_2C_2 \cdot AgNO_3$		
143–148	202[a]–209[a]	Range for heating rate 5–21.5 °C min^{-1}	[24]
171–177	200[a] or 225[b]	Instantaneous explosion or within 1 s[c]	[21]
195–200	197–202	Explosion within 5–10 s	[10]
~200	220	Details are not mentioned	[25]
–	210[a]	Explosion within 5 s	[26]
–	210, 212	Details are not mentioned	[27, 28]
	220	Instantaneous explosion	[26]

[a]Prepared in acidic solution.
[b]Prepared in neutral solution.
[c]Compared with 260 °C for MF and 383 °C for LA.

Table 12.3 Ignition temperatures of silver and copper acetylides at heating rate 1 °C s^{-1} with sample size 3 mg [12]

Sample age	Ignition temperature (°C)	
	Ag_2C_2	Cu_2C_2
3 h	155	110
24 h	158	115
3 days	156	135
8 days	157	142
2 weeks	157	163
3 weeks	157	190
6 weeks	161	184
2 months	163	175
2.5 months	169	a
7 months	170	a
16 months	167	a

[a]Explosive decomposition not observed below 200 °C.

monoxide, carbon dioxide, hydrogen, and methane. The amount of gaseous products is about one-tenth of the gaseous products that form during explosion of $Ag_2C_2 \cdot AgNO_3$ [30]. Taki later reported that 2–5 mol% of Ag_2C_2 contribute to the formation of gaseous products during explosion. The composition of these products is about 30% hydrogen, 50–60% nitrogen, 10–20% carbon dioxide, and small amounts of methane and ethane. Taki attributed the formation of gases to impurities in silver acetylide [16]

The explosion of simple Ag_2C_2 is less violent and far less noisy than explosion of $Ag_2C_2 \cdot AgNO_3$ or AgN_3 [30–32]. Eggert ascribes this to the fact that explosion of $Ag_2C_2 \cdot AgNO_3$ or AgN_3 both produce large amounts of gas while Ag_2C_2 produces only a very low amount of these products [30, 32]. Ten grams of Ag_2C_2 does not create any cavity in lead block test [29].

McCowan measured detonation velocities of silver acetylides Ag_2C_2 and $Ag_2C_2 \cdot AgNO_3$, packed in a trench $1 \times 1 \times 50$ mm cut in a perspex plate and ignited by an exploding wire. Detonation velocity of Ag_2C_2 was 1,200 m s^{-1} and 1,980 m s^{-1} for $Ag_2C_2 \cdot AgNO_3$ [10].

Simple silver acetylide is a relatively "weak" primary explosive [5]; it is less brisant than its double salt $Ag_2C_2 \cdot AgNO_3$ [2], however, has stronger explosive properties than cuprous acetylide. Urbański attributes this to an exceptionally high positive heat of formation [33].

12.1.1.3 Preparation

Silver acetylide forms by reaction of acetylene with a solution of a silver salt (nitrate, perchlorate, or chloride) in aqueous ammonia. Mixing silver nitrate with aqueous ammonia results in temporary formation of a brown precipitate which later dissolves in this solution. After the solution clears, acetylene is introduced and a gray-white precipitate forms [12]. Several authors describe that a voluminous yellow precipitate is formed initially, which then rapidly transforms into a white precipitate with a smaller volume [5, 9]. This phenomenon was discussed in the Vestin and Ralf review but the composition of the yellow precipitate has not been identified [5].

$$2\ AgNO_3 + C_2H_2 + 2\ NH_3 \longrightarrow Ag_2C_2 + 2\ NH_4NO_3$$

Details of silver acetylide preparation are described in the work of Köhn [12] or Stettbacher [9], the latter recommending to wash acetylene before introducing into the solution by passing it through four gas-washing bottles. The first contains diluted sulfuric acid which absorbs ammonia and possibly other nitrogen containing bases. The second contains potassium permanganate mixed with sulfuric acid for oxidation of phosphine and sulfane. The third contains a neutral solution of potassium permanganate and the last one pure water [9]. Using unpurified acetylene in the process of silver acetylide preparation leads to a product with a grayish or brownish tinge [29].

Simple silver acetylide does not form in aqueous potassium cyanide or potassium thiosulfate solutions [1]. Introduction of acetylene into a neutral or acidic solution of silver nitrate does not lead to the simple silver acetylide salt Ag_2C_2, but a complex acetylide $Ag_2C_2 \cdot AgNO_3$. Keiser, however, reports that it is possible to prepare Ag_2C_2 in diluted neutral solutions of silver nitrate, while in more concentrated ones the double salt $Ag_2C_2 \cdot AgNO_3$ is formed [7].

Several authors have reported that explosions can occur during some steps of silver acetylide preparation. Taki published a risk of explosion during spreading out of silver acetylide; it can explode when it is pulverized to a fine powder [16]. Vestin and Ralf reported that explosions of silver acetylide occur rather often (about every tenth preparation), especially during the evacuation (when the product is dried under vacuum) [5].

Silver acetylide may spontaneously form when acetylene comes into long-term contact with metallic silver. This may also be the case of some soldering alloys where different amounts are allowable depending on the overall composition [12].

12.1.1.4 Uses

To the best of our knowledge simple silver acetylide has never found any practical application.

12.1.2 Silver Acetylide–Silver Nitrate

$$Ag_2C_2 \cdot AgNO_3$$

12.1.2.1 Physical and Chemical Properties

The silver acetylide–silver nitrate forms fine needle and cross crystals [28] as shown on Fig. 12.1 with density 5.36 g cm^{-3} [6] or 5.369 (X-ray) [17]. The density is superior to MF and LA making $Ag_2C_2 \cdot AgNO_3$ a primary explosive with one of the highest known densities [29]. It decomposes by action of acids liberating acetylene:

$$Ag_2C_2 \cdot AgNO_3 + 3\,HCl \longrightarrow 3\,AgCl + C_2H_2 + HNO_3$$

The decomposition by action of concentrated sulfuric acid or nitric acid is gentle, unlike in the case of simple silver acetylide [9]. Silver sulfite and acetylene form by action of solutions of alkaline sulfides; decomposition takes place also by action of alkaline chloride at higher temperature; however, in this case it is rather slow and incomplete [4]. Potassium cyanide solution decomposes it in a similar way to simple silver acetylide liberating acetylene quantitatively [5]. Silver acetylide–silver nitrate is not stable in water and according to Stadler it slowly decomposes to Ag_2C_2 with part of the $AgNO_3$ dissolving in water. The composition of the precipitate corresponds to $Ag_2C_2 \cdot AgNO_3 \cdot Ag_{0.53}C_{0.52}$ after 14 h exposure in distilled water at normal temperature. However, in another section of this chapter, Stadler reported that the explosive properties of $Ag_2C_2 \cdot AgNO_3$ are not reduced by 3 days exposure to a moist environment at 90 °C, or in a moist environment with 66% carbon dioxide or by 3 days exposure to light. $Ag_2C_2 \cdot AgNO_3$ reacts with gaseous ammonia if stored above aqueous ammonia [29].

12.1.2.2 Explosive Properties

Sensitivity of silver acetylide–silver nitrate to impact and friction is lower than sensitivity of simple silver acetylide Ag_2C_2 or mercury fulminate [29]. According to Taylor and Rinkenbach the reaction conditions (especially pH) have a significant influence on sensitivity. The impact sensitivity of samples prepared from a neutral solution is higher than that of $Ag_2C_2 \cdot AgNO_3$ prepared form an acidic solution; sensitivity of $Ag_2C_2 \cdot AgNO_3$ prepared form a neutral solution is about the same as sensitivity of LA [21]. Stadler published impact sensitivity for $Ag_2C_2 \cdot AgNO_3$

12.1 Silver Acetylides

Fig. 12.1 Needle crystals of silver acetylide–silver nitrate $Ag_2C_2 \cdot AgNO_3$ by SEM

equal to 2.8–3.4 cm for 2 kg hammer in comparison with 3.2 cm for LA and 1.1 for MF [29].

Friction sensitivity of silver acetylide–silver nitrate (both prepared from acidic and neutral solutions) is published lower than the sensitivity of mercury fulminate [21]. According to our experiments sensitivity of $Ag_2C_2 \cdot AgNO_3$ is between LA and MF (see Fig. 2.19) [23]. Silver acetylide–silver nitrate is exceptionally sensitive to flame.

The ignition temperature of $Ag_2C_2 \cdot AgNO_3$ is mostly reported to be higher than for simple silver acetylide [21, 24] (see Table 12.2). The ignition temperature of the product is significantly influenced by concentration of nitric acid in the reaction mixture. It increases with nitric acid concentration from 206 °C (without HNO_3) to 280 °C (for 28% content of HNO_3). Similar behavior was observed for concentration of silver nitrate; ignition temperature increases with silver nitrate concentration from 203 °C (1% $AgNO_3$) to 282 °C (for 20% $AgNO_3$). Elaborate analysis of the measured samples was unfortunately not done before testing and it is therefore hard to relate molecular structure to sensitivity [29].

The thermal stability of silver acetylide–silver nitrate increases when some metals are present in the crystalline lattice of $Ag_2C_2 \cdot AgNO_3$. Boldyrev and Pronkin reported that the addition of cadmium nitrate into the reaction mixture during $Ag_2C_2 \cdot AgNO_3$ preparation increases the ignition temperature of $Ag_2C_2 \cdot AgNO_3$. This product is also less sensitive to ultraviolet light [26].

Stadler and Stettbacher suggested for explosive decomposition the following equation [29, 31]:

$$Ag_2C_2 \cdot AgNO_3 \longrightarrow 3\,Ag + CO_2 + CO + N_2$$

The heat of explosion is 774 kJ mol^{-1} [29]. In earlier work, Eggert reported the formation of other rather unexpected gaseous products during explosion (water, hydrogen, methane, nitrogen, and unspecified nitrogen oxides) [30]. Stadler later published that 139.4 ml gaseous products (0 °C, 101.3 kPa) form during explosion of 1 g

Table 12.4 Detonation velocity of $Ag_2C_2 \cdot AgNO_3$ compared with LA [29]

$Ag_2C_2 \cdot AgNO_3$		LA	
Density (g cm^{-3})	Detonation velocity (m s^{-1})	Density (g cm^{-3})	Detonation velocity (m s^{-1})
2.51	2,250	1.45	2,120
2.92	2,710	2.16	3,080
3.79	3,320	3.12	4,270
3.96	3,460	3.19	4,540

Table 12.5 Dependence of ignition efficiency of $Ag_2C_2 \cdot AgNO_3$ on preparation conditions and compared with LA and MF [29]

	Ignition efficiency (mg)	
Reaction conditions	Tetryl	TNT
Neutral pH, 5% cold solution $AgNO_3$	50	>150
Neutral pH, 5% hot solution $AgNO_3$	30–50	100
Acidic pH, 13% hot solution $AgNO_3$	20	50–75
LA	20	30
MF	100	120

$Ag_2C_2 \cdot AgNO_3$ with composition 38.7 mol% CO_2, 41 mol% CO, and 20.2 mol% N_2 [29]. The powder of $Ag_2C_2 \cdot AgNO_3$ measured by a lead block test is on average 136 cm^3 for 10 g sample prepared in acidic environment and 120 cm^3 for sample prepared in a neutral environment (compared with 181 cm^3 for LA) [29].

Detonation velocity of $Ag_2C_2 \cdot AgNO_3$ has been measured by Stadler. The dependence of detonation velocity on density is presented in Table 12.4 and compared with lead azide. The detonation velocity of LA is superior to that for $Ag_2C_2 \cdot AgNO_3$ at all densities [29].

The brisance of silver acetylide–silver nitrate (determined as the number of splinters) increases with increasing loading pressure and slightly exceeds that of LA pressed by 50 MPa. $Ag_2C_2 \cdot AgNO_3$ cannot be dead-pressed by this pressure level [29].

The initiating efficiency is high. It is reported as being about 4 times more powerful than MF (0.005 g for PETN, 0.07 for tetryl, and 0.25 for TNT) [13]; Stettbacher determined 0.07 g $Ag_2C_2 \cdot AgNO_3$ for tetryl (compared with 0.29 g for MF and 0.02 for SA) [9]. Stadler studied the influence of reaction conditions on ignition efficiency of $Ag_2C_2 \cdot AgNO_3$. The highest ignition efficiency was determined for $Ag_2C_2 \cdot AgNO_3$ prepared from a hot acidic solution of $AgNO_3$, lower for the product prepared from a hot neutral solution of $AgNO_3$ and the lowest for the product prepared from a cold neutral solution of $AgNO_3$. Ignition efficiency of $Ag_2C_2 \cdot AgNO_3$ prepared from a hot acidic solution is approaching that of LA [29] (Table 12.5).

12.1.2.3 Preparation

Silver acetylide–silver nitrate forms when acetylene is introduced into a neutral or slightly acidic aqueous solution of silver nitrate [5, 13, 31]. The concentration of silver nitrate solution must be lower than 10% wt. because in more concentrated solutions silver acetylide–hexanitrate forms [10, 27, 28]. Silver acetylide–silver nitrate precipitates from the solution in the form of a heavy white powder. Just as in the case of simple silver acetylide, use of unpurified acetylene results in a product with a grayish or brownish tinge [29].

$$3\ AgNO_3 + C_2H_2 \longrightarrow Ag_2C_2 \cdot AgNO_3 + 2\ HNO_3$$

The complex salt initially formed from the solution of silver nitrate may be transformed into a simple silver acetylide if the acetylene pressure at the end of precipitation exceeds some critical value. The details of the transformation are, according to Vestin and Ralf [5], described by Chavastelon who observed that introduction of acetylene into an aqueous solution containing excess of silver nitrate quantitatively gives a 1:1 complex. This complex also initially forms if excess acetylene is used. If the reaction is terminated in a reasonable time scale the precipitate is again the 1:1 complex salt. Silver acetylide, however, begins to form if the acetylene treatment continues. The change from the complex to the acetylide is not very fast and even after 40 h there may be some unconverted complex left in the reaction mixture. It is therefore important to understand that the precipitate may have a varying composition depending on the transformation extent [5].

Silver acetylide–silver nitrate also forms directly by reaction of calcium carbide with aqueous silver nitrate [4] and it even forms as a film on the surface of silver nitrate solution in an acetylene atmosphere [10]. It is also possible to prepare this substance by adding Ag_2C_2 into a stirred neutral solution of silver nitrate or by mixing Ag_2C_2 with nitric acid (50–60%) [2].

12.1.2.4 Uses

Silver acetylide–silver nitrate is reported as "probably suitable for use in primers and detonators" [2]. Several patents were published at the beginning of the twentieth century recommending use of silver acetylides in detonators [34] and as a replacement of MF in impact initiated detonators [35]. Danilov et al. reported in 2001 that silver acetylide–silver nitrate is not used in practical applications in Russia (referring to explosive applications) [25].

Silver acetylide–silver nitrate was used in light initiated flying plate generator of shock wave experiments at Sandia in 1980 [36, 37]. The use in organic synthesis has been reported by Sladkov and Ukhin [38].

12.1.3 Silver Acetylide–Silver Hexanitrate

$$Ag_2C_2 \cdot 6AgNO_3$$

Double salt of silver acetylide with six silver nitrate molecules (silver acetylide–hexanitrate) forms rhombohedral crystals if acetylene is introduced into a concentrated solution of silver nitrate (concentration higher than 25%) [28, 39, 40]. The product is very soluble in a solution of silver nitrate (mother liquor) and the solubility increases with the nitrate concentration [27, 28].

The crystal density of the product formed is 4.090 g cm^{-3} according to Guo et al. [17]; 4.79 g cm^{-3} according to Jin et al. [41], or 4.81 g cm^{-3} according to Kung-Tu Chou [40]. The crystal structure of $Ag_2C_2 \cdot 6AgNO_3$ has been studied and is reported in [39–41]. The rhombs of the acetylide–hexanitrate complex decompose in water to $Ag_2C_2 \cdot AgNO_3$ and silver nitrate. They also tend to transform to $Ag_2C_2 \cdot AgNO_3$ needles in other environments that act to reduce the silver nitrate concentration of the solution in which the rhombs exist. Hence addition of acetylene leads to $Ag_2C_2 \cdot AgNO_3$ while addition of silver nitrate leads to $Ag_2C_2 \cdot 6AgNO_3$ [28]. The complex acetylide $HgC_2 \cdot 3AgNO_3$ forms from $Ag_2C_2 \cdot 6AgNO_3$ in solution of mercuric nitrate [42].

When heated, hexanitrate first melts and then decomposes producing nitrogen oxides. The thermal decomposition is not violent as with $Ag_2C_2 \cdot AgNO_3$. The decomposition temperature is rather high and significantly exceeds that for $Ag_2C_2 \cdot AgNO_3$ (308–327 °C for hexanitrate vs. 212 °C for mononitrate) [27, 28].

According to Shaw and Fisher silver acetylide–hexanitrate is, unlike Ag_2C_2 and $Ag_2C_2 \cdot AgNO_3$, a nonexplosive substance [27]; Vestin and Ralf reported, that this compound can explode, but feebly [5].

12.1.4 Other Salts of Silver Acetylide–Silver Nitrate

Studies of acetylides have proved the existence of a large number of other silver acetylide–silver nitrate-based complexes. Pentanitrate $Ag_2C_2 \cdot 5AgNO_3$ and 5.5 nitrate hemihydrate $Ag_2C_2 \cdot 5.5AgNO_3 \cdot 0.5H_2O$ are among the more frequently mentioned ones. Guo, who also described the crystal structure of both complexes, reports a density of 4.560 g cm^{-3} for $Ag_2C_2 \cdot 5AgNO_3$ and 4.427 g cm^{-3} for $Ag_2C_2 \cdot 5.5AgNO_3 \cdot 0.5H_2O$ [17].

$Ag_2C_2 \cdot 5.5AgNO_3 \cdot 0.5H_2O$ is prepared by reaction of simple silver acetylide with a solution of silver nitrate in a water suspension at 80 °C. A violent decomposition is reported for this complex compound when it is heated. The second compound, $Ag_2C_2 \cdot 5AgNO_3$, can be prepared by reaction of Ag_2C_2 with silver trifluoroacetate. Moist simple silver acetylide reacts with a concentrated solution of silver trifluoroacetate at 80 °C and the solution is evaporated in a desiccator above P_2O_5. A gluey residue is re-dissolved in water and a small amount of silver nitrate is added at 80 °C. The resulting solution is kept at 0 °C for several days, after which a mixture of $Ag_2C_2 \cdot 5AgNO_3$ and $Ag_2C_2 \cdot 6AgNO_3$ crystallizes. The monoclinic–prismatic crystals of $Ag_2C_2 \cdot 5AgNO_3$ can be picked out from among the block-like crystals of $Ag_2C_2 \cdot 6AgNO_3$.

12.2 Cuprous Acetylide

$$Cu-C\equiv C-Cu$$

Simple cuprous acetylide was first prepared by Quetem in 1858. The preparation was based on introduction of the gases (contained acetylene) produced by passing electric sparks through liquid alcohol into a solution of cuprous chloride in aqueous ammonia [7, 8].

12.2.1 Physical and Chemical Properties

Cuprous acetylide is a red to brown-red amorphous substance. Its specific color is determined by the conditions of preparation. The substance prepared by the normal precipitation of cuprous salts by acetylene forms as a monohydrate [43]. It is interesting to note in older literature that the molecular formula of cuprous acetylide monohydrate is often written in a rather unusual form, for example $C_2H_2Cu_2O$ [44]. The water of crystallization is firmly bonded and may not be removed even at 140 °C above phosphorous oxide [43]. Older literature contains procedures for preparation of the anhydrous salt by drying the monohydrate above sulfuric acid [7].

Cuprous acetylide is not soluble in water, but quite well soluble in bases [2]. In an oxygen environment it undergoes slow decomposition, according to Klement and Köddermann-Gros [43]:

$$Cu_2C_2.H_2O + 0.5\,O_2 \longrightarrow Cu_2O + 2\,C + H_2O$$

Cuprous acetylide darkens when it is stored in the open. The change of color is accompanied by a significant increase in its sensitivity to mechanical stimuli. It is believed that both these phenomena are caused by oxidation of cuprous acetylide to cupric acetylide by oxygen in the atmosphere [45].

Potassium cyanide decomposes cuprous acetylide and an equivalent amount of acetylene is produced [46]:

$$Cu_2C_2 + 6\,KCN + 2\,H_2O \longrightarrow K_4[Cu_2(CN)_6] + 2\,KOH + C_2H_2$$

This reaction was proposed by Polyakov for quantitative analysis of copper acetylides (both cuprous and cupric). In the case of a mixture of cuprous and cupric acetylide only total acetylide content is determined. The analyzed substance is dissolved in a solution of potassium cyanide (cupric acetylide is also dissolved producing one mol of acetylene) and the acetylene is introduced into a solution of the cuprous salt. Cuprous acetylide precipitates from the solution and the excess of cuprous ions is titrated iodometrically.

Klement and Köddermann-Gros published a different course for the above-mentioned reaction [43]:

$$Cu_2C_2 \cdot H_2O + 8\,KCN + H_2O \longrightarrow 2\,K_3[Cu(CN)_4] + 2\,KOH + C_2H_2$$

In solutions of inorganic acids, cuprous acetylide hydrolyzes to acetylene and the corresponding salt. Hydrolysis rate and its course depend on the type and concentration of the acid [14]. Acetylene gas is evolved from cuprous acetylide even by action of hydrogen sulfide and potassium cyanide solution [1]. Some concentrated acids (nitric acid, sulfuric acid) and oxidizing agents (ozone, permanganate, chlorine, bromine) decompose acetylide explosively [47]. The reaction with hydrogen peroxide (30%) forms a substance with an aldehydic character [33]. Reaction with aqueous silver nitrate solution quickly leads to silver acetylide and, using a mercurous nitrate solution leads to mercurous acetylide [48].

$$Cu_2C_2 + 4\,AgNO_3 \longrightarrow Ag_2C_2 + 2\,Ag + 2\,Cu(NO_3)_2$$

$$Cu_2C_2 + 2\,Hg_2(NO_3)_2 \longrightarrow Hg_2C_2 + 2\,Hg + 2\,Cu(NO_3)_2$$

Cuprous acetylide further forms complex compounds in a similar way to silver acetylide. These cuprous complex salts form by the action of acetylene on certain cuprous salts in a neutral or slightly acidic medium [33].

12.2.2 Explosive Properties

Friction sensitivity of cuprous acetylide prepared by common procedure involving cuprous chloride (procedure published e.g., by Špičák and Šimeček [49]) is relatively low and is approaching sensitivity of PETN (Fig. 2.19) [23]. Cuprous acetylide can easily be initiated by electrostatic discharge or hot wire. The sensitivity to these two types of initiation was the reason why it was used in pyrotechnic mixtures for blasting cap fuseheads [33]. Brisance is lower than that of silver acetylide.

The ignition temperature of cuprous acetylide is reported with very wide ranges due to a strong influence of the reaction conditions on the resulting properties and the diversity of testing methods. The lowest published ignition temperature is 100 °C [44], followed by 120 °C [2, 13], 120–123 °C [33], 170 °C (in vacuum 265 °C!) [43] and the highest is in the range 260–270 C [45]. Values published by other authors are summarized in Table 12.2. Many researchers report that the ignition temperature changes as the material ages. Interestingly, some of them report a decrease of ignition temperature with time [43, 45], while others suggest an increase [12]. Morita reports that the changes are caused by atmspheric oxygen and that the ignition temperature first drops from 260–270 °C to 100 °C and later again rises above 200 °C [45]. The aging was also studied by Köhn who found the opposite trend with the substance first showing an increase in ignition temperature followed by a decrease and later by complete loss of ability to explode on heating (see Table 12.3) [12].

12.2.3 Preparation

The reaction of acetylene with an aqueous solution of cuprous chloride may yield a variety of substances depending on the reaction conditions. The most important factors are acidity and presence of other inorganic salts. Various complex salts with coordinated acetylene molecule may form (for example use of an acidic or alcohol solution results in formation of the complex salt $Cu_2Cl_2 \cdot C_2H_2$) [1].

Simple cuprous acetylide forms by reaction of gaseous acetylene introduced into the aqueous ammonia solution of the cuprous salt. Precipitates of different colors form depending on the reaction conditions.

$$Cu_2Cl_2 + C_2H_2 \xrightarrow{NH_3} Cu_2C_2 + 2\,NH_4Cl$$

Using pure cuprous chloride as a reactant leads to cuprous acetylide precipitated as a red substance. The precipitate does not form an individual compound, the ratio of C/Cu being in range 0.94–1.13 (theoretically it should be equal to 1) [50]. Klement and Köddermann-Gros analyzed the product prepared in an ammoniacal environment and found it to be 95% pure [43]. Precipitation carried out in an acidic environment produced substance which was not sensitive to impact [50].

Early scientists assumed that recovery of the acetylene from the cuprous acetylide precipitate was quantitative. However, it was soon found that the regenerated acetylene was not perfectly pure because of the addition reaction of hydrochloric acid to form vinyl chloride, and also because of the alteration of the precipitate in air to yield a cuprous compound of diacetylene [1].

In order to limit the oxidation of the cuprous salt to cupric and hence prevent the formation of cupric acetylide, a number of authors recommend the presence of reducing agents (e.g., hydroxylamine, sulfur dioxide, hydrazine sulfate, etc.) [1, 14, 33]. The applicability of certain reducing agents during the preparation of cuprous acetylide was, however, questioned by Brameld Clark and Seyfang [50]. According to these authors, the product prepared from ammoniacal solution of cuprous chloride in the presence of any of hydrazine hydrochloride, hydroxylamine hydrochloride, or formaldehyde has lower explosive properties than product prepared in the usual way. The probable reason is the formation of other by-products from the reaction (e.g, the precipitate prepared in the presence of formaldehyde has the atomic ratio of C/Cu equal to four). The product prepared in the presence of tin(II)chloride is not explosive [50].

During the preparation and processing of cuprous acetylide, attention should be paid to potential contamination by cupric acetylide. This substance is produced by a reaction of acetylene with certain copper salts which may be present as an impurity in the initial cuprous chloride. The material is unstable and highly sensitive to mechanical stimuli, exploding upon heating to 50–70 °C (see Sect. 12.3) [33].

Cuprous acetylide may also form spontaneously at places where copper or its alloys come to contact with acetylene. The acetylide does not form with any alloy containing copper but only when the content exceeds maximum allowable copper content specific for the particular alloy. The explosions of laboratory and industrial apparatuses as a result of copper acetylide formation are a subject of many reports [12].

12.2.4 Uses

Cuprous acetylide is the only acetylide which has been practically used, due to the ease with which it can be ignited by hot wire and by electric charges. It was therefore used as a component of fuse head compositions [13, 33, 51, 52]. Cuprous acetylide was also designed as a catalyst in ethynylation reactions (direct condensation of acetylene and aldehydes or ketones) [53–57].

The creation of cuprous acetylide has been suggested as a quantitative method for determination of copper in analytical chemistry [33]. Cuprous acetylide is produced in ammoniacal environments in the presence of hydroxylamine, even at very low concentrations of acetylene gas. It is on the basis of the formation of a brown precipitate of cuprous acetylide that a sensitive method for detection of acetylene was devised [58].

12.3 Cupric Acetylide

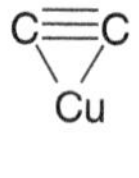

$$\begin{matrix} C\equiv C \\ \diagdown\diagup \\ Cu \end{matrix}$$

According to Fedoroff and Sheffield cupric acetylide was first prepared by Söderbaum by the introduction of acetylene into ammoniacal copper salts [2, 59].

12.3.1 Physical and Chemical Properties

Cupric acetylide is a gray-black colored substance. Depending on the method of preparation it may be produced as an amorphous substance, or as the so-called metallic, an extremely sensitive form. Cupric acetylide explosively decomposes if

12.3 Cupric Acetylide

combined with nitric acid [2] or dilute sulfuric acid, and in potassium cyanide solution it decays slowly [59]. With acids it releases acetylene [1]. With silver nitrate solution a simple silver acetylide is produced [48]. The action of potassium cyanide solution causes the release of acetylene:

$$2\ CuC_2 + 8\ KCN + 4\ H_2O \longrightarrow K_4[Cu_2(CN)_6] + 4\ KOH + 2\ C_2H_2 + (CN)_2$$

This reaction was proposed by Polyakov for determination of presence of copper acetylides (both cuprous and cupric) in mixture with other substances containing copper (see Sect. 12.2.1) [46].

12.3.2 Explosive Properties

Unlike cuprous acetylide, both forms of cupric acetylide are highly sensitive substances. Cupric acetylide may be initiated by a weak mechanical stimulus such as tapping [50]. It retains a high level of sensitivity even under water [45]. Cupric acetylide is further unstable and can explode on heating, even at temperatures as low as 50–70 °C according to Urbański [33] or at 100–120 °C according to Morita [45].

12.3.3 Preparation

Cupric acetylide is prepared by the introduction of acetylene into a solution of cupric salts. Similar to cuprous acetylide, the properties of the product (color, sensitivity) are to a large extent dependent on the conditions of the reaction [50].

$$CuCl_2 + C_2H_2 \longrightarrow CuC_2 + 2\ HCl$$

The product of the reaction forms either as an amorphous substance or as a crystalline, so-called metallic, form [50]:

- The amorphous form is a black precipitate which forms rapidly from common copper salts (chloride, sulfate, nitrate).
- The so-called metallic form is created if the reactant is cupric borate in a strongly alkaline environment, or cupric acetate in an acidic or strongly alkaline environment. These lustrous metallic looking plates form only when blowing acetylene through the solution for a long time or on standing in contact with the gas. This form is extremely sensitive to mechanical stimuli and explodes with a bright flash on a gentle tap or even on touching under solution.

When using cupric chloride as a reactant and carrying out the reaction in an alkaline environment, product properties vary significantly depending on the type of alkali. The product from an ammoniacal environment contains a significantly higher proportion of carbon (the ratio of C/Cu corresponds to 2.7–2.8 compared with a theoretical 2), while in the presence of sodium carbonate and sodium hydroxide the content of both elements is closer to theory (1.85–2.07). The acetylide precipitation is, in the latter case, practically immediate, while the rate of precipitation in ammoniacal environments depends on the concentration of ammonia; in strongly ammoniacal solutions, the precipitate appears only some time after the introduction of acetylene [50].

The composition of the product prepared in a neutral environment of cupric chloride or sulfate is close to theory (the ratio of C/Cu = 1.98). If the reactant is a cupric tartrate in an acidic environment then cuprous acetylide is the final product due to the reducing properties of tartaric acid [50]. Cupric acetylide is also produced directly by the reaction of calcium carbide in aqueous solutions with cupric salts. Calcium hydroxide and cupric hydroxide, which co-precipitate together with cupric acetylide, can be dissolved by adding dilute acetic acid [48].

Cupric acetylide (along with cuprous) may inadvertently appear on copper surfaces or alloys with high copper content when exposed to acetylene. Therefore, copper should not be used in places where it could come in contact with acetylene.

12.3.4 Uses

Because of its high sensitivity and low stability, cupric acetylide has never been used in any practical application.

12.4 Mercuric Acetylide

$$HgC_2$$

12.4.1 Physical and Chemical Properties

Mercuric acetylide is a heavy white amorphous powder with a density of 5.3 g cm^{-3} (at 16 °C) [60, 61]. Its structure was discussed by Malý and Kuča who, through X-ray analysis, determined the inter-atomic distances in the molecule HgC_2 (length of bond C≡C is 1.19 ± 0.02 Å, the length of the C–Hg bond is 2.17 ± 0.02 Å) [62].

Mercuric acetylide is insoluble in water, ethanol, and ether. Freshly produced mercuric acetylide dissolves well in ammonium acetate. It decomposes in solutions of potassium cyanide and sodium sulfide [14, 60, 61]. In inorganic acid solutions, mercuric acetylide slowly hydrolyzes to acetylene and the corresponding salt (rate depends on the type and concentration of the acid) [14]. Hydrochloric acid only has a slight effect on mercuric acetylide at normal temperatures and at higher temperatures it decomposes this acetylide to mercuric chloride and acetylene, which is partially hydrolyzed to acetaldehyde [15, 60, 61, 63]. Nitric acid dissolves acetylide, and dilute sulfuric acid causes a gradual decomposition with mainly the aldehyde as a product. Concentrated sulfuric acid, chlorine, and bromine react with mercuric acetylide explosively. Reaction with brominated water yields tetrabromoethylene C_2Br_4. With iodine dissolved in potassium iodide solution diiodoacetylene C_2I_2 is produced, or if the reaction takes place at higher temperatures, over a longer period of time in an excess of iodine, tetraiodoethylene C_2I_4 forms [61]. With silver nitrate solution it reacts quickly to form a simple silver acetylide [48].

During slow heating, mercuric acetylide slowly decomposes at 110 °C and above to mercury, carbon, and water [60, 61].

Storing mercuric acetylide in an ethanolic solution of mercuric chloride or even shortly boiling it in an aqueous solution of mercuric chloride leads to its transformation into a nonexplosive compound that has the same properties as the product of bubbling acetylene through a solution of mercuric chloride [60, 61].

12.4.2 Explosive Properties

Mercuric acetylide explosively decomposes to mercury and carbon in an amorphous form (like soot) [62]. Nothing else was found about explosive properties of this compound in an open literature.

12.4.3 Preparation

Acetylene reactions with mercury compounds are different from the analogous reaction with compounds of silver or copper. Unlike silver nitrate or cupric chloride, mercuric compounds in neutral solutions catalyze the addition of water to the acetylene molecule. The acetaldehyde and a few substituted acetylenes therefore form as a product of this reaction and form either main or side products of the action of acetylene on mercuric solutions [1].

Based on the comparison with cupric acetylide it could be assumed that mercuric acetylide will form from the reaction of acetylene with a solution of mercuric chloride. This reaction gives a white precipitate, which is not HgC_2, but some organomercury compound. To the substance prepared in a neutral environment, Keiser attributes the formula $C_2(HgCl)_2$ [15, 63], Blitz, and Mumm $(ClHg)_3C$–CHO, Biginelli CHCl=CHHgCl [1].

Mercuric acetylide HgC_2 forms when passing acetylene through Nessler reagent solution (solution of HgI_2 in KI and KOH). The product consists of a white flocculent precipitate identified by Keiser as HgC_2 [15, 63] and later confirmed by X-ray analysis [62]. However, on the basis of subsequent analysis, Fedoroff and Sheffield identified the reaction product as $3HgC_2 \cdot H_2O$ [2].

The effect of the type of mercuric reactant and composition of the reaction mixture on the yield of mercuric acetylide has been dealt with by Plimpton and Travers. Freshly precipitated mercuric oxide is suggested as the most appropriate starting material. It is dissolved in ammonia and, after ammonium carbonate addition, acetylene is introduced into the solution. The product forms as a heavy white powder. Based on the analysis of mercury in the product, the formula $3HgC_2 \cdot H_2O$ has been proposed [61]. Babko and Grebelskaya also started with mercuric oxide, but used an alkaline solution of potassium iodide as the reaction environment [14]. Introduction of acetylene into solutions of a number of other mercury salts leads to explosive compounds, probably acetylides, but unfortunately no satisfactory analysis of these compounds has been published. Exceptionally high sensitivity and brisance was observed with the precipitate formed by the introduction of acetylene into mercuric oxide solution in chloric acid [3].

Introduction of acetylene into a saturated solution of mercuric acetate in glacial acetic acid yields a substance which has an explosive character, but this is quickly lost during storage (during a week). Ferber with Romero report this substance to be mercuric acetylide and proposed the formula $2HgC_2 \cdot H_2C_2 \cdot H_2O$ [64].

An interesting compound, mercury acetylide-silver nitrate complex $HgC_2 \cdot 3AgNO_3$, forms by introducing acetylene to a concentrated solution of silver nitrate (about 30%) acidified by nitric acid and followed by addition of mercuric nitrate. The product precipitates from the solution [42].

12.4.4 Uses

Mercuric acetylide has never been employed in any practical explosive application. It has, however, found some use in analytical work because formation of the precipitate of mercuric acetylide (from an alkaline solution of mercuric iodide and potassium iodide) can be used for the quantitative determination of acetylene [1].

12.5 Mercurous Acetylide

$$Hg_2C_2 \cdot H_2O$$

Mercurous acetylide is a gray-colored substance. It crystallizes as a monohydrate insoluble in water [2, 60, 65]. Behavior during heating is described variously, some authors stating that the substance first separates out the water of crystallization and subsequently degrades [60], while others, like Burkard and Travers, describe the substance as unable to release crystal water because heating to temperatures over 100 °C leads to decomposition [65]. Mercurous acetylide is decomposed to acetylene on treatment with hydrochloric acid [65, 66]. Reaction with iodine yields mercurous iodide and tetraiodoethylene (C_2I_4) with a small proportion of diiodoacetylene (C_2I_2).

The preparation of mercurous acetylide is not very well described in the literature. The most complete seems to be the initial work of Plimpton [66] who suspended freshly prepared mercurous acetate in water (it is insoluble) and passed acetylene through such suspension. He himself, however, reported that the gray explosive substance which forms from the acetate was not analyzed. In its properties the prepared substance resembled acetylides of copper and silver.

Burkard later tried to clarify the issue of the product composition and based on the results of elemental analysis concluded that the substance is mercurous acetylide monohydrate—$Hg_2C_2 \cdot H_2O$ [65]. Another not very well described way for preparing mercurous acetylide is the reaction of cuprous acetylide in aqueous suspension with mercurous nitrate [48]. Both preparation methods must be also carried out in the absence of daylight.

12.6 Aurous Acetylide

$$Au_2C_2$$

12.6.1 Physical and Chemical Properties

Aurous acetylide is a yellow compound, it decomposes by boiling in water into its constituents (acetylene does not form), and it also decomposes by the action of hydrochloric acid (acetylene and black aurous chloride form). Cold ferric chloride or cupric sulfate does not decompose it. When the acetylide is heated very gradually, it decomposes without explosion [67].

12.6.2 Explosive Properties

Aurous acetylide is a highly sensitive substance which explodes, even by brushing with a camel's hair "brush" on filter paper [67]; Vincente et al. reported that it explodes after drying when touched by a spatula [68]. Explosion is also reported when rapidly heating it. Temperature of ignition is in range 83–157 °C (heating in the air-bath, heating rate was not specified). Solid products, gold and amorphous carbon, form when gold acetylide explodes [67].

12.6.3 Preparation

Aurous acetylide can be prepared by the reaction of acetylene with sodium aurous thiosulfate $Na_3Au(S_2O_3)_2$ in a strongly alkaline solution. The product forms as a yellow flocculent precipitate. Sodium aurous thiosulfate is prepared by reaction of auric chloride with sodium thiosulfate [67].

Aurous acetylide does not form when passing acetylene through an aqueous solution of auric chloride. Instead metallic gold precipitates. Precipitation does not occur when introducing acetylene into a solution of auric chloride alkalized by potassium hydroxide nor into aqueous or ammoniacal solutions of potassium auric cyanide [67]. Unwanted formation of aurous acetylide has been reported during preparation of some alkynylgold(I) complexes [68].

References

1. Nieuwland, J.A., Vogt, R.R.: The Chemistry of Acetylene. Reinhold Publishing Corporation, New York (1945)
2. Fedoroff, B.T., Sheffield, O.E., Kaye, S.M.: Encyclopedia of Explosives and Related Items. Picatinny Arsenal, New Jersey (1960–1983)
3. Nieuwland, J.A., Maguire, J.A.: Reactions of acetylene with acidified solutions of mercury and silver salts. J. Am. Chem. Soc. **28**, 1025–1031 (1906)
4. Gmelins Handbuch der anorganischen Chemie – Silber, pp. 270–273. Verlag Chemie, Weinheim (1973)
5. Vestin, R., Ralf, E.: Solver compounds of acetylene. Acta Chem. Scand. **3**, 101–143 (1949)
6. Redhouse, A.D., Woodward, P.: Crystallographic data for silver acetylide. Acta Crystallogr. **17**, 616–617 (1964)
7. Keiser, E.H.: The composition of the explosive copper and silver compounds of acetylene. Am. Chem. J. **14**, 285–290 (1892)
8. Quet, M.: Note sur un phénomène de polarité dans la décomposition des gaz par l'étincelle électrique, et sur les produits que l'on obtient en décomposant l'alcohol par l'étincelle électrique ou la chaleur. Comptes rendus hebdomadaires des séances de l'Académie des sciences **40**, 903–905 (1858)
9. Stettbacher, A.: Neuere Initialexplosivstoffe. Zeitschrift für das gesamte Schiess- und Sprengstoffwesen **11**, 1–4 (1916)
10. McCowan, J.D.: Decomposition of silver acetylide. Trans. Faraday Soc. **59**, 1860–1864 (1963)
11. Finch, A., Gardner, P.J., Head, A.J., Majdi, H.S.: The standard enthalpy of formation of silver acetylide. Thermochim. Acta **180**, 325–330 (1991)
12. Köhn, J.: Untersuchungen über die explosiven Eigenschaften von Silberacetylid. Amts- und Mitteilungsblatt der Bundesanstalt für Materialprüfung **8**, 57–62 (1978)
13. Krauz, C.: Technologie výbušin. Vědecko-technické nakladatelství, Praha (1950)
14. Babko, A.K., Grebelskaya, M.I.: Rastvorimost atsetilenidov medi, serebra i rtuti. Zhurnal obshchey khimii **22**, 66–76 (1952)
15. Keiser, E.H.: The metallic derivatives of acetylene. Am. Chem. J. **15**, 535–539 (1893)
16. Taki, K.: Reactive species in the explosion of silver acetylide I. Reaction with saturated hydrocarbons. Bull. Chem. Soc. Jpn. **42**, 2906–2911 (1969)
17. Guo, G.-C., Zhou, G.-D., Mak, T.C.W.: Structural variation in novel double salts of silver acetylide with silver nitrate: Fully encapsulated acetylide dianion in different polyhedral silver cages. J. Am. Chem. Soc. **121**, 3136–3141 (1999)
18. Guo, G.-C., Wang, Q.-G., Zhou, G.-D., Mak, T.C.W.: Synthesis and characterization of $Ag_2C_2 \cdot 2AgClO_4 \cdot 2H_2O$: A novel layer-type structure with the acetylide dianion functioning in a m_6-h^1,h^1:h^2,h^2:h^2,h^2 bonding mode inside an octahedral silver cage. Chem. Commun. 339–340 (1998)
19. Guo, G.-C., Zhou, G.-D., Wang, Q.-M., Mak, T.C.W.: A fully encapsulated acetylenediide in $Ag_2C_2 \cdot 8AgF$. Angew. Chem. Int. Ed. **37**, 630–632 (1998)
20. Wang, Q.-M., Mak, T.C.W.: Coexistence of differently capped trigonal prismatic $C_2@Ag_7$ cages in a triple salt of silver(I) acetylide. J. Cluster Sci. **12**, 391–398 (2001)
21. Taylor, A.C., Rinkenbach, W.H.: Sensitivities of detonating compounds to frictional impact, impact, and heat. J. Franklin Inst. **204**, 369–376 (1927)
22. Khmelnitskii, L.I.: Spravochnik po vzryvchatym veshchestvam. Voennaya ordena Lenina i ordena Suvorova Artilleriiskaya inzhenernaya akademiya imeni F. E, Dzerzhinskogo, Moskva (1962)
23. Matyáš, R., Šelešovský, J., Musil, T.: Sensitivity to friction for primary explosives. J. Hazard. Mater. **213–214**, 236–241 (2012)
24. Tammann, G., Kröger, C.: Über die Verpuffungstemperatur und Schlagempfindlichkeit von flüssigen und festen Explosivstoffen. Zeitschrift für anorganische und allgemeine Chemie **169**, 1–32 (1928)

25. Danilov, J.N., Ilyushin, M.A., Tselinskii, I.V.: Promyshlennye vzryvchatye veshchestva; chast I. Iniciiruyushchie vzryvshchatye veshchestva. Sankt-Peterburgskii gosudarstvennyi tekhnologicheskii institut, Sankt-Peterburg (2001)
26. Boldyrev, V.V., Pronkin, V.P.: Povyshenie termicheskoi ustoichivosti atsetilenida serebra vvedeniem dobavok kadmiya. Zhurnal vsesoyuznogo khimicheskogo obshchestva im. D. I. Mendeleeva **6**, 476–477 (1961)
27. Shaw, J.A., Fisher, E.: Silver acetylide compound and process of making same. US Patent 2,483,440, 1949
28. Shaw, J.A., Fisher, E.: A new acetylene silver nitrate complex. J. Am. Chem. Soc. **68**, 2745 (1946)
29. Stadler, R.: Analytische und sprengstofftechnische Untersuchungen an Azetylensilber. Zeitschrift für das gesamte Schiess- und Sprengstoffwesen **33**, 269–272, 302–305, 334–338 (1938)
30. Eggert, J.: Über Acetylensilber. Chemiker-Zeitung **42**, 199–200 (1918)
31. Stettbacher, A.: Zündsprengstoffe. Nitrocellulose **11**, 227–229 (1940)
32. Eggert, J.: Über einige Vorlesungsversuche mit Acetylen-silber. Berichte der deutschen chemischen Gesellschaft **51**, 454–456 (1918)
33. Urbański, T.: Chemistry and Technology of Explosives. PWN—Polish Scientific Publisher, Warszawa (1967)
34. Semple, J.B.: Improvements relating to the detonation of explosives. GB Patent 133,393, 1918
35. FR Patent 321,285, 1902
36. Benham, R.A.: Preliminary Experiments Using Light-Initiated High Explosive for Driving Thin Flyer Plates. Report SAND-79-1847 (1980)
37. Benham, R.A.: Initiation and Gas Expansion Model for the Light-Initiated Explosive Silver Acetylide-Silver Nitrate. Report SAND-79-1829 (1980)
38. Sladkov, A.M., Ukhin, L.Y.: Copper and silver acetylides in organic synthesis. Russ. Chem. Rev. 748–763 (1968)
39. Osterlof, J.: Crystal structure of $Ag_2C_2 \cdot 6AgNO_3$ and $2CuCl \cdot C_2H_2$. Acta Crystallogr. **7** (1954)
40. Chou, K.-T.: Crystal structure of $Ag_2C_2 \cdot 6AgNO_3$. Chem. Abstr. **60**, 2399 (1964)
41. Jin, X., Zhou, G., Wu, N., Tang, Y., Huang, H.: Structure determination and refinement of silver acetylide-silver nitrate (1:6). Chem. Abstr. **113**, 88547 (1990)
42. Shaw, J.A., Fisher, E.: Mercury acetylide-silver nitrate complex and process of making same. US Patent 2,474,869, 1949
43. Klement, R., Köddermann-Gros, E.: Die Oxydationsprodukte des Kupfer(I)-acetylides. Zeitschrift für anorganische Chemie **254**, 201–216 (1947)
44. Blochmann, R.: Ueber die Vorgänge bei der unvollständigen Verbrennung des Leuchtgases und über das Verhalten desselben in der Hitze bei Abschlufs von Luft. Justus Liebigs Annalen der Chemie **173**, 167–191 (1874)
45. Morita, S.: The effect of oxygen on the explosivity of red cuprous acetylide. Chem. Abstr. **50**, 6047 (1956)
46. Polyakov, N.N.: Opredelenie acetilenidov medi v prisutstvii drugikh soedinenii medi. Zhurnal analiticheskoi khimii **8**, 302–305 (1953)
47. Rupe, H.: Zwei Vorlesungsversuche. Journal für praktische Chemie **88**, 79–82 (1913)
48. Durand, J. F.: Doubles décompositions, en milieu aqueux, entre ses acétylures métalliques et des sels. Comptes rendus hebdomadaires des séances de l'Académie des sciences **177**, 693–695 (1923)
49. Špičák, S., Šimeček, J.: Chemie a technologie třaskavin. Vojenská technická akademie Antonína Zápotockého, Brno (1957)
50. Brameld, V.F., Clark, M.T., Seyfang, A.P.: Copper acetylides. J. Soc. Chem. Ind. **66**, 346–353 (1947)
51. Rintoul, W.: Explosives. Rep. Prog. Appl. Chem. **5**, 523–565 (1920)
52. Burrows, L.A., Lawson, W.E.: Electric blasting initiator. US Patent 2,086,531, 1937
53. Reppe, W.: Äthinylierung. Justus Liebigs Annalen der Chemie **596**, 6–11 (1955)

54. Reppe, W., Keyssner, E.: Verfahren zur Herstellung von Alkoholen der Acetylenreihe. DE Patent 726,714, 1937
55. Reppe, W., Keyssner, E.: Production of alkinols. US Patent 2,232,867, 1941
56. Reppe, W., Steinhofer, A., Spaenig, H., Locker, K.: Production of alkinols. US Patent 2,300,969, 1942
57. Reppe, W., Steinhofer, A., Trieschmann, H.-G.: Verfahren zur Ausführung katalytischer Umsetzungen. DE Patent 734,881, 1939
58. Pietsch, E., Kotowski, A.: Über den Nachweis sehr geringer Mengen von Acetylen. Angew. Chem. **44**, 309–312 (1931)
59. Söderbaum, H.G.: Ueber die Einwirkung des Acetylens auf Cuprisalze. Berichte der deutschen chemischen Gesellschaft **30**, 814–815 (1897)
60. Bagal, L.I.: Khimiya i tekhnologiya iniciiruyushchikh vzryvchatykh veshchestv. Mashinostroenie, Moskva (1975)
61. Plimpton, R.T., Travers, M.W.: Metallic derivates of acetylene. I. Mercuric acetylide. J. Chem. Soc. 264–269 (1894)
62. Malý, J., Kuča, L.: Struktura a rozpad acetylidu rtuti. Chemické listy 47 (1953)
63. Keiser, E.H.: Metallic derivates of acetylene. J. Chem. Soc. 61 (1894)
64. Ferber, E., Römer, E.: Über einige neue Acetylen-Quecksilber-Komplexsalze. Journal für praktische Chemie 277–283 (1934)
65. Burkard, E., Travers, W.T.: The action of acetylene on the acetates of mercury. J. Chem. Soc. **81**, 1270–1272 (1902)
66. Plimpton, R.T.: Metallic derivates of acetylene. Proc. Chem. Soc. **8**, 109–111 (1892)
67. Mathews, J.A., Watters, L.L.: The carbide of gold. J. Am. Chem. Soc. **22**, 108–111 (1900)
68. Vicente, J., Chicote, M.-T., Abrisqueta, M.-D., Jones, P.G.: New neutral and anionic alkynylgold(I) complexes *via* new synthetic methods. Crystal and molecular structures of [(PPh$_3$)$_2$N][Au(CCCH$_2$OH)$_2$], [Au(CCSiMe$_3$)(CN^{t}Bu)], and [Au(CCR)PR$'_3$] (R$'$=Cyclohexyl, R=CH$_2$Cl, CH$_2$Br; R$'$=Ph, R=SiMe$_3$, ^{t}Bu). Organometallics **16**, 5628–5636 (1997)

Chapter 13
Other Substances

In this chapter, we have decided to summarize substances that are not very well described in open literature. There are three groups of substances that we consider interesting enough to mention here—salts of nitramines, organophosphates, and hydrazine nitrates.

13.1 Salts of Nitramines

Primary nitramines are substances with acidic hydrogen and which therefore easily form salts. Since many nitramines are high explosives it is logical that their salts, especially those of heavy metals, were those investigated. Only a small number of such substances, however, fall into a group of primary explosives.

The heavy metal salts of methylnitramine (MNA) have been reported by Davis [1] as "primary explosives which had not been extensively investigated." Searching more recent literature revealed that there has not been much work published since the time of World War II.

Silver and lead salts of methylenedinitramine (MEDINA) have, according to Urbański, sufficient initiating properties to be used practically. The sensitivity to impact and the ignition temperatures of both salts are summarized in Table 13.1 [2].

Ethylenedinitramine (EDNA) is a powerful secondary explosive patented by Hale [3]. The two acidic hydrogens of the nitramino group can be relatively easily replaced with metal ions. Every once in a while speculation arises regarding the applicability of EDNA salts for use as primary explosives. The sensitivity of the metallic salts is, however, relatively low. The cupric salt of ethylenedinitramine is less sensitive than both LA and LS. It follows from Table 13.2 that it is comparable to secondary explosives. Ignition of CuEDNA occurs at 196 °C, DSC onset is at 162 °C (10 °C min^{-1}) [4]. Cupric, ferrous, lead, and potassium salts of EDNA are more sensitive then RDX but not sensitive enough to qualify for use as primary explosives. Impact sensitivities obtained by Blatt [5] and summarized in Fedoroff, Shefield, and Kaye's [6] are listed in Table 13.3.

Table 13.1 Sensitivity and ignition temperatures of methylenedinitramine salts [2]

Salt	Impact sensitivity h_{50}; 2 kg weight (cm)	Ignition temperature (°C)
Ag	10	195
Pb	12	213

Table 13.2 Sensitivity of CuEDNA compared with LA and LS [4]

Compound	Impact sensitivity—h_{50}; Ball & Disc test (cm)	Friction sensitivity (m s^{-1})^a	Sensitivity to electric discharge	
			Standard test (mJ)	Test for sensitive expl. (No.7) (µJ)
CuEDNA	>40	>3.7	>40	>40
LA^b	19	1.2	–	2.5
LS^b	11	1.8	–	7

^{a}Emery paper friction test; values are pendulum velocities required to produce a 50 % probability of ignition.
^{b}Service material RD 1343 for LA (precipitated from sodium carboxymethylcellulose/sodium hydroxide) and RD 1303 for LS. Emery paper friction test; values are velocity of pendulum required to produce a 50 % probability of ignition.

Table 13.3 Impact sensitivity of metallic salts of EDNA [5, 6]

Compound	CuEDNA	FeEDNA	K$_2$EDNA	PbEDNA	RDX
Impact sensitivity (cm/5 kg hammer)	17	21	43	30	51

The heavy metallic salts (including the cupric salt) of ethylenedinitramine can be easily prepared by reacting ethylenediamine with the corresponding sulfate via the sodium salt. In the case of copper, the product precipitates from the solution in the form of a blue/green solid which tends to form compact material after drying in air overnight. Such product can be reduced to a powder by gentle grinding in toluene using a glass rod. The final powdery product is obtained after filtering off the toluene and drying under vacuum [4]. The same method of preparation has been reported for the lead salt [7].

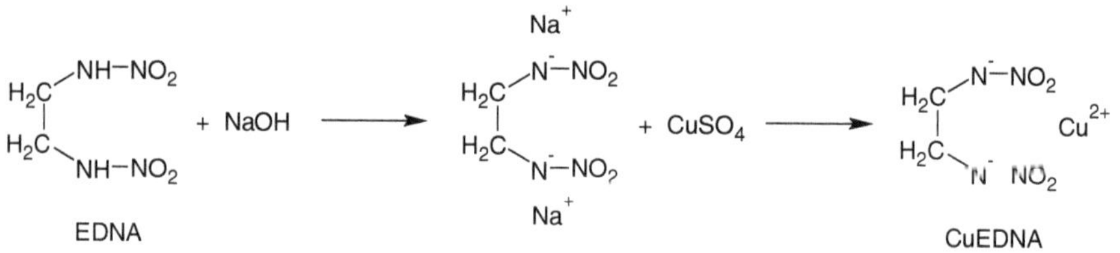

$$\text{EDNA} \xrightarrow{+ \text{NaOH}} \text{Na salt} \xrightarrow{+ \text{CuSO}_4} \text{CuEDNA}$$

The cupric salt of ethylenedinitramine has been studied as a potential green primary explosive replacing LA. Low sensitivity, however, excluded it from such a use [4].

Another relatively recently investigated group of primary nitramines are salts of 4,6-diazido-2-nitramino-1,3,5-triazine (DANT). These substances have been

13.1 Salts of Nitramines

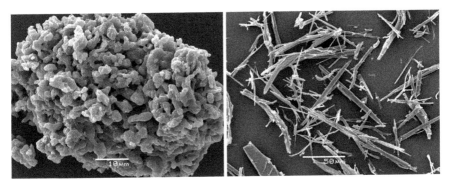

Fig. 13.1 SEM photographs of cesium (*left*) and rubidium (*right*) salts of DANT (kindly provided by Tomáš Musil)

Table 13.4 Sensitivity and DSC data of rubidium and cesium salts [8]

Compound	Impact sensitivity Ball drop (mJ)	Friction sensitivity Small BAM (g)		Onset DSC, 20 °C min^{-1} (°C)
		No-fire level	Low fire level	
Rb salt	12	850	900	200
Cs salt	13	175	200	182
LS	25	40	50	290

Table 13.5 Performance of rubidium and cesium salts (witness plate test) [8, 9]

Compound	Loading pressure (MPa)	Density (g cm^{-3})	Average dent in Al block (mm)
Rb salt	69	1.810	0.8
Cs salt	69	2.026	0.5
LA (RD 1333)	69	3.119	37.3

examined while searching for new potential green primary explosives. From the published results, it seems that the most promising are the cesium and rubidium salts of DANT. SEM photographs of rubidium and cesium salts are shown in Fig. 13.1.

The sensitivity of the cesium and rubidium salts of 4,6-diazido-2-nitramino-1,3,5-triazine has been published by Hirlinger and Bichay and is reprinted in Table 13.4 [8].

The performance of these two salts was measured by the plate dent test. Both salts have a performance which is too low for use as a lead azide replacement (as follows from Table 13.5) [8, 9].

Preparation of DANT has been briefly described by Hirlinger and Bichay [8] Fronabarger et al. [11] and more recently by Musil et al. in detail [10]. Cyanuric chloride is the starting material that is aminated at low temperature to yield 2-amino-4,6-dichloro-1,3,5-triazine which in turn reacts with sodium azide to form 2-amino-4,6-diazido-1,3,5-triazine. Finally, the product is nitrated at low temperature with 100 % nitric acid to produce the final product, 4,6-diazido-2-nitramino-1,3,5-triazine. The alkali salts are readily formed by reaction with the appropriate hydroxides in methanol [8, 10, 11].

The rubidium and cesium salts did not perform well in trials for LA replacement and were therefore proposed as LS replacements. Using less electropositive metals may yield candidates for lead azide replacement [8].

13.2 Organophosphates

An interesting and relatively unexpected group of organophosphate compounds has recently been reported. Two candidates have been mentioned: 4,4,6,6,8,8-hexaazido-1,3,5,7,2λ^5,4λ^5,6λ^5,8λ^5-tetrazatetraphosphocine-2,2-diamine (DAHA) and 7,7,9,9-tetraazido-1,4-dinitro-1,4,6,8,10-pentaaza-5λ^5,7λ^5,9λ^5-triphospha-spiro[4.5]deca-5,7,9-triene (ENTA) [12]. Surprisingly, there is nothing reported in the literature about diazido, dinitro, or dinitramino derivates of DAHA in position 2 instead of the amino groups nor about diazido or dinitramino derivates of ENTA in position 1,4 instead of the nitro groups.

DAHA   ENTA

13.2 Organophosphates

Table 13.6 Sensitivity and DSC data of ENTA [12]

Compound	Impact sensitivity Ball drop (mJ)	Friction sensitivity Small BAM (g)		DSC, 20 °C min^{-1} (°C)	
		No-fire level	Low fire level	Onset	Peak
ENTA	–	300	400	235	266
LS	25	40	50	290	305

Table 13.7 Results of plate dent test of ENTA [12]

Compound	Loading pressure (MPa)	Density (g cm^{-3})	Average dent in Al block (mm)
ENTA	69	1.756	1.3
LA (RD 1333)	69	3.119	37.3

ENTA is described as being a yellow or white [12] or colorless [13] crystalline solid. Density of ENTA is 1.83 g cm^{-3} [12] and melting point 144–145 °C [12, 13]. DAHA forms as a white crystalline material with melting point 72–74 °C [14].

The sensitivity of ENTA (Table 13.6) and its performance measured by witness plate test (Table 13.7) compared to LA and LS indicate its low sensitivity but unfortunately also its low performance. DAHA is sensitive to hot wire and successfully initiates secondary explosives (RDX, CL-20) in a detonator (MK-1) [23]. DAHA is a thermally stable compound despite its low melting point (72–74 °C). It decomposes violently at 230 °C (DSC, onset) [14].

Bichay and Hirlinger [12] reported that the detonation ability of ENTA in the above-mentioned test strongly depends on synthesis conditions. The ENTA sample reported in the earlier tables failed to detonate (1.3 mm dent) whereas other tested samples produced detonation (values for those samples were not found). The authors expected that higher purity and larger crystals would improve the detonation characteristics of ENTA.

The synthesis of ENTA begins with the commercially available starting material hexachlorocyclotriphosphazene which reacts with 2 equivalents of ethylenediamine to yield 7,7,9,9 tetrachloro-1,4,6,8,10-pentaaza-5λ^5,7λ^5,9λ^5-triphosphaspiro[4.5]deca-5,7,9-triene-1,4-diamine (a compound very susceptible to hydrolysis). This compound can be re-crystallized from methylene chloride or hexane. The second step is nucleophilic azide displacement of the residual chloro groups with lithium or sodium azide in dry acetone. 7,7,9,9-Tetraazido-1,4,6,8,10-pentaaza-5λ^5,7λ^5,9λ^5-triphosphaspiro-[4.5]deca-5,7,9-triene-1,4-diamine forms a clear gum which slowly crystallizes under high vacuum. Nitration of 7,7,9,9-tetraazido-1,4,6,8,10-pentaaza-5λ^5,7λ^5,9λ^5-triphosphaspiro[4.5]deca-5,7,9-triene-1,4-diamine is the last step in ENTA synthesis. Nitronium tetrafluoroborate in acetonitrile is necessary because 100 % nitric acid in acetanhydride does not yield any product. More suitable usage of *N*,*N'*-dinitroethylendiamine (EDNA) or its dilithium or disodium salt instead of ethylenediamine is not possible, probably because of the poor nucleophilicity of the nitramine group [12, 13].

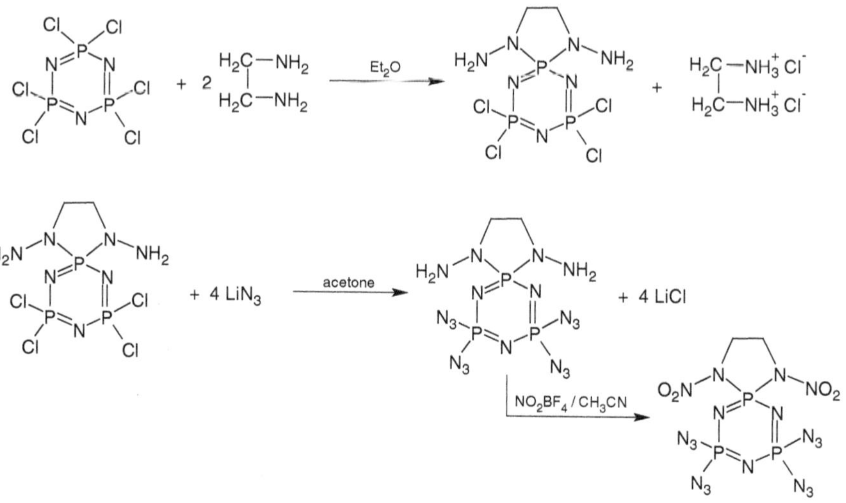

[Reaction scheme: hexachlorocyclotriphosphazene + 2 H₂C(NH₂)–H₂C(NH₂) (ethylenediamine) in Et₂O → 1,1-ethylenediamino-3,3,5,5-tetrachlorocyclotriphosphazene + 2 H₂C-NH₃⁺Cl⁻ / H₂C-NH₃⁺Cl⁻]

[Reaction scheme: above intermediate + 4 LiN₃ in acetone → 1,1-ethylenediamino-3,3,5,5-tetraazidocyclotriphosphazene + 4 LiCl; then NO₂BF₄/CH₃CN → ENTA (with N–NO₂ groups)]

ENTA

DAHA is prepared from readily available octachlorocyclotetraphosphazene. It is aminated with ammonium hydroxide in diethylether at low temperature, forming the intermediate 1,1-diamino-3,3,5,5,7,7-hexachlorocyclotetraphosphazene. DAHA forms in the second step by nucleophilic azide displacement of the residual chloro groups with sodium azide in acetone [14, 24].

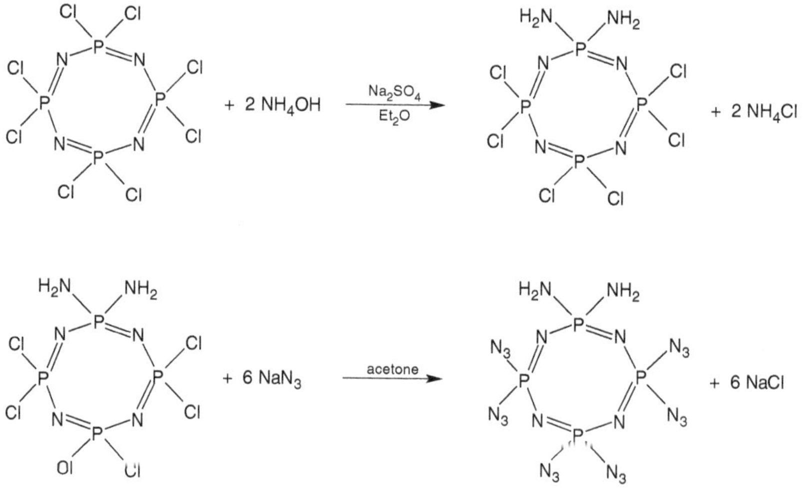

[Reaction scheme: octachlorocyclotetraphosphazene + 2 NH₄OH, Na₂SO₄/Et₂O → 1,1-diamino-hexachlorocyclotetraphosphazene + 2 NH₄Cl]

[Reaction scheme: above + 6 NaN₃, acetone → DAHA + 6 NaCl]

DAHA

Both substances have been proposed as "green candidates" for new primary explosives [12, 14, 23]. Details of the results, except the ones in Table 13.7, are not known.

13.3 Hydrazine Complexes

Table 13.8 Form and color of metallic hydrazine nitrates [16]

Formula	Color, form
[Cd(N$_2$H$_4$)$_3$](NO$_3$)$_2$[a]	White crystalline powder
[Co(N$_2$H$_4$)$_3$](NO$_3$)$_2$[a]	Brown fine crystals[a]
[Mn(N$_2$H$_4$)$_2$](NO$_3$)$_2$	Yellow-brown powder
[Ni(N$_2$H$_4$)$_3$](NO$_3$)$_2$	Rose violet colored fine crystals
[Zn(N$_2$H$_4$)$_3$](NO$_3$)$_2$[a]	White crystalline powder

[a]The same colors and shapes have been published by Lao [17] for products containing two hydrazine ligands.

13.3 Hydrazine Complexes

With bivalent metals, hydrazine nitrate forms coordination compounds that may be represented by the general formula Me[(N$_2$H$_4$)$_{2 \text{ or } 3}$](NO$_3$)$_2$ [6]. The two most often mentioned substances are nickel hydrazine nitrate (NHN) and cobalt hydrazine nitrate (CoNH).

These compounds are not typical primary explosives as they have sensitivity and explosive properties in between that of primary and secondary explosives [15]. The amount of information about these compounds is rather limited and in some cases quite inconsistent.

The description of crystal type and color of several metallic hydrazine nitrates are summarized in Table 13.8. The density of NHN is 2.07 g cm^{-3}. Nickel and cobalt hydrazine nitrate are practically insoluble in common organic solvents (e.g., ethanol, acetone) and sparingly soluble in water (0.47 g l^{-1} at 30 °C, 2.31 g l^{-1} at 70 °C). NHN burns upon the action of 96 % sulfuric acid, but weak acid makes it decompose more gently; CoHN explodes in contact with concentrated sulfuric acid. NHN mildly reacts with a 10 % solution of sodium hydroxide and this can be used for producing a chemical decomposition. NHN is stable on exposure to sunlight and it hardly reacts at all with common metals. Cobalt hydrazine nitrate is unstable and decomposes slowly, even in the presence of traces of moisture, changing in color from brown to greenish [16, 17].

Nickel hydrazine nitrate is not easily detonated by flame or heat in charges of a few grams even when mixed with lead styphnate, picrate, or RDX. These mixtures were loaded in a blasting cap and initiated but the best that has been observed was a deflagration [6, 16]. Other results have been obtained by Talawar et al. [18]. They made a research of the applicability of NHN in detonators. NHN is ignitable by the flame of a match, a safety fuse, or a hot wire. Whether it develops detonation strongly depends on its confinement and its density.

Impact sensitivity and ignition temperature for several metallic hydrazine nitrates are summarized in Table 13.9.

The detonation velocity of NHN is 3,600 m s^{-1} at 0.8 g cm^{-3} and 6,900–7,000 m s^{-1} at 1.7 g cm^{-3} [16, 21]. To achieve the density of 1.7 g cm^{-3} NHN was pressed by 150 MPa, which shows its good resistance to dead pressing. The dependency of the detonation velocity on the diameter of the charge at density 0.62 g cm^{-3} is summarized in Médard and Barlot paper [16].

Table 13.9 Metallic hydrazine nitrates: ignition temperatures and sensitivity to impact and friction

Compound	Impact sensitivity		Friction sensitivity (N)	Ignition temperature (°C)	References
	As reported	Recalculated to energy (J)			
$[Cd(N_2H_4)_3](NO_3)_2$[a]	~1.5 m/10 kg	147	–	236–245	[16]
$[Co(N_2H_4)_3](NO_3)_2$[a]	~1.75 m/2 kg	34	–	206–211	[16]
	59 cm/2 kg (h_{50})	11.6 (h_{50})	>60	–	[19]
$[Mn(N_2H_4)_2](NO_3)_2$	Not sensitive		–	150	[16]
$[Ni(N_2H_4)_3](NO_3)_2$	1.4 m/2 kg	27.5	–	212–215	[16]
	84 cm/2 kg (h_{50})	16.5	>10	219 (210 onset)	[20]
	21 cm/0.4 kg (h_0) lower than MF	0.82	–	–	[21]
			–	260–270	[17]
	96 cm/2 kg (h_{50})	18.8 (h_{50})	10	219	[18]
	–		72 (F_{50})	232[b]	Our data
$[Zn(N_2H_4)_3](NO_3)_2$[a]	>3 m/10 kg	>294	–	310	[16]

[a]The same values of impact sensitivities and ignition temperatures are reported by [17] for products containing two hydrazine ligands.
[b]Heating rate 5 °C min^{-1} (DTA).

Metallic hydrazine nitrates is prepared by the reaction of the relevant nitrate with hydrazine in water or alcohol [16, 21]. The product is isolated from the reaction mixture as a precipitate [6]. The preparation of NHN according to the procedure of Zhu [21] gives 95 % yield and leads to fine powder shown in Fig 13.2.

$$Ni(NO_3)_2 + 3\,N_2H_4 \longrightarrow Ni[(N_2H_4)_3](NO_3)_2$$

Detailed preparation methods for NHN and CoHN have been described by Chhabra et al. [20]. They used diluted aqueous solutions (4 and 8 %) of nickel nitrate preheated to 65 °C and simultaneously mixed it with a concentrated solution (>80 %) of hydrazine hydrate. The reaction gave a 90 % yield of NHN. When using cobalt nitrate the yield was 85 % of CoHN. Reducing the temperature to 25 °C led to yields around 90 % for both complexes; however, the products obtained at lower temperatures were amorphous and did not flow as well as the granular ones produced at higher temperatures.

According to Chinese authors [17] researchers outside China use 30 % nickel nitrate and a 40 % solution of hydrazine hydrate. Using solutions of such a high concentration leads to the formation of a light pink powder with unsuitable properties. To obtain good product, the Chinese use lower concentrations in both nickel nitrate (5–10 %) and hydrazine hydrate (10–15 %). The reaction temperature is 50–70 °C and within 30–40 min yields about 80 % of product in the form of spherical crystals of NHN with density 1.0 g cm^{-3}.

Information about the process is only available to a limited extent. It is believed that an aqueous solution of nickel salt is placed into the reactor and hydrazine is slowly added. Most of the time nickel nitrate is used, but some trials using nickel acetate and sodium nitrate have also been found. Acidity of the nickel solution in

Fig. 13.2 SEM photograph of nickel hydrazinium nitrate (NHN) (kindly provided by Tom Musil)

the reactor is adjusted by addition of acetic acid which is done before the addition of the hydrazine. The initiating efficiency of NHN to RDX is 0.17 g and to PETN it is 0.13 g (compared to 0.2 g of MF) [17].

Of all the hydrazine nitrate salts only nickel hydrazine nitrate has been suggested for practical applications. Although it was first reported more than 50 years ago it still has not found wider application [22]. Despite the earlier information that nickel hydrazine nitrate is not reliably initiated in blasting caps [6, 16] it has been recently suggested as a primary explosive that is suitable as a replacement for lead azide as an intermediate charge in commercial detonators [15, 22]. Other authors recommended NHN as a complete replacement for lead styphnate and a partial replacement of lead azide in detonators [18].

One analog of NHN is a substance with the nitrate group replaced by azide—nickel hydrazine azide (NHA). Only very limited information is available, including a preparation procedure which is practically the same as in the case of NHN. Aqueous solutions of the nickel salt and sodium azide are placed in a stirred reactor, acidity is adjusted by addition of acetic acid, the mixture is heated to 50–70 °C, and an aqueous solution of hydrazine hydrate is slowly added. NHA has green polycrystalline form with crystal density 2.12 g cm^{-1}. Average particle size of product with good fluidity is around 80 μm and bulk density of such powder is 0.57–0.65 g cm^{-3}. The thermal decomposition is a two-step process which starts at 165–206 °C (heating rates from 10 to 30 °C min^{-1}). The second step starts at 206–400 °C. NHA does not become dead pressed by 70 MPa, because it is less sensitive to impact but more to friction than PETN [19].

References

1. Davis, T.L.: The Chemistry of Powder and Explosives. Wiley, New York (1943)
2. Urbański, T.: Chemistry and Technology of Explosives. PWN—Polish Scientific Publisher, Warszawa (1967)
3. Hale, G.C.: Explosive. US Patent 2,011,578, 1934

4. Millar, R.W.: Lead-free Initiator Materials for Small Electro-explosive Devices for Medium Caliber Munitions. Report QinetiQ/FST/CR032702/1.1, 2003
5. Blatt: Compilation of Data on Organic Explosives, Sensitivities to Impact of K, Fe, Pb and Cu Salts. Report, Office of Scientific Research and Development, 1944
6. Fedoroff, B.T., Sheffield, O.E., Kaye, S.M.: Encyclopedia of Explosives and Related Items. Picatinny Arsenal, New Jersey (1960–1983)
7. Franz, A.O.: Explosive compound, process of making same and a composition thereof. US Patent 2,708,623, 1955
8. Hirlinger, J.M., Bichay, M.: New Primary Explosives Development for Medium Caliber Stab Detonators. Report SERDP PP-1364, US Army ARDEC, 2004
9. Hirlinger, J., Fronabarger, J., Williams, M., Armstrong, K., Cramer, R.J.: Lead azide replacement program. In: Proceedings of NDIA, Fuze Conference, Seattle, Washington, April 5–7, 2005
10. Musil, T., Matyáš, R., Lyčka, A., Růžička, A.: 4,6-Diazido-N-nitro-1,3,5-triazine-2-amine (DANT)—synthesis and characterization. Propellants Explosives Pyrotechnics (2012) 37, 275–281 (2012)
11. Fronabarger, J., Sitzmann, M.E., Williams, M.D.: Method for azidotriazole, nitrosoguanazine, and related compounds. US Patent 7,375,221 B1, 2008
12. Bichay, M., Hirlinger, J.M.: New Primary Explosives Development for Medium Caliber Stab Detonators. Report SERDP PP-1364, US Army ARDEC, 2004
13. Dave, P.R., Forohar, F., Axenrod, T., Bedford, C.D., Chaykovsky, M., Rho, M.-K., Gilardi, R., George, C.: Novel spiro substituted cyclotriphosphazenes incorporating ethylenedinitramine units. Phosphorus Sulfur Silicon 90, 175–184 (1994)
14. Forohar, F., Dave, P.R., Iyer, S.: Substituted cyclotetraphosphazene compound and method of producing the same. US Patent 6,232,479, 2001
15. Hariharanath, B., Chandrabhanu, K.S., Rajendran, A.G., Ravindran, M., Kartha, C.B.: Detonator using nickel hydrazine nitrate as primary explosive. Defence Sci. J. 56, 383–389 (2006)
16. Médard, L., Barlot, J.: Prépatarion et propriétés des nitrates complexes des métaux bivalentes et de l'hydrazine et d'un styphnate complexe analogue. Mémorial des poudres 34, 159–166 (1952)
17. Lao, Y.-L.: Chemistry and Technology of Primary Explosives. Beijing Institute of Technology Press, Beijing (1997)
18. Talawar, M.B., Agrawal, A.P., Chhabra, J.S., Ghatak, C.K., Asthana, S.N., Rao, K.U.B.: Studies on nickel hydrazinium nitrate (NHN) and bis-(5-nitro-$2H$ tetrazolato-N^2)tetraamino cobalt(III) perchlorate (BNCP): Potential lead-free advanced primary explosives. J. Sci. Ind. Res. 63, 677–681 (2004)
19. Zhu, S., Wang, Z., Li, Y., Zhang, L., Ye, Y.: Performance of NHA and its application. In: Schelling, F.J. (ed.) Proceedings of 35th International Pyrotechnics Seminar, pp. 201–206, Fort Collins, Colorado (2008)
20. Chhabra, J.S., Talawar, M.B., Makashir, P.S., Asthana, S.N., Haridwar, S.: Synthesis, characterization and thermal studies of (Ni/Co) metal salts of hydrazine: Potential initiatory compounds. J. Hazard. Mater. A99, 225–239 (2003)
21. Zhu, S., Wu, Y., Zhang, W., Mu, J.: Evaluation of a new primary explosive: Nickel hydrazine nitrate (NHN) complex. Propellants Explosives Pyrotechnics 22, 317–320 (1997)
22. Thangadurai, S., Kartha, K.P.S., Sharma, D.R., Shukla, S.K.: Review of some newly synthesized high energetic materials. Sci. Technol. Energetic Mater. 65, 215–226 (2004)
23. Hirlinger, J.M., Cheng, G.: Investigation of alternative energetic compositions for small electro-explosive devices for medium caliber ammunition, in, US Army armaments research WP-1307-FR-01, Picatinny Arsenal (2004)
24. Fincham, J.K., Shaws, R.A.: Phosphorus-nitrogen compounds. Part 62. The reactions of 2,2-diamino-4,4,6,6,8,8-hexachloro- and 2,6-diamino-2,4,4,6,8,8- hexachloro-cyclotetraphosphazatetraene with sodium methoxide in methanol. The first example of amino group migration in the tetramer system. ^{31}P, ^1H and ^{13}C nuclear magnetic resonance spectroscopic investigations of the products, Phosphorus, Sulfur and Silicon 47, 109–117 (1990)

Index

A
Acetylides, 303
Alkaline fulminates, 63
2-Amino-4,6-dichloro-1,3,5-triazine, 328
7-Amino-4,6-dinitrobenzofuroxan, 168
Aminoguanidine, 195
5-Amino-1H-tetrazole (5-ATZ), 194
Aminotetrazole salts, 194
1-Amino-1-(tetrazol-5-yldiazenyl)guanidin
 (GNGT, tetrazene), 4, 6, 26, 189
Ammonium fulminate, 63
Antimony sulfide, 5
Antimony triazide, 110
Arsenic triazide, 110
Aurous acetylide, 321
Azides, 71
Azides, organic, 110
Azidoformamidine, 195
5-Azido 1-methyltetrazole, 222
5-Azido-1-picryltetrazole, 222
5-Azidotetrazoles, 209
 organic derivatives, 222
1-Azido-2,4,6-trinitrobenzene, 111
Azoimide (hydrazoic acid), 71
5,5′-Azotetrazole salts, 212

B
Barium styphnate, 149
Benzo[d][1,2,3]oxadiazole, 157
Benzotrifuroxan (BTF), 167
Bis(furoxano)-2-nitrophenol (BFNP), 179
Bis(2-hydroperoxypropane)peroxide, 255
Bismuth triazides, 110
Black powder, 5

C
Cadmium azide, 106, 109
Cadmium fulminate, 63
5-Carboxamidotetrazolatopentaaminecobalt
 (III) perchlorate, 233
CCI Clean Fire, 7
Charge diameter, 20
(3-(Chloromethyl)pentamethylhexoxonane),
 271
5-Chlorotetrazole salts, 207
Cobalt hydrazine nitrate (CoNH), 331
Cobalt perchlorate complexes, 228
Compacting pressure, 13
Confinement, 20
Copper azides, 96
 undesired formation, 102
Copper(II) tetrakis-[1,5-
 cyclopentamethylenetetrazol]
 perchlorate (CuL_4), 241
Corazol, 238
CuEDNA, 325
Cupric acetylide, 316
Cupric azide, 97
Cupric salt of 5-nitrotetrazole, 201
Cuprous acetylide, 313
Cuprous azide, 97
 impact sensitivity, 28
Cuprous salt of 5-nitrotetrazole, 203
1-Cyanoformimidic acid hydrazide, 232
5-Cyanotetrazole, 231
Cyanuric triazide, 111
1,5-Cyclopentamethylenetetrazole
 complexes, 238

D
Dead pressing, 11
Deflagrating substances/mixtures (priming mixtures, primes), 2
Deflagration, 2
Deflagration to detonation transition (DDT), 2, 227
Density, 11, 13
Detonants, 2, 79
Detonating substances (primary explosives), 2
Detonators, 1
Diacetone diperoxide (tetramethyltetroxane; DADP), 256
5,7-Diamino-4,6-dinitrobenzofuroxan (CL-14), 167
Diaminohexachlorocyclotetraphosphazene, 330
1,5-Diaminotetrazole ferrous perchlorate complex (DFeP), 245
4,6-Diazido-2-nitramino-1,3,5-triazine (DANT), 326
2-Diazo-4,6-dinitrophenol (dinol, diazol, DDNP, DDNPh, DADNPh), 157
Diazophenols, 157
Dihydroperoxide 2,2-dihydroperoxypropane, 255
4,6-Dinitrobenzofuroxan (4,6-DNBF), 168
Dinitroguanidines, 250
Dinitroresorcine (DNR), 133
Dinol (DDNP), 4, 7
Disilver 5-amino-1H-tetrazolium perchlorate Ag$_2$(ATZ)ClO4, 196

E
Ease of ignition (Bickford fuse) test, 32
Electrostatic discharge, sensitivity, 31
Environmental hazards, 6
Ethylene dinitramine (EDNA), 325
Ethylene tetrazylazide, 222
Explosive train, 1

F
Ferrous 5-nitrotetrazole, 199
Flame sensitivity, 32
Formonitrile oxide, 38
Friction initiation/sensitivity, 29
Fulminates, 37
    blasting caps, Alfred Nobel, 55
Fulminic acid, 37
Furoxan, 167

G
Glass powder, 5
Gold fulminate, 63
Green initiating substances, 6
Gum Arabic, 5

H
Heavy metals, 7
Hexaazidotetrazatetraphosphocine-2,2-diamine (DAHA), 328
Hexamethylene triperoxide diamine (diazahexaoxabicyclotetradecane, HMTD), 275
Hg free, 7
High-voltage detonator (HVD), 235
1-(1H-Tetrazol-5-yl)guanidinium nitrate, 219
Hydrazine nitrate, 331
7-Hydroxylamino-4,6-dinitro-4,7-dihydrobenzofuroxan, 175

I
Igniters, 1
Ignition, reliable, 4
Impact sensitivity, 25
Initiating efficiency, 13
Initiating substances (primaries), 1
α-Isocyanilic acid, 37

K
KDNBF, 173, 176

L
LDDS, 148
Lead azide, 2, 4, 31, 74–89
Lead (IV) azide, 88
Lead dinitroresorcinol, 133, 135
Lead hypophosphite nitrate, 5
Lead mononitroresorcinol, 131
Lead picrate, 132
Lead salt of trinitroresorcine, 5
Lead styphnate (LS), 4, 6, 86, 138
    basic, 145
    double salts, 148
Lead tricinate, 6

M
Mercuric acetylide, 318
Mercuric azide, 105, 109

Mercuric salt of 5-nitrotetrazole, Hg(NT)$_2$, 201
Mercurous acetylide, 320
Mercurous azide, 105, 108
Mercury fulminate (MF), 2, 4, 39–59
Metafulminuric acid, 37
Metallic hydrazine nitrates, 331
Methylenedinitramine (MEDINA), 325
Methyl-5-nitrotetrazoles, 221

N
Nickel hydrazine nitrate, 331
Nitramines, 325
Nitrogen dibromideiodide, 294
Nitrogen halides, 289
Nitrogen tribromide, 294
Nitrogen trichloride, 289
Nitrogen triiodide, 295
    sensitivity, 23
Nitroglycerine, initiation, 55–56
5-Nitro-1H-tetrazole (HNT), 197
5-Nitro-2-picryltetrazole, 222
5-Nitrotetrazolato-N^2-ferrate coordination complexes, 248
5-Nitrotetrazole, cupric salt, 201
    cuprous salt, 203
    mercuric salt, 201
    organic derivatives, 221
    silver salt 201
Nitro-2-(2,4,6-trinitrofenyl)tetrazole, 221
5-(N-Nitramino)tetrazoles, 188
NONTOX, 7

O
Organic azides, 110
Organic peroxides, 255
Organometallic azides, 111
Organophosphates, 328
Output charges, 2
2-Oxa-3,4-diazabicyclo[3.2.2]nonatetraene, 157
Oxidizers, 5

P
PbNATNR, 148
Pentaamine(1,5-cyclopentamethylene-tetrazolato-N^3) cobalt(III) perchlorate (PAC), 238, 251
Pentaamine(5-cyano-2H-tetrazolato-N^2)cobalt (III) perchlorate (CP), 228
Pentaamine(5-nitro-2H-tetrazolato-N^2) cobalt (III) perchlorate (NKT), 237

Peroxides of acetone, 255
Picramic acid, 157, 162
Picric acid, 131
5-Picrylaminotetrazole, 218
Polynitrophenols, 131
Potassium chlorates, 5
Potassium fulminate, 63
Potassium 7-hydroxylamino-4,6-dinitro-4,7-dihydrobenzofuroxan, 175
Predetonation, 32
Primary explosives, 2, 4
    explosive properties, 11
Primer, 1
Priming compositions (priming mixtures), 5

R
Reinforcing cap, 20

S
Secondary explosives, initiation, 22
Sensitivity, 23
    electrostatic discharge, 31
    flame, 32
    friction, 29
Sensitizers, priming compositions, 4
Silver acetylide, 303
Silver acetylide-hexanitrate, 312
Silver acetylide-silver nitrate, 308
Silver azide (SA), 4, 89
    explosive properties, 92
    preparation, 93
    sensitivity, 91
    uses, 96
Silver fulminate, 59
Silver salt of 5-nitrotetrazole, 201
Simple initiating impulse (SII), 1
SINOXID, 7
SINTOX, 7
Smokeless powder, 5
Sodium fulminate, 63
Specific surface, 15, 19
Surface mining blasting, 25

T
TAAT (4,4′,6,6′-tetra(azido)azo-1,3,5-triazine), 116
TAHT (4,4′,6,6′-tetra(azido)hydrazo-1,3,5-triazine), 116
Tetraacetone tetraperoxide (octamethyloctaoxacyclododecane, TeATeP), 274

Tetraaminecarbonatocobalt(III) nitrate (CTCN), 242
Tetraamine-cis-bis(5-nitro-2H-tetrazolato-N^2) cobalt(III) perchlorate (BNCP), 241
Tetraazido-1,4-benzoquinone (TeAzQ), 121
Tetraazido-1,4-dinitropentaazatriphosphaspiro [4.5]deca-5,7,9-triene (ENTA), 328
Tetrachloropentaazatriphosphaspiro[4.5] decatriene-1,4-diamine, 329
Tetramethylene diperoxide dicarbamide (tetraoxatetraazacyclotetradecanedione, TMDD), 280
Tetrazene (GNGT), 4, 6, 26, 189
Tetrazoles, 187
  organic substituent, 217
Thallium fulminate, 63

Thallous azide, 105
Theoretical maximum density (TMD), 13
Toxicity, 6
Triacetone triperoxide (hexamethylhexaoxonane, TATP), 257, 262
Triazide (cyanuric triazide), 111
Triazidotrinitrobenzene (TATNB, TNTAB), 111, 118
Tricinate, 6
Trinitroresorcine, 138

W
WINCLEAN, 7
Wollfenstein's superoxide (TATP), 257, 262